# Additive Manufacturing of Emerging Materials

Bandar AlMangour
Editor

# Additive Manufacturing of Emerging Materials

Springer

*Editor*
Bandar AlMangour
School of Engineering and Applied
Sciences
Harvard University
Cambridge, MA, USA

ISBN 978-3-030-06282-8 ISBN 978-3-319-91713-9 (eBook)
https://doi.org/10.1007/978-3-319-91713-9


Softcover re-print of the Hardcover 1st edition 2019


Printed on acid-free paper

This Springer imprint is published by the registered company Springer Nature Switzerland AG.
The registered company address is: Gewerbestrasse 11, 6330 Cham, Switzerland

# Contents

**Additive Manufacturing of In Situ Metal Matrix Composites** . . . . . . . . . 1
Taban Larimian and Tushar Borkar

**Optimization of Electrical Discharge Machining of Titanium Alloy (Ti6Al4V) by Grey Relational Analysis Based Firefly Algorithm** . . . . . . . . . . . . . . . . . . . . . . . . . . . . . . . . . . . . . . 29
Anshuman Kumar Sahu and Siba Sankar Mahapatra

**Laser-Based Additive Manufacturing of Lightweight Metal Matrix Composites** . . . . . . . . . . . . . . . . . . . . . . . . . . . . . . 55
Eskandar Fereiduni, Mostafa Yakout, and Mohamed Elbestawi

**Process-Structure-Property Relationships in Additively Manufactured Metal Matrix Composites** . . . . . . . . . . . . . . . . . . . . . . . . 111
Eskandar Fereiduni and Mohamed Elbestawi

**Additive Manufacturing of Titanium Alloys for Biomedical Applications** . . . . . . . . . . . . . . . . . . . . . . . . . . . . . . . . . . . . . . . . . . . . . 179
Lai-Chang Zhang and Yujing Liu

**Corrosion Behaviors of Additive Manufactured Titanium Alloys** . . . . . . 197
Lai-Chang Zhang and Peng Qin

**Effect of Process Parameters of Fused Deposition Modeling and Vapour Smoothing on Surface Properties of ABS Replicas for Biomedical Applications** . . . . . . . . . . . . . . . . . . . . . . . . . . . . . . . . . . 227
Jasgurpreet Singh Chohan, Rupinder Singh, and Kamaljit Singh Boparai

**Development of Rapid Tooling Using Fused Deposition Modeling** . . . . . 251
Kamaljit Singh Boparai and Rupinder Singh

**Development of ABS-Graphene Blended Feedstock Filament for FDM Process** . . . . . . . . . . 279
Gurleen Singh Sandhu and Rupinder Singh

**Investigate the Effects of the Laser Cladding Parameters on the Microstructure, Phases Formation, Mechanical and Corrosion Properties of Metallic Glasses Coatings for Biomedical Implant Application** . . . . . . . . . . 299
Mahmoud Z. Ibrahim, Ahmed A. D. Sarhan, M. O. Shaikh, T. Y. Kuo, Farazila Yusuf, and M. Hamdi

**Fabrication of PLA-HAp-CS Based Biocompatible and Biodegradable Feedstock Filament Using Twin Screw Extrusion** . . . . . . . . . . 325
Nishant Ranjan, Rupinder Singh, I. P. S Ahuja, and Jatenderpal Singh

**Index** . . . . . . . . . . 347

# Additive Manufacturing of In Situ Metal Matrix Composites

**Taban Larimian and Tushar Borkar**

## 1 Introduction

Additive manufacturing (AM), also known as three-dimensional printing has recently emerged as a subject of intense worldwide attention. Additive manufacturing is the process of building three-dimensional (3D) objects in a layer form using CAD data. Over the past three decades additive manufacturing has advanced from creating basic models or rapid prototyping into near net shape processes and gradually taking over the conventional methods in production of complex shaped objects [1]. Additive manufacturing is also beneficial in terms of decreasing the lead time and reducing the cost of a production of small number of parts [2]. Advances in the additive manufacturing technology have made it possible to manufacture complex-shaped metal components strong enough for real engineering applications [3–16]. While laser additive manufacturing is becoming more and more important in the context of advanced manufacturing for the future, most of the current efforts are focusing on optimizing the required parameters for processing well-matured alloys from powder feedstock to achieve reproducible properties, comparable to, or better than, their conventionally processed counterparts. However, laser additive manufacturing or processing also opens up a new horizon in terms of processing novel alloys and composites that are difficult to process using conventional techniques. According to the Wohlers Report 2015, the current additive manufacturing market of $4.1 billion is expected to reach $21.2 in 5 years [17]. An additive manufacturing system takes a computer generated geometric model as its input and builds the geometry by depositing the constituent materials precisely in a layer-by-layer fashion [1, 17, 18]. There are more than ten AM techniques that have been developed so far, which include electron- beam melting (EBM), selective

T. Larimian · T. Borkar (✉)
Mechanical Engineering Department, Cleveland State University, Cleveland, OH, USA
e-mail: T.Borkar@csuohio.edu

 
B. AlMangour (ed.), *Additive Manufacturing of Emerging Materials*,
https://doi.org/10.1007/978-3-319-91713-9_1

laser melting (SLM), stereolithography (SLA), fused deposition modeling (FDM), and digital light processing (DLP), etc. [1, 17–24]. Because of the utilization of the layer-by-layer process, additive manufacturing systems are capable of creating geometrically complex prototypes and products efficiently in small to medium quantities. It is thus best suited for applications requiring complex, high value, time-sensitive, and customized products such as automobile and aerospace parts (i.e. complex designs), broken part replacement (i.e. time-sensitive), and medical implants (i.e. highly customized) such as replacement hip joints. This chapter mainly focused on processing of in situ metal matrix composites via laser additive manufacturing processes such as laser engineered net shaping (LENS)- directed energy deposition type AM process and selective laser melting (SLM)- powder bed fusion type AM process.

The LENS™ process, in recent years, is having a strong impact on the rapid prototyping, small scale fabrication, and repair of complex parts in industry. Originally, the LENS™ process was developed at Sandia National Laboratories and subsequently commercialized by Optomec Design Company. LENS™ process uses focused laser beam for melting the metallic powders and 3-D CAD files for designing the solid 3-D object prior to fabrication. LENS™ is a freeform additive manufacturing (AM) process for near-net shaping of nearly fully dense, homogeneous bulk materials. The process begins with a computer aided design (CAD) file of a 3-dimensional (3D) component, which is sliced into a series of layers with a predetermined layer spacing/thickness on the order of 0.01 in (0.25 mm). Each layer contains a tool path, which is followed by the multi-axis stage, while pre-alloyed or blends of elemental powders are injected into a melt pool produced by a high-powered laser. The process continues via the sequential deposition of layers to develop the overall 3D shape of the component. Most of the worldwide research and development activities related to additive manufacturing, or 3D printing as it is often referred to, of metallic systems have focused on a rather limited set of alloys, such as: stainless steels, certain nickel base superalloys, and conventional titanium alloys. Furthermore, these alloys have all been processed using pre-alloyed powder feedstock with efforts directed to optimizing feedstock characteristics and deposition parameters to achieve additively manufactured components with properties comparable to wrought or conventional thermo-mechanically processed materials. While these efforts are critical in order to establish 3D printing/additive manufacturing as viable technologies for future manufacturing, the potential of these technologies is underexploited for the development and processing of in situ metal matrix composites via additive manufacturing processes.

Controllable microstructure by altering the processing parameters is another advantage that laser additive manufacturing offers. Processing parameters such as energy density, hatch spacing, powder feed rate scanning speed and strategy have tremendous effect on mechanical properties, residual stress and microstructural characteristics such as morphology and grain size [25]. The importance of processing parameters is due to their impact on the melt pool size, cooling rate, thermal gradient and consequently the residual stresses [25]. Two of the most important parameters are laser power and scanning speed. Higher energy along with lower scanning speed will result in more incident energy which results in lower cooling rate and consequently a coarser microstructure. According to equation bellow, the energy density can be altered by changing the scanning speed, more energy equals more melted material [26].

$$E_\rho = \frac{P}{\pi r^2} \frac{2r}{v} \frac{2r}{s}$$

$E_\rho$ = energy density (J/m$^2$).
P = laser power (W).
v = scanning speed (m/s).
s = scan line spacing (m), and.
r = beam radius (m).

Despite the increase in melting, increasing the laser power has one main disadvantage of causing balling defect and dross formation in melt pools and might result in rougher surface finish [27].

## 2 Metal Matrix Composites

There is an increasing demand from aerospace and automotive sectors to develop new advanced materials which contribute to weight savings, improve energy efficiency, withstand harsh structural loadings, and enhanced tribological performances. To attain such special attributes, the material should possess high specific strength, elastic modulus, and stiffness additional to enhanced functional characteristics. Traditionally, metals and alloys were unable to fulfill these demands. Fortunately, development of metal matrix composites (MMCs), a perfect response and promising material, cater to these requirements. Tailorable characteristic and improved properties of MMCs, such as higher specific stiffness and strength, excellent wear resistance, controlled coefficient of thermal expansion, higher fatigue resistance, and better stability at elevated temperature, are now quickly replacing existing metals or their alloys. MMCs comprise of combined properties of metals as matrices (ductility and toughness), and ceramic reinforcements (high modulus and strength) in the form of continuous or discontinuous fibers, whiskers, or particulates. The reason for increasing demand of metal matrix composites for structural applications is mainly due to their exceptional mechanical and physical properties as well as inexpensive reinforcements and isotropic properties [28]. Metal matrix composites have also played a role in reducing the weight in aerospace structures by decreasing the alloy density [29]. The commonly used reinforcement materials are oxides, nitrides and carbides, whereas titanium, magnesium, aluminum, copper and nickel alloys are used as matrix materials [28, 30]. These metal matrix composites are broadly classified as ex-situ and in-situ metal matrix composites based on the nature of reinforcement formation during their fabrication process. During these techniques, ceramic reinforcements were externally added prior to composite fabrication into the matrix material (which may be in molten or powder form). The main disadvantages of these conventional ex-situ metal matrix composites involve size limitation of reinforcing phase, which is nothing but the starting powder size, interfacial reaction between the reinforcement and matrix, and poor wettability between the reinforcements and the matrix due to surface contamination of reinforcements. The physical and mechanical properties of metal matrix composites are primarily governed by size and volume fraction of reinforcement as well as the nature of the matrix reinforcement interface. The uniform dispersion of fine and thermally stable ceramic particulates in the metal matrix is desirable for achieving optimum mechanical properties of metal matrix composites. This leads to the development of novel in situ metal matrix composites in which precipitates are synthesized in metallic matrix by a chemical reaction between elements or between elements and compounds during the composite fabrication. The in-situ metal matrix composites exhibit many advantages over ex-situ metal matrix composites: (a) In-situ formed reinforcements are thermodynamically stable and lead to less degradation at elevated temperatures; (b) Strong interfacial bonding between the matrix and the reinforcement due to clean matrix-reinforcement interface due to absence of any interfacial reaction between matrix and reinforcement; and (c) Better physical and mechanical properties due to homogeneous dispersion of fine-scale reinforcements into matrix.

In situ formation of reinforcement is the promising fabrication route for processing metal matrix composites in terms of both technical and economic considerations. In situ metal matrix composites have better control over physical and mechanical properties due to their greater control on the size and level of reinforcements, as well as the matrix-reinforcement interface. Mechanical and physical properties of metal matrix composites are mainly governed by properties of the matrix, dispersion of the reinforcement, interfacial bonding between matrix and reinforcement, and finally the processing method. Various processing routes have been developed due to the great potential and widespread applications of these in situ metal matrix composites that involve matrix materials (titanium, nickel, aluminum and copper) and in-situ reinforcements (carbides, nitrides, and borides). Laser Additive processing of in situ metal matrix composites opens up new horizon to process multifunctional monolithic metal matrix composites that are difficult to process via traditional manufacturing processing such as laser cladding. These MMCs have been processed either by in situ reaction between elemental blend powders or by in situ reaction between elemental blend powders and reactive gases (nitrogen, oxygen, etc.) during laser additive processing. The following section discusses few examples of in situ metal matrix composites processed via laser additive manufacturing processes.

## 2.1 In-Situ Reactions Between Elemental Blend Powders

### 2.1.1 Aluminum Matrix Composites: Al-$Fe_2O_3$

As mentioned before, Selective laser melting is a laser based additive manufacturing process which is not only capable of producing fully dense three dimensional parts, but can also be used for achieving in-situ reactions between powder particles [31]. Compare to aluminum and it's alloys, Aluminum metal matrix reinforced with alumina and silicon carbide has better properties in terms of higher strength, lighter weight, thermal conductivity, better wear resistance, and stiffness. Therefore aluminum matrix composites are widely used in the aerospace and automotive industries [32, 33]. The mechanical properties of aluminum alloys can be easily tailored by varying the processing conditions such as solidification rate, is another advantage [34, 35, 52]. As compared to conventional processing routes additive manufacturing provides processing flexibility, which helps in achieving refine and uniform distribution of reinforcement (alumina and iron) within aluminum matrix leading towards better mechanical performance [31]. Sasan et al. have processed mixture of aluminum and $Fe_2O_3$ to obtain in situ $Al_2O_3$ and $Fe_3Al$ within aluminum matrix [31]. $Fe_2O_3$ is an inexpensive reinforcement which can reduce the amount of energy required for the selective laser melting of aluminum matrix composites [36]. The following stoichiometric reaction reveals the fact that laser produces extra heat and activates thermite interactions during SLM processing: $8Al + 3Fe_2O_3 \rightarrow 2Fe_3Al + 3Al_2O_3$ + heat, in which Al reduces the $Fe_2O_3$ [37]. The extra heat produced during the in-situ reaction not only helps to decrease the energy input required to melt powders but also helps in consolidation of the composites powder. Also, Melting pool which plays an important role in process optimization can be extended in comparison to pure aluminum [36]. Figure 1a, b show the effect of $Fe_2O_3$ content on the melt pool size for the parts fabricated with the same SLM processing parameters (laser energy and scanning speed).

The Al/5wt.%$Fe_2O_3$ composites create smaller molten pool (Fig. 1a) as compared to Al/15wt.%$Fe_2O_3$ composites (Fig. 1b) mainly due to the higher heat released from

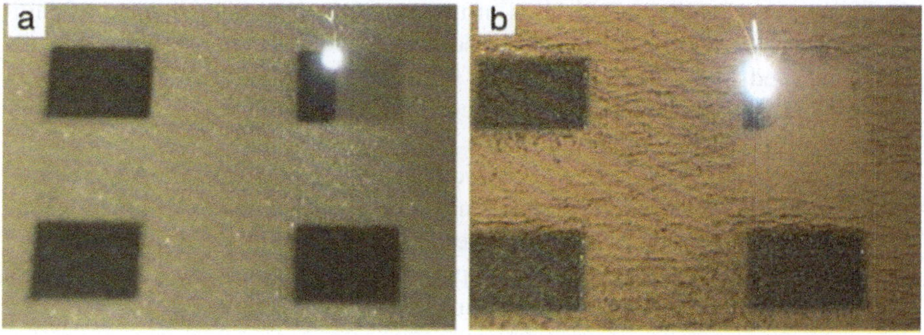


**Fig. 1** The visual appearance of parts being fabricated from (**a**) Al/5wt.%$Fe_2O_3$ and (**b**) Al/15wt.% $Fe_2O_3$ powder mixture using $P = 82$ W and $v = 0.33$ m/s. (Reprinted with permission from [36])

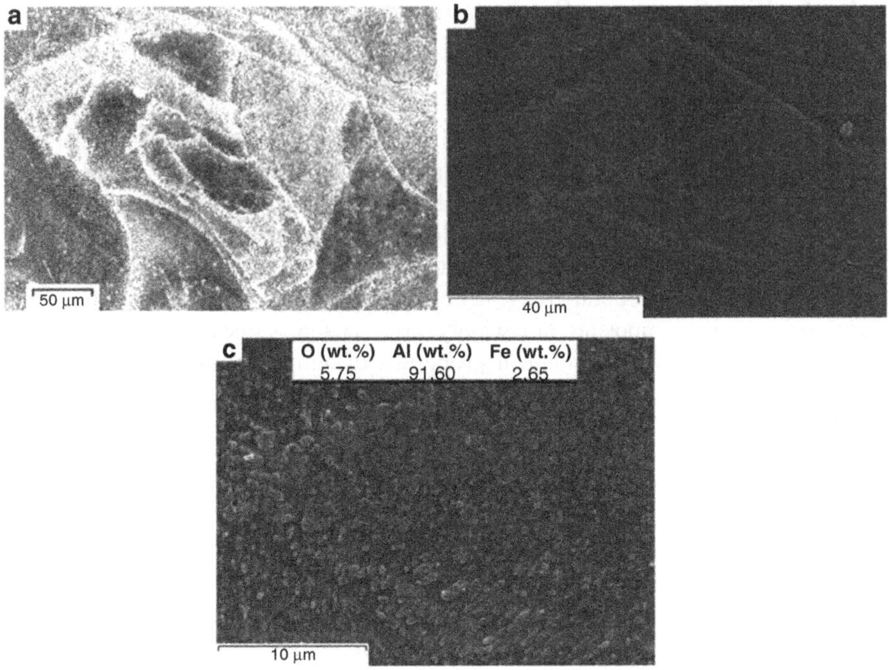


**Fig. 2** Typical microstructural views of the SLM part fabricated from Al/5 wt% $Fe_2O_3$ mixture; (**a**) Optical microscopy image showing random grain structure, (**b**) SEM overall view of grain structure and (**c**) previous image with higher magnification showing fine particles formed inside grains and boundaries (overall chemical composition was acquired by EDS). (Reprinted with permission from [31])

the mixture with higher $Fe_2O_3$ content [36]. Selective laser melting process is capable of producing net-shape aluminum matrix composite parts with uniform properties [31]. The content of $Fe_2O_3$ has significant influence on hardness, density and surface roughness since it can manipulate the amount of energy released in the reaction and subsequently the surface profile. The amount of $Fe_2O_3$ does not have significant effect on the densification of the composites but microhardness significantly increases with increases in $Fe_2O_3$ content. The optical microscopy image of a SLM processed Al-$Fe_2O_3$ composites are shown in Fig. 2a, exhibiting random grain structure [31]. Scanning electron microscope (SEM) images shows uniform distribution of fine particles within grains and grain boundaries (Fig. 2b, c). The microstructure shows uniform distribution as well as excellent interfacial bonding of very fine particles (50–2000 nm) within Al matrix which leads to improvement of the mechanical properties of these composites. X-ray diffraction pattern (Fig. 3) confirms the presence of $Al_2O_3$ in these composites which confirms the formation of in situ precipitates during the SLM processing of Al-$Fe_2O_3$ composites.

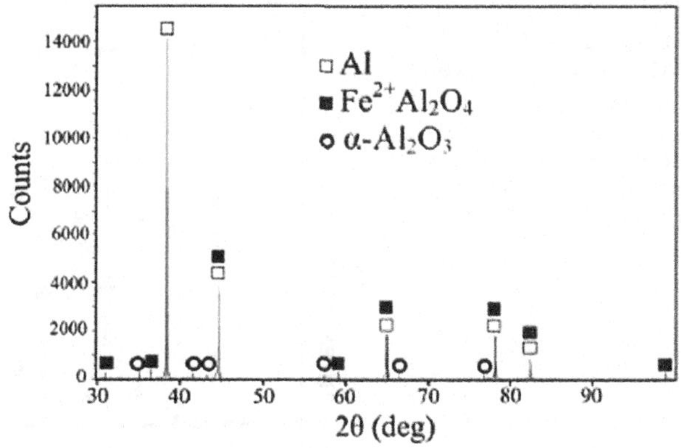


**Fig. 3** The XRD pattern of the SLM sample shown in Fig. 2. (Reprinted with permission from [31])

### 2.1.2 $AlSi_{10}Mg$-SiC

Selecting the proper reinforcement for metal matrix gives the opprtunity to design MMCs with better strength and stiffness. Low hardness and poor wear performance of alliuminum alloys hinder their use in high temperature surface engineering applications. One way to improve their hardness and wear propeerties is to reinforce them with hard precipitates such as SiC, $Al_2O_3$, and $TiB_2$, due to their good corrosion resistance and high hardness [38–41]. Alluminum matrix composites are widely utilized in aerospace and automative indstries, such as use of fiber reinforced alluminum metal composites by Toyota for their diesel engine pistone or using 25 vol%.SiC particulate as reinforcement in 6061 aluminum matrix composite for aerospace applications. In-situ $Al_4SiC_4$ + SiC composites have been fabricated via selective laser melting (SLM) of $AlSi_{10}Mg$ and SiC powder mixture. Major draw backs in SLM processing of aluminum matrix composites are the low wettability between the reinforcing particles and the Al matrix, low laser energy absorption by aluminum and it's propensity to react with oxygen at high temperatures [42–45]. Fe Cheng et al. have processed in situ $Al_4SiC_4$+ SiC reinforced Al matrix composites via SLM process in order to investigate the effect of starting SiC particle size on densification, microstructure, in-situ reaction, and tribological properties of these composites. Figure 4 shows the XRD patterns of SLM processed $Al_4SiC_4$+ SiC/ Al composites fabricatedusing different size SiC particles. These results clearly shows the effect of SiC particles significantly affects the in situ reaction between Al and SiC during SLM processing. The larger SiC particles leads to weaker $Al_4SiC_4$ peaks whereas stronger intensity observed for the composites processed using finer SiC particles. This observation clearly suggests that finer SiC particles enhance the in situ reaction during SLM processing which leads to increase in volume fraction of $Al_4SiC_4$. Reinforcemnet in Al matrix. Table 1 shows the variation of $2\theta$ position and the Al peak intensities in SLM processed $Al_4SiC_4$+ SiC/ Al composites using different SiC particle sizes. This clearly shows that intensity of Al peaks decreases with reduction in silicon carbide particles size, whereas FWHM increases with decrease in SiC particle size [46, 47]. Since aluminum has a relatively low

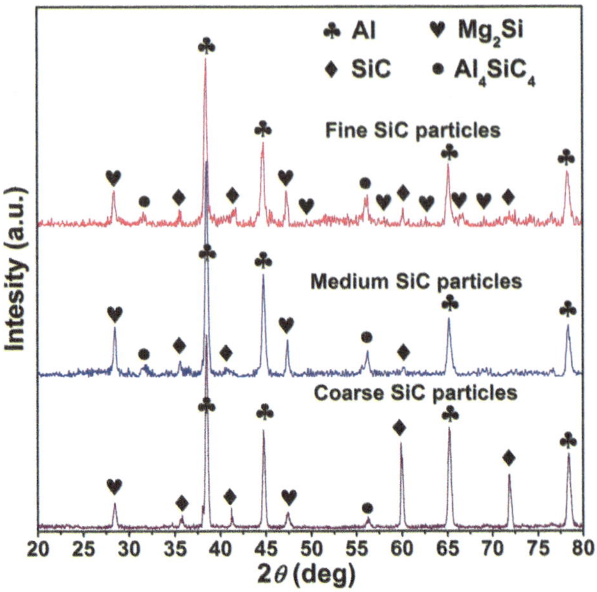


**Fig. 4** XRD spectra of SLM-processed ($Al_4SiC_4$ + SiC)/Al hybrid reinforced composites with the variation of particle sizes of the starting SiC powder. (Reprinted with permission from [46])

**Table 1** Variation of $2\theta$ locations and intensities of the Al diffraction peaks in SLM-processed (Al4SiC4 + SiC)/Al hybrid reinforced composites with different SiC particle sizes

| Sample | $2\theta$ (°) | Intensity | FWHM | $2\theta$ (°) | Intensity | $2\theta$ (°) | Intensity |
|---|---|---|---|---|---|---|---|
| Standard | 38.47 | | | 44.72 | | 65.10 | |
| $D_{50}$ = 50 μm | 38.52 | 628 | 0.302 | 44.74 | 322 | 65.28 | 317 |
| $D_{50}$ = 15 μm | 38.52 | 492 | 0.356 | 44.74 | 239 | 65.18 | 135 |
| $D_{50}$ = 5 μm | 38.50 | 231 | 0.387 | 44.82 | 116 | 65.16 | 86 |

Adapted from [46]

absorption rate for laser energy, addition of high energy absorptive- silicon carbide particles will help in overall increase in energy absorption rate that leads to form sufficient liquid and more stable molten pools [45, 48]. Furthermore, increase in molten pool's temperature due to increase in the energy absorption rate will result in a better liquid-solid wettability [46]. Size of SiC particles has an effect on density of aluminum matrix composites, since the specific surface area between melt of aluminum and silicon carbide changes with the SiC particle size. The specific surface area between the reinforcement and matrix is limited in the case of coarser SiC particles, consequently the wettability between SiC particles and aluminum matrix decreases leading towards lowering the density SLM processed Al composite. Fine SiC particles help in accelerating the in-situ reaction between matrix and reinforcements which increases their density as well as mechanical properties, improving the density of Al composites [5, 46]. Decreasing the size of SiC particle also enhances the Wear performance and micro hardness of the samples due to higher density and homogenous microstructure. When the SiC is too coarse, less amount of particles

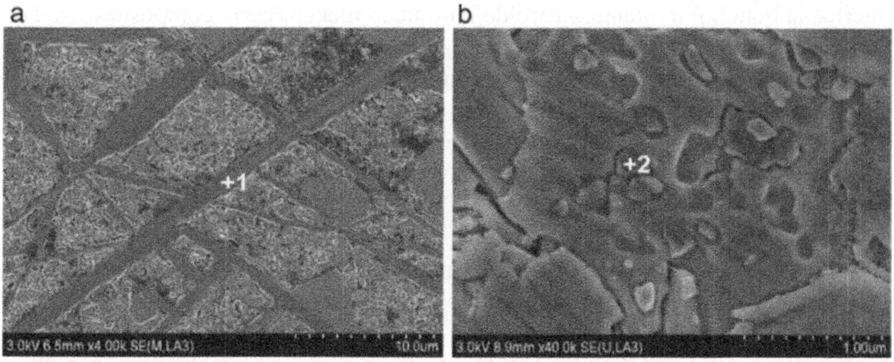


**Fig. 5** FE-SEM images showing the typical morphologies of in-situ reinforcements and the EDX results showing the chemical compositions in SLM-processed (Al4SiC4 + SiC)/Al hybrid reinforced composites using the fine SiC particles: (**a**) plate-like reinforcement; (**b**) particle-shaped reinforcement. (Reprinted with permission from [46])

melt during the SLM process, and the interface between SiC particle and Aluminum is insufficient, consequently the in situ reaction between Al and SiC is hindered. Furthermore, residual irregular-shaped SiC particles result in microstructural heterogeneity and a significant decrease in densification level [46]. When the SiC particles are finer, the average size of the remaining SiC particles is smaller and the composite is more homogenous. Two different microstructures of Al4SiC4 that are created from the in situ reaction between aluminum and SiC micron-sized plate-like structure and nearly nano-sized particle structure shown in Fig. 5 [46]. Process of forming Al4SiC4 phase during the in-situ reaction between Al and SiC, includes creation of irregular nuclei on the interface of Al and SiC. As the in-situ reaction progresses, these nuclei begin to grow and shape plate-like structures. Homogeneous dispersion of plate-like Al4SiC4 has a crucial role in enhancing the mechanical properties of the SLM processed parts [46].

### 2.1.3 Nickel Matrix Composites: Ni-Ti-C

Nickel and nickel base superalloys are widely used in aerospace applications (aircraft jet engines, and land base turbines) as well as in land-based applications, including the petrochemical and nuclear energy sectors, due to their excellent properties such as high resistance to corrosion as well as fatigue and low thermal expansion. Titanium carbide (TiC) has a very high hardness (2859–3200 HV), high melting point (3160 °C), low density (4.93 g/cm$^3$), and high mechanical strength, but it is very brittle and cannot be used as a monolithic ceramic [49–54]. Therefore, titanium carbide reinforced nickel matrix composites are considered as a good candidate for high temperature refractory, abrasive, and structural as well as surface engineering applications. Also, in contrast to most other metals, nickel exhibits a low wetting angle with titanium carbide, which leads to significant improvements in the interfacial bonding in titanium carbide reinforced nickel matrix composites. Ni-TiC composites exhibit a good balance of properties combining the ductility and toughness of the nickel matrix with the high strength and modulus of the TiC reinforcement making these composites promising candidates for high temperature structural applications. Furthermore, TiC reinforced nickel matrix composites can also be considered as potential replacements for WC-Co based wear-resistant materials for surface engineering applications. Ni-TiC-C composites exhibited excellent wear resistant properties due to the presence of the graphitic phase, which plays an important role as an in situ solid lubricant during friction. Using laser engineered net shaping (LENS™), novel monolithic composites based on Ni-Ti-C have been developed that combine properties such as solid lubrication (e.g. graphite), high hardness (e.g. TiC), and high fracture toughness (e.g. nickel), for structural as well as surface engineering applications. These multifunctional, monolithic composites are needed in industrial applications, such as drilling components (wear band, stabilizer, drill collar, etc.), tunnel boring, and land base turbines. Laser Engineered Net Shaping (LENS™) provides *processing flexibility*, e.g. in situ composites and functionally graded materials, that traditional laser cladding/hard facing techniques cannot provide. A new class of Ni-Ti-C based metal matrix composites has been developed using the laser engineered net shaping (LENS™) process. These composites consist of an in situ formed and homogeneously distributed titanium carbide (TiC) phase reinforcing the nickel matrix. Additionally, by tailoring the C/Ti ratio in these composites, an additional graphitic phase can also be engineered into the microstructure. The following four types of composites have been processed via LENS™ process in order to investigate the effect of excess graphite on microstructure, microhardness, and tribological properties of these composites: Ni-10Ti-5C, Ni-10Ti-10C, Ni-7Ti-20C, and Ni-3Ti-20C. 3D characterization of laser-deposited in situ Ni-Ti-C based metal matrix composites, reveals homogeneously distributed primary and eutectic titanium carbide precipitates as well as a graphitic phase encompassing the primary carbides, within a nickel matrix. The morphology and spatial distribution of these phases in three dimensions reveals that the eutectic carbides form a network linked by primary carbides or graphitic nodules at the nodes, suggesting interesting insights into the sequence of phase evolution. These three-phase Ni-TiC-C composites exhibit excellent tribological properties, in terms of an extremely low coefficient of friction while maintaining a relatively high hardness. Backscattered SEM images of Ni-Ti-C composites with varying C/Ti ratio is shown in Fig. 6. Ni-10Ti-5C (Fig. 6a) composite exhibits only fine needle like eutectic TiC precipitates whereas Ni-10Ti-10C (Fig. 6b) composite shows the presence of both fine needle like eutectic as well as cuboidal primary TiC precipitates. As carbon to titanium ratio increases both Ni-7Ti-20C (Fig. 6c) and Ni-3Ti-20C (Fig. 6d) composites show the presence of black graphite phase along with dark grey TiC precipitates. Ni-7Ti-20C exhibits larger primary TiC precipitates than that of Ni-3Ti-20C mainly because of higher atomic percentage of titanium. The 3D microstructure of Ni-10Ti-10C and Ni-3Ti-20C composites is shown in Fig. 7, where primary (cuboidal shaped) and eutectic (plate-shaped) TiC reinforcements in a

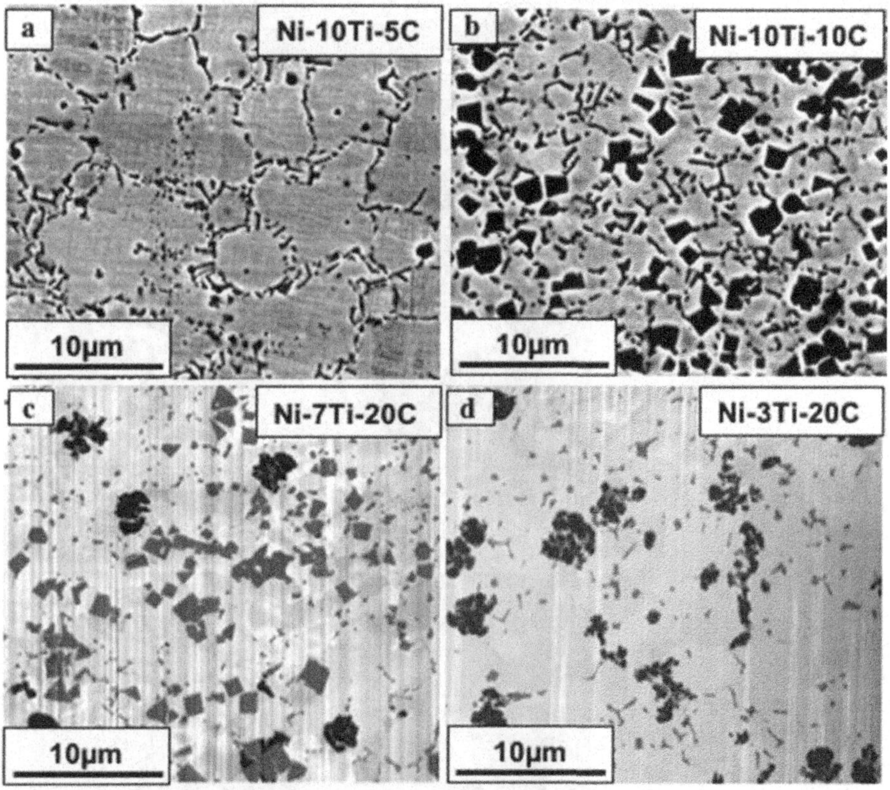


**Fig. 6** Backscatter SEM images of LENS deposited (**a**) Ni-10Ti-5C, (**b**) Ni-10Ti-10C, (**c**) Ni-7Ti-20C, and (**d**) Ni-3Ti-20C composites. (Modified from [4])

Ni matrix have been reconstructed. 3D microstructural characterization of Ni-Ti-C composites revealed:

(a) Different morphologies of primary and eutectic TiC precipitates: primary TiC precipitates exhibit a cuboidal morphology whereas eutectic TiC precipitates appear to exhibit a plate-like morphology.
(b) Primary Graphite phase engulfs TiC precipitates.
(c) Primary TiC precipitates act as heterogeneous nucleation sites for the eutectic precipitates during solidification leading to a carbide network formation.

3D characterization leads to novel understanding of the sequence of phase evolution during solidification for these complex metal matrix composites. The connectivity between the carbide precipitates is nearly impossible to visualize based on the 2D SEM images shown in Fig. 6. The distinction between the cuboidal primary TiC precipitates and the plate or needle-shaped eutectic TiC precipitates is more difficult in Ni-3Ti-20C composites because the primary precipitates are smaller in size as clearly shown in the 3D reconstruction. The Vickers microhardness values for the Ni-Ti-C composites and pure Ni have been listed in Table 2. Comparing these

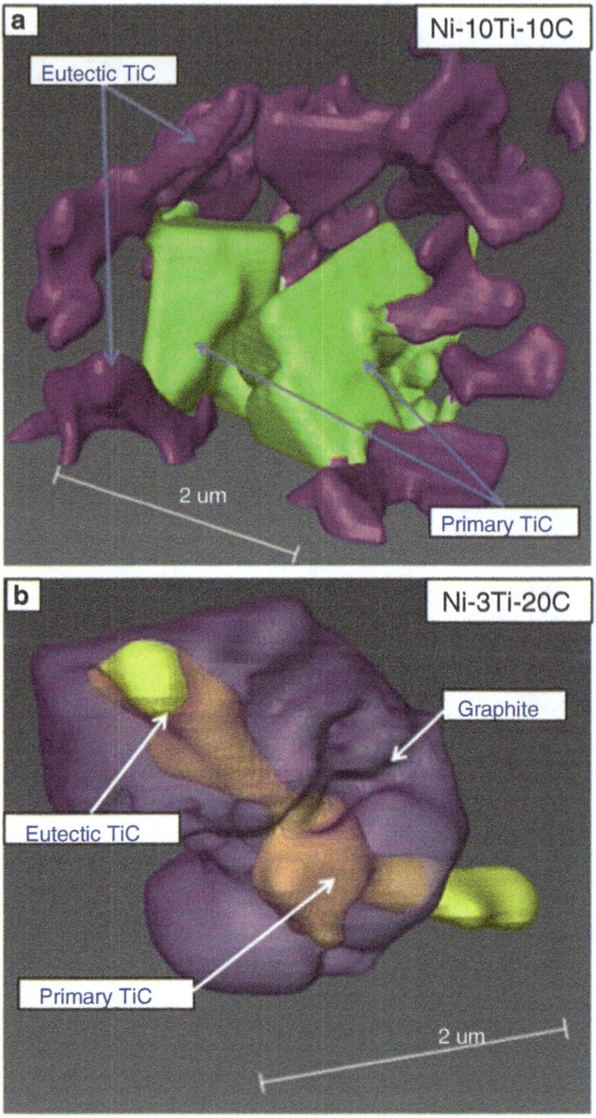


**Fig. 7** 3D reconstruction of LENS deposited (**a**) Ni-10Ti-5C and (**b**) Ni-3Ti-20 composites. (Modified from [4])

**Table 2** Microhardness of Ni-Ti-C composites

| Sample | Hardness (HV) |
|---|---|
| Pure Ni | 165 ± 6 |
| Ni-3Ti-20C | 240 ± 6 |
| Ni-7Ti-20C | 290 ± 7 |
| Ni-10Ti-10C | 370 ± 10 |
| Ni-10Ti-5C | 265 ± 2 |

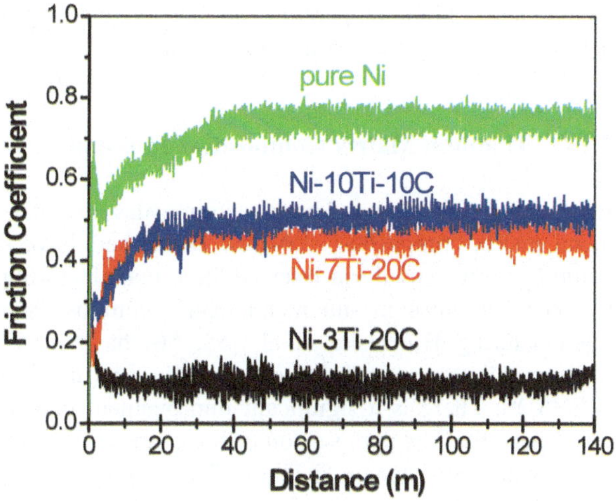


**Fig. 8** Steady state friction coefficient as a function of sliding distance up to 140 m for LENS deposited pure Nickel, Ni-10Ti-10C, Ni-7Ti-20C, and Ni-3Ti-20C composites. (Modified from [4])

microhardness values, the Ni-10Ti-10C composite exhibited a substantially higher hardness of 370 VHN as compared to 165 VHN for the LENS™ deposited pure Ni and all other Ni-Ti-C composites. These microhardness values clearly show a trend of decreasing hardness as a function of increasing C/Ti ratio in the composite. The Ni-10Ti-10C composites exhibit highest hardness mainly due to the presence of high volume fraction of titanium carbides (both primary and eutectic) as compared to pure nickel as well as other Ni-Ti-C composites. Figure 8 shows Friction coefficient versus distance plot LENS™ deposited Ni-10Ti-10C, Ni-3Ti-20C, Ni-7Ti-20C composites and pure nickel. It is clear that the graphite and TiC phases in the composite were beneficial towards reducing the friction coefficient with respect to the pure nickel sample. While the presence of TiC reduces the coefficient of friction, as observed in case of the Ni-10Ti-10C composite, the presence of the lubricious graphitic phase can play a more dominant role in reducing the friction for these composites. This is evident from the friction curves for the Ni-7Ti-20C and Ni-3Ti-20C composites. The friction coefficient in case of Ni-7Ti-20C is marginally lower as compared to Ni-10Ti-10C due to the presence of the graphitic phase in the former. However, the most promising composite appears to be the Ni-3Ti-20C composite, which exhibits a drastic reduction in friction coefficient (~0.2) when compared to any of the other composites, mainly due to the presence of a substantial fraction of the graphitic phase as well as TiC precipitates. All the Ni-Ti-C composites exhibit very high microhardness as well as excellent tribological properties when compared to pure nickel. The Ni-10Ti-10C composite exhibits highest microhardness due to the presence of primary as well as eutectic TiC precipitates, whereas the Ni-3Ti-20C composite exhibits excellent tribological properties due the presence of solid lubricious graphite phase along with TiC precipitates. Thus, these Ni-Ti-C composites, especially the Ni-3Ti-20C composite, appear to be promising materials for high temperature surface engineering applications requiring high hardness with improved solid lubrication.

### 2.1.4 Titanium Matrix Composites: Ti-B and Ti64-BN

Even though in situ metal matrix composites have several advantages over ex situ metal matrix composites, very few researchers have studied in situ metal matrix composites processed via laser additive manufacturing. This section will give the overview of those in situ metal matrix composites processed via laser additive manufacturing Himanshu et al. [55, 56] have performed in situ nitridation of titanium alloys using laser gas nitriding in the laser engineered net shaping (LENS™). There is a significant improvement in microhardness as well as wear resistance observed in those composite coatings due to the presence of TiN and $Ti_2N$ along with unreacted $\alpha$-Ti phase. These nitrided precipitates exhibited dendritic morphology and uniformly distributed within $\alpha$-Ti matrix. Das et al. [57, 58] have processed TiB + TiN reinforced composite coatings using Ti-6Al-4 V and hexagonal boron nitride (h-BN) powders via LENS™ process. TiB-TiN reinforced Ti64 coatings have been processed with varying BN wt% as well as laser power. The TiN phase exhibits coarse dendritic and needle-like structure whereas TiB phase exhibits fine rod-like structure (Fig. 9). During laser processing BN decomposes and react with molten titanium which give rise to TiB and TiN precipitates. These composite coatings have shown significant improvement in fracture toughness as well as tribological properties as compared to Ti-6Al-4 V alloy. Hooyar et al. [59] have processed Ti-TiB composites via selective laser melting process using milled $Ti$-$TiB_2$ powders. SEM images of both cross-sectional and longitudinal sections of SLM processed Ti-TiB composites are shown in Fig. 10. The TiB precipitates exhibit needle-shape morphology and are uniformly distributed within titanium matrix. The Ti-TiB composites have shown significant improvement in microhardness (402 HV) and compressive yield strength (1103 MPa) as compared to SLM-processed CP-Ti (261 HV and 560 MPa). These superior mechanical properties have mainly attributed due to the strengthening and hardening effects of the TiB particles and grain refinement of the $\alpha$-Ti matrix. Vamsi et al. [60] have deposited $ZrO_2$ coatings on titanium substrate using Zr powder via LENS process in Ar + $O_2$ inert atmosphere. Laser oxidized Zr coatings have shown enhancement in osteoblast cell adhesion as well as in the wear resistance due to the presence of monoclinic as well as tetragonal $ZrO_2$. Ti alloy-TiB composites combine the high strength and stiffness of the borides with the toughness and damage tolerance of the Ti-alloy matrix. These composites have been extensively researched since they offer attractive properties, such as: high stiffness, enhanced elevated temperature strength, good creep performance, fatigue resistance and wear resistance. Since the boride reinforcement was formed as a consequence of a chemical reaction in such composites, a homogeneous dispersion consisting of refined scale borides results. Furthermore, the boride phase that forms in these in situ composites is in thermodynamic equilibrium with the matrix. Unlike reinforcement phases added from external

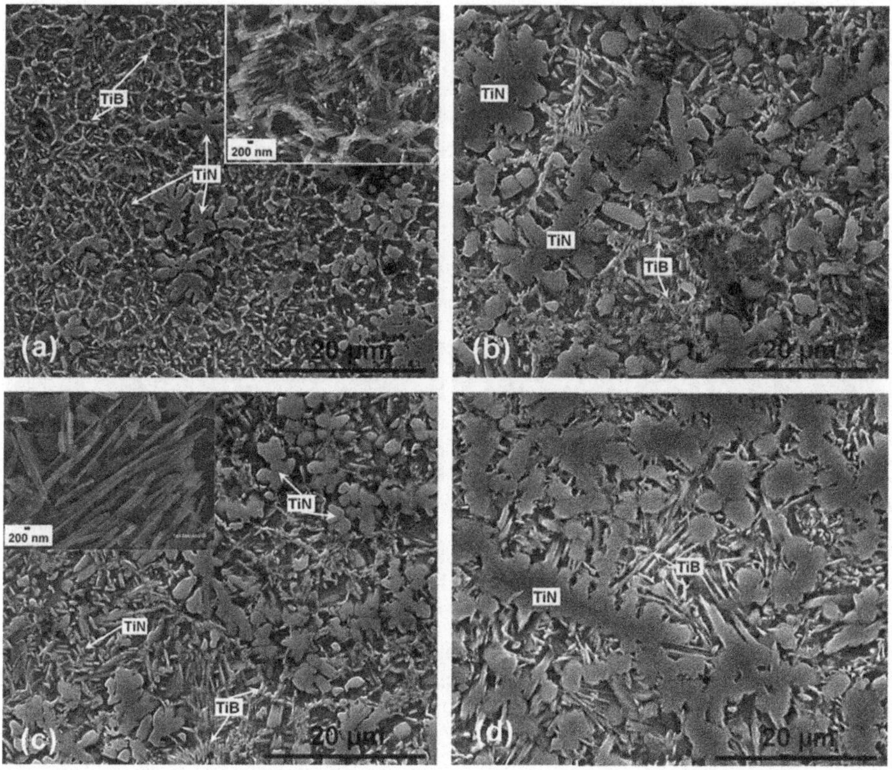


**Fig. 9** Typical high-magnification SEM microstructures showing Ti–BN reactions products in laser-processed, BN-reinforced Ti6Al4V alloy composite coatings: (**a**) 5BN-300/20 (38 J $mm^{-2}$); (**b**) 15BN-300/20 (38 J $mm^{-2}$), (**c**) 5BN-400/10 (102 J $mm^{-2}$), (**d**) 15BN-400/10 (102 J $mm^{-2}$). Inset shows scale of TiB nanorods in respective coatings. (Reprinted with permission from [57])

sources, in situ composites consist of contaminant-free boride-matrix interfaces, which are significantly stronger. Banerjee et al. have processed Ti-TiB composites via the LENS™ process from a blend of pure elemental Ti and B powders while Ti alloy-TiB composites were deposited from a blend of pre-alloyed Ti-6Al-4 V and elemental boron [61–64]. The homogeneous dispersion of the fine TiB precipitates achieved in laser deposited Ti alloy-TiB composites deposited using LENS process was expected to significantly strengthen the $\alpha$-Ti or $\alpha/\beta$ Ti-6Al-4 V matrix and consequently holds promise from the viewpoint of enhanced high temperature mechanical properties such as creep as well as wear resistance of these composites. Figure 11 shows the microstructure of LENS™ deposited Ti-6Al-4 V-TiB composites. The needle-shaped TiB precipitates are uniformly distributed with titanium matrix. In addition, since these were in situ composites, the reinforcing TiB precipitates were thermodynamically stable at high temperatures, chemically compatible with the matrix, as well as bonded strongly to the matrix due to the clean interface between the matrix and the reinforcement. Sonia et al. have processed *in situ* Ti-Nb-

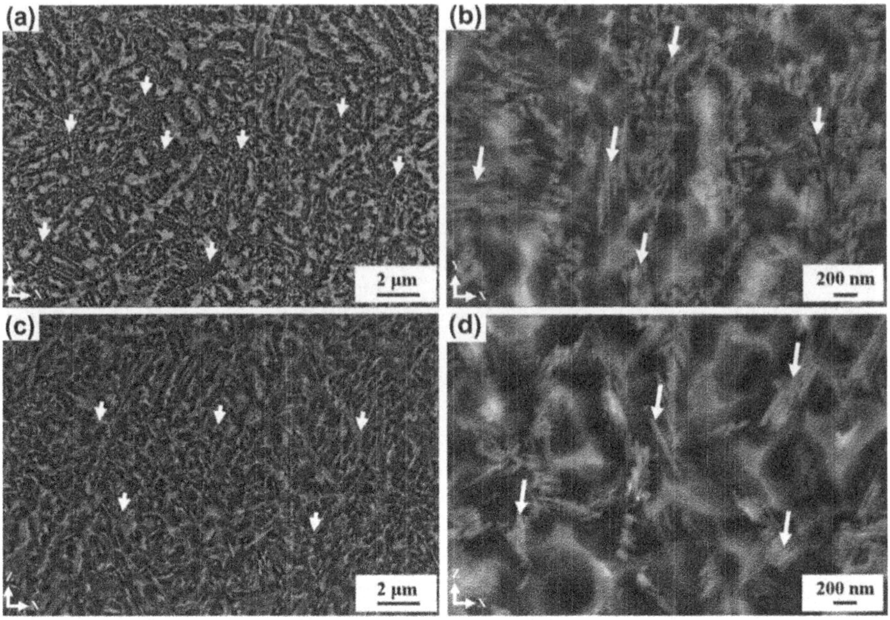


**Fig. 10** SEM images for the microstructures of the SLM-produced Ti–TiB composite at different magnifications: (**a**, **b**) cross-sectional views and (**c**, **d**) longitudinal views showing needle-shape TiB particles within the Ti matrix. White arrows indicate TiB particles. (Reprinted with permission from [59])

Zr-Ta-TiB composites via LENS™ technique from a blend of pure elemental titanium (Ti), niobium (Nb), zirconium (Zr), and tantalum (Ta) powders mixed with titanium diboride ($TiB_2$) powders [47, 48]. The microstructure of the LENS™--deposited TNZT+2B alloy at different magnifications is shown in the backscatter SEM images in Fig. 12. Two types of TiB precipitates have been observed within titanium matrix. The coarser precipitates have hexagonal-shaped geometry whereas eutectic TiB precipitates exhibit fine needle-shape morphology. TNZT-TiB composites have shown significant improvement in wear resistance as compared to TNZT alloys.

## 2.2 *In Situ Reaction Between Elemental Blend Powders and Reactive Gases*

### 2.2.1 Ti-Mo-N

Titanium alloys are attractive candidates for structural, marine, aerospace, biomedical (such as in dental and orthopedic as bone implants) and other industrial applications due to their excellent strength to weight ratio, ductility and formability,

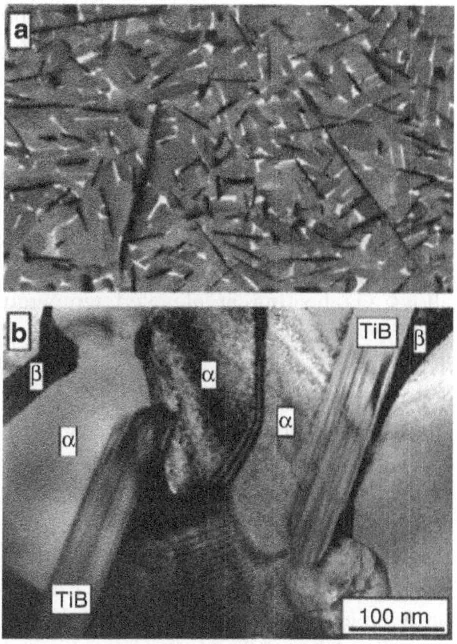


**Fig. 11** (**a**) Backscatter SEM micrograph showing the overall microstructure of the LENS™ deposited Ti-6Al-4 V–TiB composite with TiB precipitates distributed within an $\alpha + \beta$ matrix. (**b**) Bright-field TEM micrograph showing two TiB precipitates in the Ti-6Al-4 V matrix. (Reprinted with permission from [64])

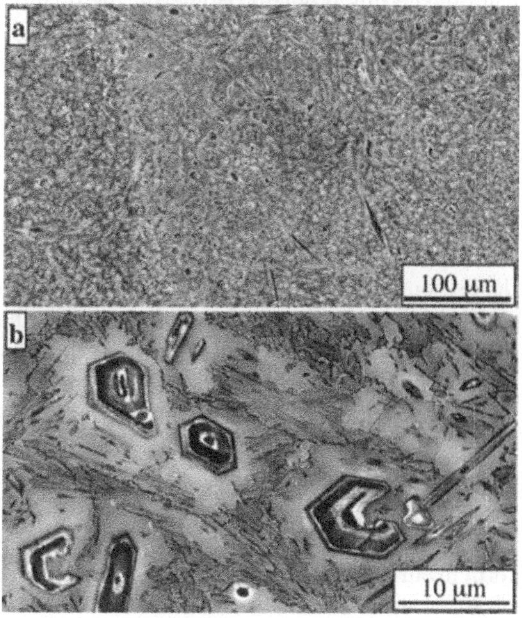


**Fig. 12** (**a**) Lower magnification and (**b**) higher magnification backscatter SEM images of LENS™ as-deposited TNZT+2B alloy composites showing both coarser primary borides exhibiting contrast within the same boride precipitate as well as finer scale eutectic borides. (Reprinted with permission from [62])

corrosion resistance and biocompatible properties. However, titanium alloys suffer from rather poor surface hardness and wear resistance properties. One way of improving the hardness as well as tribological properties of titanium alloys is by reinforcing the soft matrix with hard precipitates, such as titanium nitrides, carbides, and borides. These reinforcements can be introduced either via direct incorporation of such hard compounds in the matrix during processing, or via in situ reaction with solid or gaseous precursors, resulting the formation of hard precipitates as well as via surface engineering techniques such as nitriding. Focusing on nitride reinforced titanium alloys, the most commonly used technique is surface nitridation of these alloys via heating at elevated temperatures in a flowing nitrogen atmosphere. Another approach that has been employed is the direct introduction of $\delta$-TiN and TiB particles during laser deposition of Ti-6Al-4 V. The forming processing approach, nitridation in a gaseous atmosphere, has been applied to the case of $\beta$ Ti-Mo alloys, resulting in the formation of a continuous surface $\delta$-TiN layer and a sub-surface microstructure consisting of laths (or plates) of either the same $\delta$-TiN phase or a nitrogen rich $\alpha$(Ti, N) solid solution phase, dispersed within the $\beta$ matrix. While such surface nitridation via heating in a nitrogen atmosphere is a simple and inexpensive way to achieve a case-hardened layer, the time required and the depth of penetration are rather limited and it is difficult to introduce hard nitrides (or other nitrogen enriched hard phases) within the bulk of the material. The direct introduction of nitride and boride particles during laser deposition of titanium alloy powders can obviate this problem. However, the quality of such an interface between such externally introduced reinforcements and the alloy matrix can be rather difficult to control and therefore ideally a reinforcement created as a product of an *in-situ* reaction offers the advantage of a thermodynamically stable and clean interface with the matrix. Furthermore, by employing an *in-situ* nitride (or nitrogen enriched hard $\alpha$ phase) formation reaction, a more uniform and homogeneous distribution of the nitride phase can be potentially achieved throughout the matrix. *In situ* nitridation during laser deposition of titanium-molybdenum alloys from elemental powder blends has been achieved by introducing the reactive nitrogen gas during the deposition process. Ti-Mo-N alloys have been deposited using the laser engineered net shaping (LENS™) process and resulted in the formation of a hard $\alpha$(Ti,N) phase, exhibiting a dendritic morphology, distributed within a $\beta$(Ti-Mo) matrix with fine scale transformed $\alpha$ precipitates. Varying the composition of the Ar + $N_2$ gas employed during laser deposition permits a systematic increase in the nitrogen content of the as-deposited Ti-Mo-N alloy. Interestingly, the addition of nitrogen, which stabilizes the $\alpha$ phase in Ti, changes the solidification pathway and the consequent sequence of phase evolution in these alloys. The nitrogen enriched hcp $\alpha$(Ti,N) phase has higher c/a ratio, exhibits an equiaxed morphology, and tends to form in clusters separated by ribs of the molybdenum (Mo) rich $\beta$ phase. The Ti-Mo-N alloys also exhibit a substantial enhancement in microhardness due to the formation of these $\alpha$(Ti,N) phase, combining it with the desirable properties of the $\beta$-Ti matrix, such as excellent ductility, toughness, and formability. All Ti-Mo-N alloys were deposited inside a glove box of LENS™ machine with controlled gas atmosphere. Pure Ar (Alloy 1), 25% $N_2$–75% Ar (Alloy 2), 50% $N_2$–50%Ar (Alloy 3) and

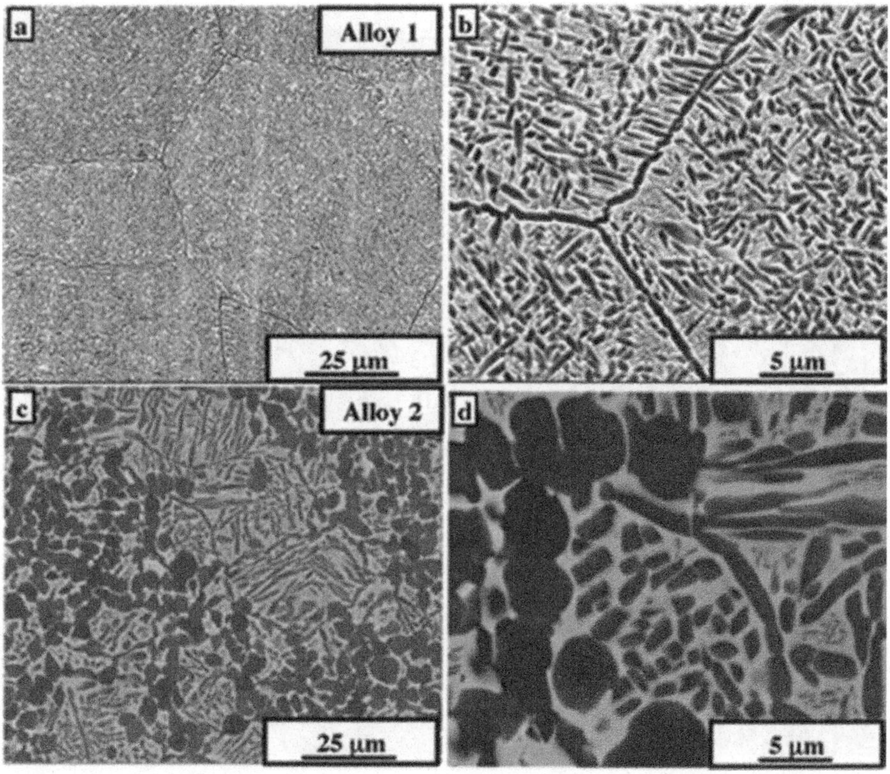


**Fig. 13** (**a**) Low and (**b**) high magnification backscatter SEM image of LENS deposited Alloy 1. (**c**) Low and (**d**) high magnification backscatter SEM image of LENS deposited Alloy 2. (Reprinted with permission from [5])

75% $N_2$–25% Ar (Alloy 4) were the atmospheres used for fabrication of deposits. A series of backscattered SEM images recorded from the Ti-Mo-N alloys are shown in Figs. 13 and 14. Figure 13a, b show the microstructure for the binary Ti-Mo alloy deposited in 100% argon (Ar) atmosphere. The nominal composition of this alloy measured using energy dispersive spectroscopy (EDS) in the SEM is Ti-10wt%Mo. The microstructure primarily consists of uniformly distributed fine scale precipitates of a second phase, presumably $\alpha$ precipitates, exhibiting a bimodal size distribution within the $\beta$-matrix. This bimodal size distribution of $\alpha$ precipitates can be rationalized on the basis of a two-step decomposition of the $\beta$-matrix, wherein the coarser $\alpha$ precipitates presumably result from the deposition of the layer during LENS™ processing, while solid-state re-heating of the same layer, when the subsequent layer is being deposited on the top, results in the secondary decomposition of $\beta$ forming the finer scale $\alpha$. The inherent rapid cooling rates involved in laser deposition, also contribute to such non-equilibrium $\beta$ decomposition processes. Thus, while during laser deposition of a layer, the phase evolution follows the sequence liquid to $\beta$ to $\beta + \alpha$, the inherently rapid cooling rates involved do not result in an

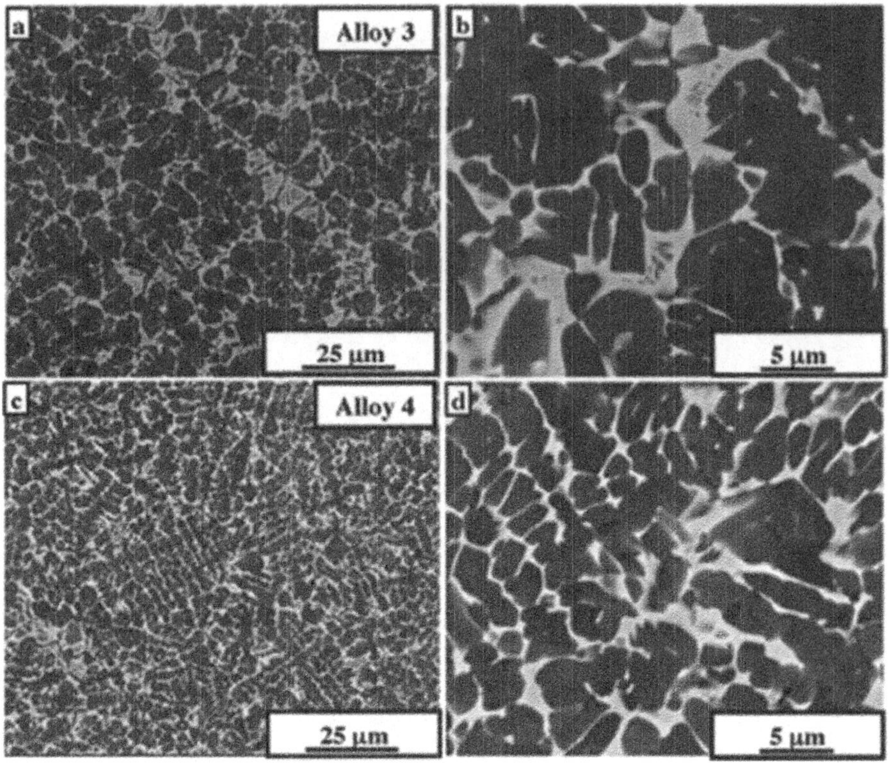


**Fig. 14** (**a**) Low and (**b**) high magnification backscatter SEM image of LENS deposited Alloy 3. (**c**) Low and (**d**) high magnification backscatter SEM image of LENS deposited Alloy 4. (Reprinted with permission from [5])

equilibrium composition of the $\beta$ matrix. Therefore, on subsequent re-heating of the same layer when a layer on top is being deposited, there is a further decomposition of the retained $\beta$ matrix to form fine scale secondary $\alpha$ precipitates. The aspect ratio of this finer scale $\alpha$ is substantially larger than those of the coarser $\alpha$ precipitates. Figure 13c, d show the microstructure of the Ti-Mo-N alloy deposited using a 75% Ar-25% $N_2$ atmosphere in the LENS™ glovebox as well as the center gas purge and the powder carrier gas. The microstructure clearly exhibits two precipitate phases having rather different morphologies. The coarser precipitates appear to exhibit an equiaxed or globular morphology with curved interfaces separating them from the $\beta$ matrix, suggesting that these interfaces are presumably incoherent in nature. In contrast, the finer scale precipitates exhibit a sharper faceted morphology suggesting that the precipitate/matrix interface is likely to be semi-coherent. The identity of these two types of precipitates cannot be determined based solely on the backscatter SEM evidence presented in Fig. 13c, d. However, the morphology and contrast of the finer scale precipitates suggests that the $\alpha$ precipitates exhibited smaller aspect ratios when compared with those observed in case of the binary Ti-10Mo alloy deposited under a pure Ar atmosphere (Fig. 13a, b). As the $N_2$ content in the Ar + $N_2$ mixture used in LENS™ deposition increases, the size and volume fraction of the coarser equiaxed second phase precipitates increases significantly in the microstructure, as revealed in Fig. 14. Backscattered scanning electron microscope images of Ti-10Mo alloys LENS™ deposited under a 50% Ar-50%$N_2$ mixture are shown in Fig. 14a, b while Fig. 14c, d correspond to an alloy deposited under a 25% Ar-75% $N_2$ mixture. There is also a corresponding decrease in the $\beta$ volume fraction with increasing $N_2$ content in the gaseous atmosphere used for deposition. Thus, the primary microstructural influence of introducing nitrogen during laser deposition appears to be the formation of a novel equiaxed precipitates within the $\beta$ matrix together with finer scale $\alpha$ precipitates, with the volume fraction of the equiaxed precipitates increasing with increasing nitrogen content in the alloy.

The Vickers microhardness values for all the four alloys, 1–4, have been plotted in Fig. 15. A systematic increase in the microhardness values, with increasing nitrogen content in the alloys can be clearly observed. The binary Ti-Mo alloy, LENS™ deposited under a pure Ar atmosphere, exhibits an average hardness of ~500 HV. The increase in microhardness of the Ti-Mo-N alloys, with increasing $N_2$ content in the reactive atmosphere during LENS™ deposition, can be attributed to the formation of the $\alpha$-$TiN_{0.3}$ random solid solution hexagonal close packed (hcp) phase. For the highest nitrogen containing alloy 4, the microhardness value is ~800 HV, which is ~60% higher as compared with the LENS™ deposited binary Ti-10wt%Mo (alloy 1).

In situ nitridation of Ti-10wt%Mo alloys has been achieved by the introduction of reactive nitrogen gas during the laser deposition (LENS™) of these alloys from elemental powder blends. The nitrogen content in these laser deposited alloys has been tuned via changing the ratio of argon to nitrogen used in LENS™ deposition. The enrichment of these alloys with nitrogen results in the formation of primary precipitates of the $\alpha$(Ti,N) phase within the $\beta$ matrix. The higher c/a ratio of these hcp $\alpha$(Ti,N) precipitates, coupled with the Mo enrichment in the $\beta$ matrix, results in a

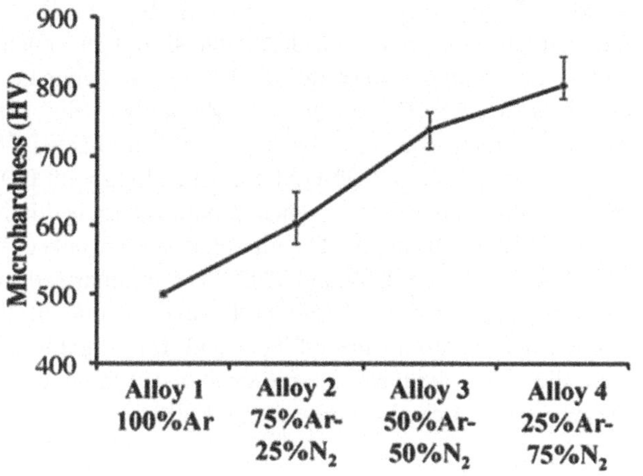


**Fig. 15** A plot showing the variation in Vickers microhardness values for LENS deposited Ti-Mo-N alloys (Alloy 1 to 4). (Reprinted with permission from [5])

substantially large misfit, consequently leading to a loss of precipitate/matrix coherency during growth and coarsening. This results in the $\alpha$(Ti,N) precipitates adopting an equiaxed morphology, and they tend to aggregate into clusters separated by thin Mo-rich $\beta$ ribs. These $\alpha$(Ti,N) precipitates increase the microhardness of the alloy to a substantial degree. Additional fine scale secondary $\alpha$ precipitates, exhibiting a lath or plate-like morphology are also formed within the retained $\beta$ matrix of these Ti-Mo-N alloys. The ability to introduce controlled volume fractions of the hard nitrogen enriched $\alpha$(Ti,N) phase can be very useful in tailoring the local microhardness and consequently wear resistance of these alloys. Furthermore, using the LENS™ process it is possible to grade the nitrogen content within the same alloy and thus process a compositionally-graded microstructure with systematically varying properties.

### 2.2.2 Ti64/TNZT-N

Among different metallic alloys, titanium alloys have wide application in industries from orthopedic implants in biomedical to turbine gas engines in aerospace because of their high strength to weight ratio, formability and corrosion resistance [5]. However, they have poor wear resistance. In order to overcome this, titanium alloys need surface treatment such as surface nitridation to improve their wear performance [58]. During surface nitriding, the chemically inert layer is produced in a temperature between (700–1100 °C) in a nitrogen enriched atmosphere [65–74]. Toughness and damage resistance of titanium alloys significantly increases when reinforcement with carbides, nitrides or borides and become potential candidates for industrial utilization [5, 75]. In situ gas reaction in comparison to externally introduced reinforcement has the advantage of mechanically and thermodynamically stability a well as more uniform distribution of reinforcement particles through the alloy matrix [62, 74]. Hamid et al. have processed in situ nitridation of titanium alloys (TNZT and Ti64 alloys) via LENS™ process. The backscattered SEM images of Ti64 and nitrided Ti64 alloys are shown in Fig. 16. The Ti64 alloys exhibit typical $\alpha$-$\beta$ microstructure whereas nitrided Ti64 shows presence of 2 types of equiaxed-$\alpha$ laths with smaller aspect ratio. The typical dendritic microstructures have been observed in LENS™ deposited TNZT alloys (Fig. 17a) whereas darker dendritic TiN/$TiN_2$ precipitates have been observed in nitrided TNZT composites (Fig. 17b). The cross-sectional STEM-HADF image along with EDS elemental map of nitrided TNZT is shown in Fig. 17c, where $\beta$ matrix enriched in Nb and Ta and $\alpha$ precipitates are deficient in Nb and Ta. The microhardness results of the LENS™ deposited Ti64, TNZT, and nitrided Ti64 and TNZT are summarized in Fig. 18. Ti64 has a higher hardness compared to Ti-Nb-Zr-Ta mainly due to the presence of mixed $\alpha$-$\beta$ microstructure. Both nitrided Ti64 and TNZT exhibits higher microhardness than base alloys mainly due to the presence of TiN/$Ti_2N$ precipitates in Ti-Nb-Zr-Ta, and the nitrogen-enriched $\alpha$ phase in Ti64 alloys.

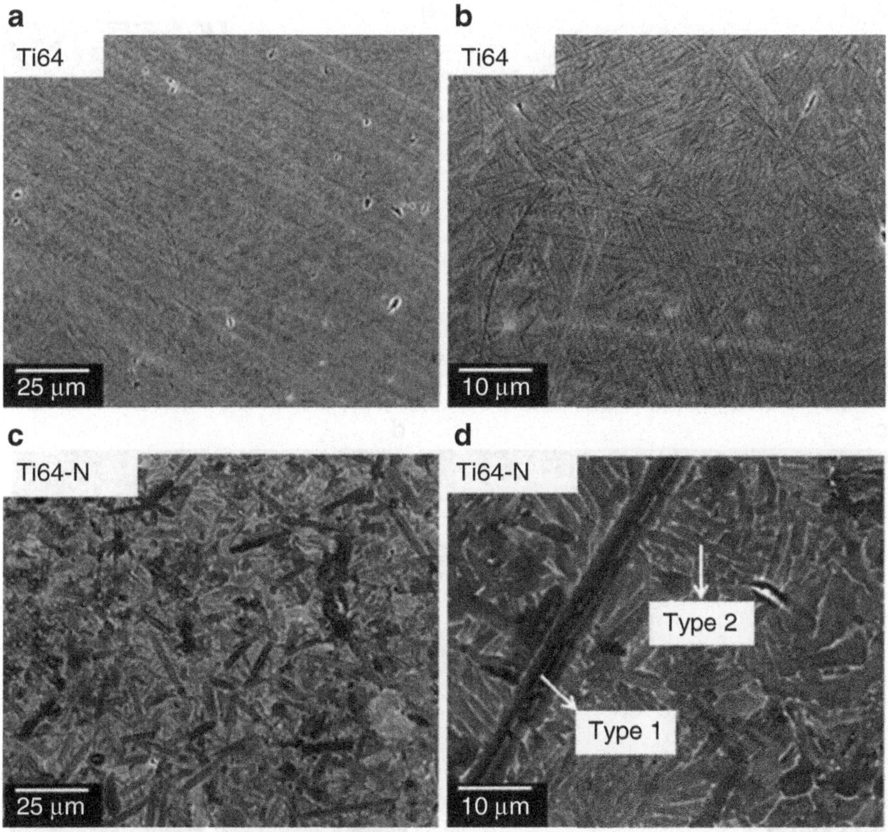


**Fig. 16** (**a**) Low- and (**b**) high-magnification backscatter SEM images of LENS deposited Ti64 alloy. (**c**) Low- and (**d**) high-magnification backscatter SEM images of LENS deposited nitrided Ti64 (Ti64-N) alloy. (Reprinted with permission from [74])

# 3 Summary

Laser additive manufacturing of in situ metal matrix composites offers various advantages as compared to the conventional processing and also has capability to produce near-net shape components. This chapter provided the overview of laser additive manufacturing of in situ metal matrix. Broadly, two types of in situ reactions have been discussed: in situ reaction between elemental blend powder and in situ reaction between elemental blend powder and reactive gases. In situ Aluminum matrix composites processed via selective laser melting (SLM) process have shown significant improvement in hardness, tribological and mechanical properties as compared to their base alloy. Laser additively manufactured in situ nickel-titanium-graphite composites have shown significant improvement in microhardness as well as tribological properties as compared to pure nickel and is potential candidate for

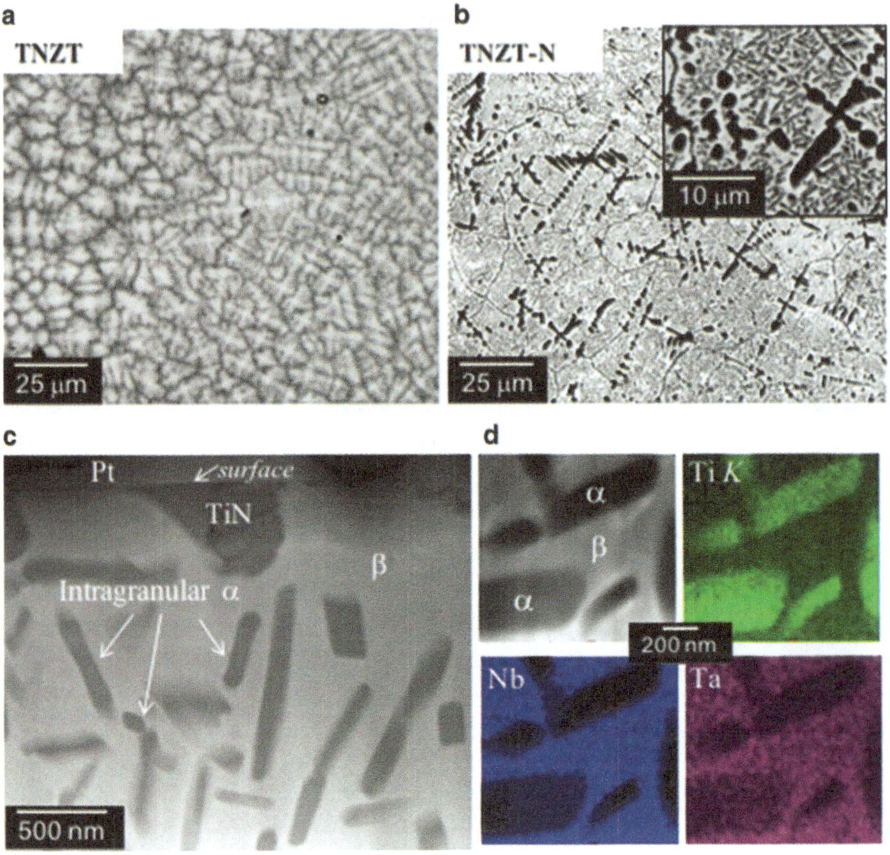


**Fig. 17** Low-magnification backscatter SEM images of LENS deposited (**a**) TNZT and (**b**) nitrided TNZT (TNZT-N) alloys. (**c**) Cross-sectional HAADF-STEM image of nitrided TNZT alloy with corresponding titanium, niobium and tantalum EDS maps. (Reprinted with permission from [74])

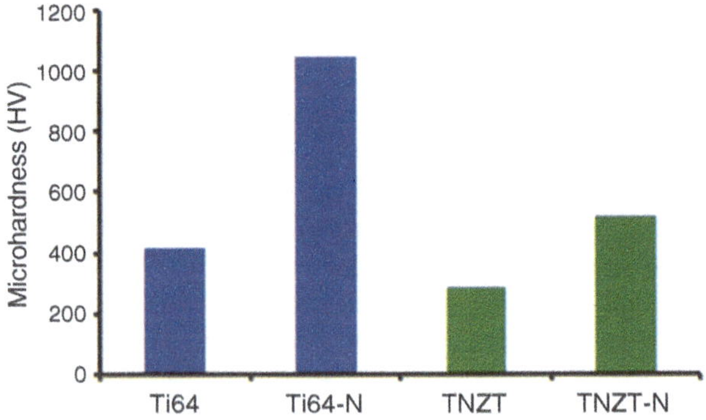


**Fig. 18** Plot showing the variation in Vickers microhardness values for LENS deposited Ti64, Ti64-N, TNZT, and TNZT-N alloys

high temperature surface engineering applications. In situ titanium alloys composites have shown improvement in wear properties and ideal for biomedical applications.

## References

1. I. Gibson, D. W. Rosen, and B. Stucker, *Additive manufacturing technologies: Rapid prototyping to direct digital manufacturing*. 2010.
2. N. Guo and M. C. Leu, "Additive manufacturing: Technology, applications and research needs," *Frontiers of Mechanical Engineering*, vol. 8, no. 3. pp. 215–243, 2013.
3. T. Borkar *et al.*, "A combinatorial assessment of AlxCrCuFeNi2 (0," *Acta Mater.*, vol. 116, pp. 63–76, 2016.
4. T. Borkar *et al.*, "Laser-deposited in situ TiC-reinforced nickel matrix composites: 3D microstructure and tribological properties," *JOM*, vol. 66, no. 6, pp. 935–942, 2014.
5. T. Borkar, S. Gopagoni, S. Nag, J. Y. Hwang, P. C. Collins, and R. Banerjee, "In situ nitridation of titanium-molybdenum alloys during laser deposition," *J. Mater. Sci.*, vol. 47, no. 20, pp. 7157–7166, 2012.
6. T. Borkar, R. Conteri, X. Chen, R. V. Ramanujan, and R. Banerjee, "Laser additive processing of functionally-graded Fe–Si–B–Cu–Nb soft magnetic materials," *Mater. Manuf. Process.*, vol. 32, no. 14, pp. 1581–1587, 2017.
7. C. V. Mikler *et al.*, "Laser Additive Manufacturing of Magnetic Materials," *JOM*, vol. 69, no. 3, pp. 532–543, 2017.
8. W. J. Sames, F. A. List, S. Pannala, R. R. Dehoff, and S. S. Babu, "The metallurgy and processing science of metal additive manufacturing," *International Materials Reviews*, vol. 61, no. 5. pp. 315–360, 2016.
9. W. E. Frazier, "Metal additive manufacturing: A review," *Journal of Materials Engineering and Performance*, vol. 23, no. 6. pp. 1917–1928, 2014.
10. J. C. Zhao, "Combinatorial approaches as effective tools in the study of phase diagrams and composition-structure-property relationships," *Prog. Mater. Sci.*, vol. 51, no. 5, pp. 557–631, 2006.
11. S. Curtarolo, G. L. W. Hart, M. B. Nardelli, N. Mingo, S. Sanvito, and O. Levy, "The high-throughput highway to computational materials design," *Nat. Mater.*, vol. 12, no. 3, pp. 191–201, 2013.
12. D. D. Gu, W. Meiners, K. Wissenbach, and R. Poprawe, "Laser additive manufacturing of metallic components: materials, processes and mechanisms," *Int. Mater. Rev.*, vol. 57, no. 3, pp. 133–164, 2012.
13. P. A. Kobryn and S. L. Semiatin, "The laser additive manufacture of Ti-6Al-4V," *JOM*, vol. 53, no. 9, pp. 40–42, 2001.
14. Y. J. Liang, D. Liu, and H. M. Wang, "Microstructure and mechanical behavior of commercial purity Ti/Ti-6Al-2Zr-1Mo-1V structurally graded material fabricated by laser additive manufacturing," *Scr. Mater.*, vol. 74, pp. 80–83, 2014.
15. D. C. Hofmann *et al.*, "Compositionally graded metals: A new frontier of additive manufacturing," *J. Mater. Res.*, vol. 29, no. 17, pp. 1899–1910, 2014.
16. D. C. Hofmann *et al.*, "Developing gradient metal alloys through radial deposition additive manufacturing," *Sci. Rep.*, vol. 4, 2014.
17. T. Wohlers and T. Caffrey, *Wohlers Report 2015: 3D Printing and Additive Manufacturing State of the Industry Annual Worldwide Progress Report*. 2014.
18. E. C. Santos, M. Shiomi, K. Osakada, and T. Laoui, "Rapid manufacturing of metal components by laser forming," *Int. J. Mach. Tools Manuf.*, vol. 46, no. 12–13, pp. 1459–1468, 2006.
19. P. Heinl, A. Rottmair, C. Körner, and R. F. Singer, "Cellular titanium by selective electron beam melting," *Adv. Eng. Mater.*, vol. 9, no. 5, pp. 360–364, 2007.

20. J.-P. Kruth, P. Mercelis, J. Vaerenbergh, L. Froyen, and M. Rombouts, "Binding mechanisms in selective laser sintering and selective laser melting," *Rapid Prototyp. J.*, vol. 11, no. 1, pp. 26–36, 2005.
21. I. Zein, D. W. Hutmacher, K. C. Tan, and S. H. Teoh, "Fused deposition modeling of novel scaffold architectures for tissue engineering applications," *Biomaterials*, vol. 23, no. 4, pp. 1169–1185, 2002.
22. F. Abe, K. Osakada, M. Shiomi, K. Uematsu, and M. Matsumoto, "The manufacturing of hard tools from metallic powders by selective laser melting," *J. Mater. Process. Technol.*, vol. 111, no. 1–3, pp. 210–213, 2001.
23. B. Vandenbroucke and J. Kruth, "Selective laser melting of biocompatible metals for rapid manufacturing of medical parts," *Rapid Prototyp. J.*, vol. 13, no. 4, pp. 196–203, 2007.
24. L. Thijs, F. Verhaeghe, T. Craeghs, J. Van Humbeeck, and J. P. Kruth, "A study of the microstructural evolution during selective laser melting of Ti-6Al-4V," *Acta Mater.*, vol. 58, no. 9, pp. 3303–3312, 2010.
25. N. Shamsaei, A. Yadollahi, L. Bian, and S. M. Thompson, "An overview of Direct Laser Deposition for additive manufacturing; Part II: Mechanical behavior, process parameter optimization and control," *Addit. Manuf.*, vol. 8, pp. 12–35, 2015.
26. R. Morgan, C. J. Sutcliffe, and W. O'Neill, "Density analysis of direct metal laser re-melted 316L stainless steel cubic primitives," *J. Mater. Sci.*, vol. 39, no. 4, pp. 1195–1205, 2004.
27. J. P. Kruth, G. Levy, F. Klocke, and T. H. C. Childs, "Consolidation phenomena in laser and powder-bed based layered manufacturing," *CIRP Ann. - Manuf. Technol.*, vol. 56, no. 2, pp. 730–759, 2007.
28. I. A. Ibrahim, F. A. Mohamed, and E. J. Lavernia, "Particulate reinforced metal matrix composites - a review," *Journal of Materials Science*, vol. 26, no. 5. pp. 1137–1156, 1991.
29. A. P. Divecha, S. G. Fishman, and S. D. Karmarkar, "Silicon Carbide Reinforced Aluminum—A Formable Composite," *JOM J. Miner. Met. Mater. Soc.*, vol. 33, no. 9, pp. 12–17, 1981.
30. P. Sahoo and M. J. Koczak, "Microstructure-property relationships of in situ reacted TiC/AlCu metal matrix composites," *Mater. Sci. Eng. A*, vol. 131, no. 1, pp. 69–76, 1991.
31. S. Dadbakhsh and L. Hao, "In situ formation of particle reinforced Al matrix composite by selective laser melting of Al/Fe2O3 powder mixture," *Adv. Eng. Mater.*, vol. 14, no. 1–2, pp. 45–48, 2012.
32. Z. Zhong and N. P. Hung, "Grinding of alumina/aluminum composites," *J. Mater. Process. Technol.*, vol. 123, no. 1, pp. 13–17, 2002.
33. S. J. Zhu and T. Iizuka, "Fabrication and mechanical behavior of Al matrix composites reinforced with porous ceramic of in situ grown whisker framework," *Mater. Sci. Eng. A*, vol. 354, no. 1–2, pp. 306–314, 2003.
34. E. J. Lavernia, J. D. Ayers, and T. S. Srivatsan, "Rapid solidification processing with specific application to aluminium alloys," *Int. Mater. Rev.*, vol. 37, no. 1, pp. 1–44, 1992.
35. E. J. Lavernia and T. S. Srivatsan, "The rapid solidification processing of materials: Science, principles, technology, advances, and applications," *Journal of Materials Science*, vol. 45, no. 2. pp. 287–325, 2010.
36. S. Dadbakhsh, L. Hao, P. G. E. Jerrard, and D. Z. Zhang, "Experimental investigation on selective laser melting behaviour and processing windows of in situ reacted Al/Fe 2O 3 powder mixture," *Powder Technol.*, vol. 231, pp. 112–121, 2012.
37. R. H. Fan, H. L. Lü, K. N. Sun, W. X. Wang, and X. B. Yi, "Kinetics of thermite reaction in Al-Fe2O3 system," *Thermochim. Acta*, vol. 440, no. 2, pp. 129–131, 2006.
38. M. Das, V. K. Balla, D. Basu, S. Bose, and A. Bandyopadhyay, "Laser processing of SiC-particle-reinforced coating on titanium," *Scr. Mater.*, vol. 63, no. 4, pp. 438–441, 2010.
39. R. Anandkumar, A. Almeida, and R. Vilar, "Wear behavior of Al-12Si/TiB2 coatings produced by laser cladding," *Surf. Coatings Technol.*, vol. 205, no. 13–14, pp. 3824–3832, 2011.
40. J. M. Torralba, C. E. Da Costa, and F. Velasco, "P/M aluminum matrix composites: An overview," *Journal of Materials Processing Technology*, vol. 133, no. 1–2, pp. 203–206, 2003.

41. D. Gu, F. Chang, and D. Dai, "Selective Laser Melting Additive Manufacturing of Novel Aluminum Based Composites With Multiple Reinforcing Phases," *J. Manuf. Sci. Eng.*, vol. 137, no. 2, p. 21010, 2015.
42. X. S. Cong, P. Shen, Y. Wang, and Q. Jiang, "Wetting of polycrystalline SiC by molten Al and Al-Si alloys," *Appl. Surf. Sci.*, vol. 317, pp. 140–146, 2014.
43. C. Xue and J. K. Yu, "Enhanced thermal transfer and bending strength of SiC/Al composite with controlled interfacial reaction," *Mater. Des.*, vol. 53, pp. 74–78, 2014.
44. A. Simchi and D. Godlinski, "Effect of SiC particles on the laser sintering of Al-7Si-0.3Mg alloy," *Scr. Mater.*, vol. 59, no. 2, pp. 199–202, 2008.
45. E. Louvis, P. Fox, and C. J. Sutcliffe, "Selective laser melting of aluminium components," *J. Mater. Process. Technol.*, vol. 211, no. 2, pp. 275–284, 2011.
46. F. Chang, D. Gu, D. Dai, and P. Yuan, "Selective laser melting of in-situ Al4SiC4 + SiC hybrid reinforced Al matrix composites: Influence of starting SiC particle size," *Surf. Coatings Technol.*, vol. 272, pp. 15–24, 2015.
47. A. L. Patterson, "The scherrer formula for X-ray particle size determination," *Phys. Rev.*, vol. 56, no. 10, pp. 978–982, 1939.
48. J. P. Kruth, X. Wang, T. Laoui, and L. Froyen, "Lasers and materials in selective laser sintering," *Assem. Autom.*, vol. 23, no. 4, pp. 357–371, 2003.
49. S. Gopagoni *et al.*, "Microstructural evolution in laser deposited nickel-titanium-carbon in situ metal matrix composites," *J. Alloys Compd.*, vol. 509, no. 4, pp. 1255–1260, 2011.
50. Z. de Liu, J. Tian, B. Li, and L. ping Zhao, "Microstructure and mechanical behaviors of in situ TiC particulates reinforced Ni matrix composites," *Mater. Sci. Eng. A*, vol. 527, no. 16–17, pp. 3898–3903, 2010.
51. D. Strz�ciwilk, Z. Wokulski, and P. Tkacz, "Microstructure of TiC crystals obtained from high temperature nickel solution," *J. Alloys Compd.*, vol. 350, no. 1–2, pp. 256–263, 2003.
52. G. Xiao, Q. Fan, M. Gu, Z. Wang, and Z. Jin, "Dissolution-precipitation mechanism of self-propagating high-temperature synthesis of TiC-Ni cermet," *Mater. Sci. Eng. A*, vol. 382, no. 1–2, pp. 132–140, 2004.
53. Y. Li, P. Bai, Y. Wang, J. Hu, and Z. Guo, "Effect of TiC content on Ni/TiC composites by direct laser fabrication," *Mater. Des.*, vol. 30, no. 4, pp. 1409–1412, 2009.
54. D. Strzeciwilk, P. Tkacz, and Z. Wokulski, "Transmission electron microscope studies of TiC crystals," *Cryst. Res. Technol.*, vol. 35, no. 11–12, pp. 1295–1303, 2000.
55. H. Sahasrabudhe, J. Soderlind, and A. Bandyopadhyay, "In Situ Nitridation of Titanium Using Lens™," *Biomaterials Science: Processing, Properties and Applications V: Ceramic Transactions, Volume 254*, pp. 149–159, 2015.
56. H. Sahasrabudhe, J. Soderlind, and A. Bandyopadhyay, "Laser processing of in situ TiN/Ti composite coating on titanium," *J. Mech. Behav. Biomed. Mater.*, vol. 53, pp. 239–249, 2016.
57. M. Das, V. K. Balla, D. Basu, I. Manna, T. S. Sampath Kumar, and A. Bandyopadhyay, "Laser processing of in situ synthesized TiB-TiN-reinforced Ti6Al4V alloy coatings," *Scr. Mater.*, vol. 66, no. 8, pp. 578–581, 2012.
58. M. Das *et al.*, "In situ synthesized TiB-TiN reinforced Ti6Al4V alloy composite coatings: Microstructure, tribological and in-vitro biocompatibility," *J. Mech. Behav. Biomed. Mater.*, vol. 29, pp. 259–271, 2014.
59. H. Attar, M. Bönisch, M. Calin, L. C. Zhang, S. Scudino, and J. Eckert, "Selective laser melting of in situ titanium-titanium boride composites: Processing, microstructure and mechanical properties," *Acta Mater.*, vol. 76, pp. 13–22, 2014.
60. V. K. Balla, W. Xue, S. Bose, and A. Bandyopadhyay, "Laser-assisted Zr/ZrO2 coating on Ti for load-bearing implants," *Acta Biomater.*, vol. 5, no. 7, pp. 2800–2809, 2009.
61. R. Banerjee, P. C. Collins, and H. L. Fraser, "Laser Deposition of In Situ Ti – TiB Composites," *Adv. Eng. Mater.*, vol. 4, no. 11, pp. 847–851, 2002.
62. S. Samuel, S. Nag, T. W. Scharf, and R. Banerjee, "Wear resistance of laser-deposited boride reinforced Ti-Nb-Zr-Ta alloy composites for orthopedic implants," *Mater. Sci. Eng. C*, vol. 28, no. 3, pp. 414–420, 2008.

63. S. Nag, S. Samuel, A. Puthucode, and R. Banerjee, "Characterization of novel borides in Ti-Nb-Zr-Ta + 2B metal-matrix composites," *Mater. Charact.*, vol. 60, no. 2, pp. 106–113, 2009.
64. A. Genç, R. Banerjee, D. Hill, and H. L. Fraser, "Structure of TiB precipitates in laser deposited in situ, Ti-6Al-4V-TiB composites," *Mater. Lett.*, vol. 60, no. 7, pp. 859–863, 2006.
65. D. G. Bansal, O. L. Eryilmaz, and P. J. Blau, "Surface engineering to improve the durability and lubricity of Ti-6Al-4V alloy," *Wear*, vol. 271, no. 9–10, pp. 2006–2015, 2011.
66. M. Geetha, A. K. Singh, R. Asokamani, and A. K. Gogia, "Ti based biomaterials, the ultimate choice for orthopaedic implants - A review," *Progress in Materials Science*, vol. 54, no. 3. pp. 397–425, 2009.
67. A. Liu, "Transcendental Data: Toward a Cultural History and Aesthetics of the New Encoded Discourse," *Crit. Inq.*, vol. 31, no. Autumn, pp. 49–84, 2004.
68. R. A. Buchanan, E. D. Rigney, and J. M. Williams, "Ion implantation of surgical Ti???6Al??? 4V for improved resistance to wear???accelerated corrosion," *J. Biomed. Mater. Res.*, vol. 21, no. 3, pp. 355–366, 1987.
69. C. B. Johansson, J. Lausmaa, T. Röstlund, and P. Thomsen, "Commercially pure titanium and Ti6AI4V implants with and without nitrogen-ion implantation: surface characterization and quantitative studies in rabbit cortical bone," *J. Mater. Sci. Mater. Med.*, vol. 4, no. 2, pp. 132–141, 1993.
70. C. Hu, H. Xin, L. M. Watson, and T. N. Baker, "Analysis of the phases developed by laser nitriding Ti-6Al-4V alloys," *Acta Mater.*, vol. 45, no. 10, pp. 4311–4322, 1997.
71. A. Czyrska-Filemonowicz *et al.*, "Transmission electron microscopy and atomic force microscopy characterisation of titanium-base alloys nitrided under glow discharge," *Acta Mater.*, vol. 53, no. 16, pp. 4367–4377, 2005.
72. A. Zhecheva, W. Sha, S. Malinov, and A. Long, "Enhancing the microstructure and properties of titanium alloys through nitriding and other surface engineering methods," *Surface and Coatings Technology*, vol. 200, no. 7. pp. 2192–2207, 2005.
73. M. Nakai *et al.*, "Surface hardening of biomedical Ti-29Nb-13Ta-4.6Zr and Ti-6Al-4V ELI by gas nitriding," *Mater. Sci. Eng. A*, vol. 486, no. 1–2, pp. 193–201, 2008.
74. H. Mohseni, P. Nandwana, A. Tsoi, R. Banerjee, and T. W. Scharf, "In situ nitrided titanium alloys: Microstructural evolution during solidification and wear," *Acta Mater.*, vol. 83, pp. 61–74, 2015.
75. R. Banerjee, P. C. Collins, A. Genç, and H. L. Fraser, "Direct laser deposition of in situ Ti-6Al-4V-TiB composites," *Mater. Sci. Eng. A*, vol. 358, no. 1–2, pp. 343–349, 2003.

# Optimization of Electrical Discharge Machining of Titanium Alloy (Ti6Al4V) by Grey Relational Analysis Based Firefly Algorithm

**Anshuman Kumar Sahu and Siba Sankar Mahapatra**

## 1 Introduction

Electrical discharge machining (EDM) is a non-traditional machining process widely used in the machining of difficult to machine materials where other conventional machining process are unable to perform. The EDM process is extensively use to produce complex, intrinsic cavity in difficult to machine materials to be use in aerospace, biomedical, automobile, die and mold making industries. In EDM, both work piece and tool electrode are immerse inside a dielectric medium. Here both work piece and tool electrode are electrically conducting in nature. When voltage is supply to the EDM circuit, spark is produce between the electrode gap and very high temperature of around 10,000 °C is generated. Due to the generation of such a high temperature, tiny amount of material melt and evaporate and produce crater on the work piece surface, which lead to material removal from the work piece surface. As the EDM tool electrode shape is mirror in the work piece, the shape of tool is the essential parameter along with tool electrode material, design and process of manufacturing. Other conventional and non-conventional machining process does the production of EDM electrode, therefore, the cost of production of EDM electrodes account for more than 50% of the cost of the final product [1]. To produce complex, intrinsic shape with excellent surface finish in the work piece material, pushed the production of EDM tool electrode by conventional machining process into more expensive and time taking, which increased the cost of the final product. Therefore, an alternative process i.e. additive manufacturing (AM) technology provide the direct fabrication of the EDM electrode. AM process consists of a group of manufacturing techniques that yield direct production of EDM tool electrode from three-dimensional CAD model, especially for complex, intrinsic shape of

A. K. Sahu (✉) · S. S. Mahapatra
Department of Mechanical Engineering, National Institute of Technology,
Rourkela, Odisha, India



B. AlMangour (ed.), *Additive Manufacturing of Emerging Materials*,
https://doi.org/10.1007/978-3-319-91713-9_2

tool electrode to reduce production time and processing cost. It is also called as rapid proto typing (RP) process. Selective laser sintering (SLS) is the most suitable AM process used for the preparation of EDM tool electrode, that reduce the tool production time and total production cost of the final product. The SLS process produce parts layer by layer by selectively melting and fused lose power particles by the help of laser beam. The main difficulty of the production of EDM electrode by the SLS process is the selection of the appropriate material for tool. So that, it can be easily prepared by the SLS process as well as contain the properties of EDM tool electrode [2]. The SLS process is the most successful AM process, which directly generate very complex, intrinsic and accurate parts by using wide range of powder materials, increasing its applications in the fields of tooling industries.

Titanium and it's alloys are having high corrosion resistance, high strength to weight ratio and can be use in a wide range of temperature variation. Titanium alloy (Ti6Al4V) is highly strong, lightweight, durable and long lasting. Therefore, Ti-alloy is use in biomedical, automobile, aerospace, military and chemical industries. In the medical field, large amount of product, which required complex intrinsic cavity with excellent surface finish, can be manufactured by EDM process, which are used in prosthesis, surgery devices and tissue engineering. Titanium and it's alloys are also difficult to machine by conventional machining process due to very low thermal conductivity, low elastic modulus and chemical reactivity with the tool materials along with premature failure of tools. Therefore, nontraditional machining like EDM is used for the machining of Ti-alloy to produce intrinsic and complex shape with excellent surface finish to meet its desired applications in different fields as stated above [3, 4].

Therefore, to consider the ease of manufacturing EDM electrode by SLS process and to meet the required properties of EDM tool electrode, composite tool electrode of aluminium, silicon and magnesium (i.e. AlSi10Mg) is prepared by SLS process. To study the performance of this AlSi10Mg composite electrode prepared via SLS process Ti-alloy (Ti6Al4V) is taken as work piece material with commercially available EDM 30 oil as dielectric fluid during EDM process and the performance of this EDM electrode is compared with the conventional copper and graphite tool electrodes. To reduce the number of experiment, Taguchi's $L_{18}$ orthogonal array is used. The process parameters like open circuit voltage (V), discharge current (Ip), pulse-on-time (Ton), duty cycle ($\tau$) and types of tool are varied during the EDM process. The effect of these process parameters on material removal rate (MRR), tool wear rate (TWR) and average surface roughness (Ra) are studied.

In the EDM process, multiple conflicting performance characteristics need to be optimize in order to maximize productivity of the process. For example, material removal rate need to be maximize whereas tool wear rate and average surface roughness are minimize for improving the EDM process. Therefore, proper selection of the various process parameters is an important issue in EDM process. From recent literature, it is found that different optimizations techniques have been used for the optimization of the EDM process to enhance the performance of the processes. Various techniques used for the purpose include Grey Relational Analysis (GRA), Satisfaction Function and Distance Based Approach, Utility Concept and Quantum

Behaved Particle Swarm Optimization (QPSO), VIKOR based harmony search algorithm, Fuzzy TOPSIS [3, 5–8]. In this work, to find out the best parametric combination that can simultaneously optimize three performance measures, grey relational analysis (GRA) based Firefly algorithm (FA) is used and to get optimal parametric setting to get best output responses of EDM process. Here, GRA method used to convert the multi responses into single response i.e. grey relational grade (GRG) index and Firefly algorithm is used to find the optimum parametric setting.

## 2 Literature Review

The following section highlights the investigations outcome of past research work on preparation of EDM tool electrode by AM process like SLS process and the performance of these EDM tool electrodes during machining by taking different types of tool electrode composition and different work piece.

Czelusniak et al. have prepared EDM tool electrodes of metal matrix of CuNi combine with Mo, $TiB_2$ and $ZrB_2$ by SLS for EDM of AISI H13 tool steel. The composite electrodes given better performance than the SLS Cu powder electrode. TiB2–CuNi achieved the best overall EDM performance, removing more material from the work piece and presenting lowest electrode wear. ZrB2–CuNi, electrode gave superior finishing performance and Mo–CuNi gave poorest [9]. Again, in another work, Czelusniak et al. have prepared ZrB2-CuNi electrode by SLS process and used as EDM tool electrode. They have studied the effect of SLS process parameters on porosity of electrodes, surface appearance and interiority. ZrB2-CuNi electrodes have much superior performance than SLS copper powder electrodes, but inferior to the solid copper electrodes [2].

Durr et al. have focused on the direct production of EDM tool electrode by SLS process. They have manufactured Bronze-Nickel with Copper Phosphite composite tool electrode by SLS process. They have taken alloyed cold work steel X210Cr12 and C45 Steel as their work piece materials to study the performance of the EDM electrodes prepared via SLS process. They observed that relative electrode wear increases with increase in porosity contain of the electrodes and pulse-on-time and relative electrode wear is more in X210Cr12 as compare to C45 Steel [10].

Zhao et al. have prepared EDM electrode by SLS process by taking steel and phosphate with polyester as binder. They have study the performance of these EDM electrodes by taking 45 steel as work piece material during EDM. In this work, it is found that electrode wear increases with increase in pores and holes in electrodes. Similarly, with increase in pulse-on-time and peak current, electrode wear rate decreases. Likewise, the surface roughness of the EDM machined surface decreases with increase in pulse-off-time [11].

Tang et al. have used the abrading tool of corundum and silicon carbide to prepared graphite electrode by rapid prototyping (RP) process and then the graphite electrode is use to produce steel mold by EDM process [12]. Similarly, Ding et al. have prepared two different types of EDM electrode by RP process, like copper electrode by electroforming and graphite electrode by abrading process. They have used these EDM electrodes to form steel mold [13].

Arthur et al. have used rapid prototyping (RP) process to prepare the EDM electrode. They have prepared EDM electrode from epoxy with silver paint and copper coating. They have investigated the performance of these RP EDM electrodes by taking hardened tool steel. They found that the tool electrodes are rupture during machining where coating thickness is less than 180 μm. These electrodes can be used for semi-roughing or finishing operations [1].

In recent past, different optimization techniques are used for optimization of the manufacturing process. Sahu et al. have used grey relational analysis (GRA) method to optimize the EDM process during machining of Ti-alloy and 316L stainless steel by taking copper as tool electrode in EDM oil dielectric medium [3]. Similarly, Rahul et al. have used satisfaction function and distance-based approach for optimization of EDM process during machining of Inconel 718. They have combine satisfaction function, distance-based approach and Taguchi's philosophy to optimize the EDM process [5]. Mohanty et al. have used utility concept and quantum behaved particle swarm optimization (QPSO) to optimize the EDM process during machining of Inconel 718 work piece material by taking three diffrerent types of tool electrodes like brass, copper and graphite [6]. Sahu et al. have used Taguchi based VIKOR method combine with Harmony search algorithm to optimize the surface finish parameters during electrical discharge coating process of tungsten carbide layer on 1040 stainless steel work piece. They have used copper-tungsten composite electrode prepared via powder metallurgy route for the EDC process by taking EDM 30 oil as dielectric fluid [7]. Dewangan et al. have used Fuzzy TOPSIS to optimize the surface integrity and dimensional accuracy during EDM of AISI P20 tool steel [8]. Datta and Mahapatra have used grey based Taguchi method to optimize wire EDM process of D2 tool steel [14]. Raja et al. have used Firefly algorithm to optimize the EDM process of hardened die steel [15]. Similarly, Shukla and Singh have used firefly algorithm to optimize electrical discharge machining and abrasive water jet machining [16]. Mishra et al. have study the effect of external perimeter on flexural strength of fused deposition modelling build parts and used Firefly algorithm to obtain the best parametric setting for improve flexural strength [17].

Critical review of the past research study suggests a good number of works are done on EDM process and different optimization techniques are used to obtain best parametric setting to enhance the EDM machining process. However, analysis of machining parameters by taking different types tool electrodes and also the use of RP process like SLS to prepare the tool electrode are less. Furthermore, it is also observe that less number of work have been done to machine a relatively low conductivity material like Titanium alloy (Ti6Al4V), which is extensively use in aerospace and biomedical industries. Numerous studies are reported to obtain optimal parametric setting by different researchers in different manufacturing process. However, it is observed that no attempt is applied in the combination of grey relational analysis (GRA) in combination with Firefly algorithm (FA) during the EDM process to obtain optimal parametric setting to enhance the EDM performance. Therefore, in this work, grey relational analysis based Firefly algorithm (FA) used to optimize the

EDM process during machining of Titanium alloy by taking three different types of tool electrodes like AlSi10Mg SLS electrode, copper and graphite.

# 3 Methodology

## *3.1 Grey Relational Analysis (GRA)*

Julong Deng first time proposed the grey system theory (GST) in 1982. In his work, the researcher has proposed the grey relational analysis (GRA) to find out the relationship between the variables by using grey degrees [18, 19]. In GRA, the experimental data i.e. the performance measures are first normalized into the range of 0–1. This process is termed as grey relational generation. Then, grey relational coefficient is calculated by considering the desired and actual experimental dada. The overall grey relational grade is calculated by averaging the grey relational coefficient of all the responses. The GRA method converts a multi response optimization problem into a single response optimization problem by considering the objective function as overall grey relational grade. The detailed procedure of the grey relational analysis (GRA) method is explained as follows [3, 14].

1. Calculate the scale value ($Y_{ij}$) of the observations.

$$\text{For Lower is the better, } Y_{ij} = \frac{y_{ij}^{max} - y_{ij}}{y_j^{max} - y_j^{min}} \tag{1}$$

$$\text{For Higher is the better, } Y_{ij} = \frac{y_{ij} - y_{ij}^{min}}{y_j^{max} - y_j^{min}} \tag{2}$$

Where $y_{ij}$ = observed responses of the $i^{th}$ number of experiment in the $j^{th}$ response.

$y_j^{max}$ = maximum value of the $j^{th}$ response.
$y_j^{min}$ = minimum value of the $j^{th}$ response.

2. Calculate the Grey relational co-efficient ($\gamma_{ij}$).

$$\gamma_{ij} = \frac{\left(\Delta_j^{min} + \xi\Delta_j^{max}\right)}{\left(\Delta_{ij} + \Delta_j^{max}\right)} \tag{3}$$

Where, $\Delta_{ij} = |1 - Y_{ij}|$

$$y_j^{min} = \min\left(\Delta_{1j}, \Delta_{2j}, \ldots\ldots\ldots, \Delta_{mj}\right) \quad y_j^{max} = \max\left(\Delta_{1j}, \Delta_{2j}, \ldots\ldots\ldots, \Delta_{mj}\right)$$

$$\xi \in [0, 1], \xi = 0.5$$

where $\Delta_{ij}$= deviation sequence, which is absolute of the difference between the ideal value or reference sequence (one) and the scale value ($Y_{ij}$) and $\xi$ = distinguishing coefficient, $\xi \in [0, 1]$, $\xi = 0.5$.

3. Grey relational grade (GRG)

$$\mathrm{GRG} = \sum_{j=1}^{p} w_j \gamma_{ij} \tag{4}$$

$w_j$ = weightage of the each performance measures. Here equal weightage of 0.33 has been taken for each performance measures.

Where, $\sum_{j=1}^{p} w_j = 1$

## 3.2 *Firefly Algorithm (FA)*

Xin-She Yang has developed a new nature inspired optimization algorithm called Firefly algorithm (FA). The FA algorithm is inspired from the rhythmic flashlight of fireflies. The flashing light is developed by the bioluminescence process. Two important characteristics of this flashes are to attract mating partners and attracting of potential prey. The Firefly attracted by other fireflies according to higher flash light intensity. The light intensity I at a distance r decreases as the distance r increases like $I \propto 1/r^2$. The attraction among fireflies is local or global according to the absorbing coefficient. The fireflies are subdivided into subgroups due to the intensity of light of the neighboring fireflies. So, each subgroup is swarm around a local mode. The flashing light has been formulated such that the objective function can be optimized. This algorithm can be used for multimodal optimization problems [15–17, 20–23].

The Firefly algorithm is based on three rules.

1. All fireflies are unisex and one firefly is attracted to other brighter fireflies regardless of their sex.
2. Attractiveness is proportional to the brightness of the fireflies. The less bright firefly will move towards the brighter firefly. The attractiveness of fireflies is proportional to the brightness and attractiveness decreases with increase in the distance between the two fireflies. The firefly will move randomly when no brighter firefly is available.
3. The brightness of fireflies can be determined by the landscape of the objective function.

The light intensity (I) varies with the distance (r) between the fireflies can be denoted as in Eq. (5).

$$I(r) = I_0 e^{-\gamma r^2} \tag{5}$$

Where

$I_0$ = light intensity of the source.
$\gamma$ = light absorption coefficient that controls the decrease in the light intensity.

The attractiveness ($\beta$) can be described by a monotonically decreasing function of the distance (r) between any two fireflies as in Eq. (6).

$$\beta(r) = \beta_0 e^{-\gamma r^2} \tag{6}$$

Where $\beta_0$= maximum attractiveness at r = 0.

The distance between two fireflies i and j at position $x_i$ and $x_j$ can be represented as in Eq. (7).

$$r_{ij} = |x_i - x_j| = \sqrt{\sum_{k=1}^{d} \left(x_{i,k} - x_{j,k}\right)^2} \tag{7}$$

Where $x_{i,k}$ and $x_{j,k}$ are the $k^{th}$ components of the spatial coordinates $x_i$ and $x_j$ of the $i^{th}$ and $j^{th}$ fireflies respectively.

d = number of dimensions.

The movement of the firefly i is determined by the formula of Eq. (8).

$$x_i = x_i + \beta_0 e^{-\gamma r^2} \left(x_j - x_i\right) + \alpha\left(\text{rand} - \frac{1}{2}\right) \tag{8}$$

In Eq. (8), the first term denotes the present position of the firefly i, the second term denote the attractiveness of the fireflies and the last terms denotes the free movement of firefly during the absent of brighter firefly. In most of the cases, the randomization $\alpha$ is taken between 0 and 1. Similarly, the light absorption coefficient $\gamma$ taken between 0.1 and 10. The flowchart of FA algorithm is given in Fig. 1 as follows.

## 4 Materials and Methods

To reduce the cost of manufacturing of complex EDM electrode, rapid tooling (RT) technique is used, which is an additive manufacturing (AM) process. In this work, AM process like selective laser sintering (SLS) process use to manufacture EDM electrode. Aluminium, silicon and magnesium (i.e. AlSi10Mg) powder is used to manufacture EDM tool electrode by SLS process. The chemical composition of the

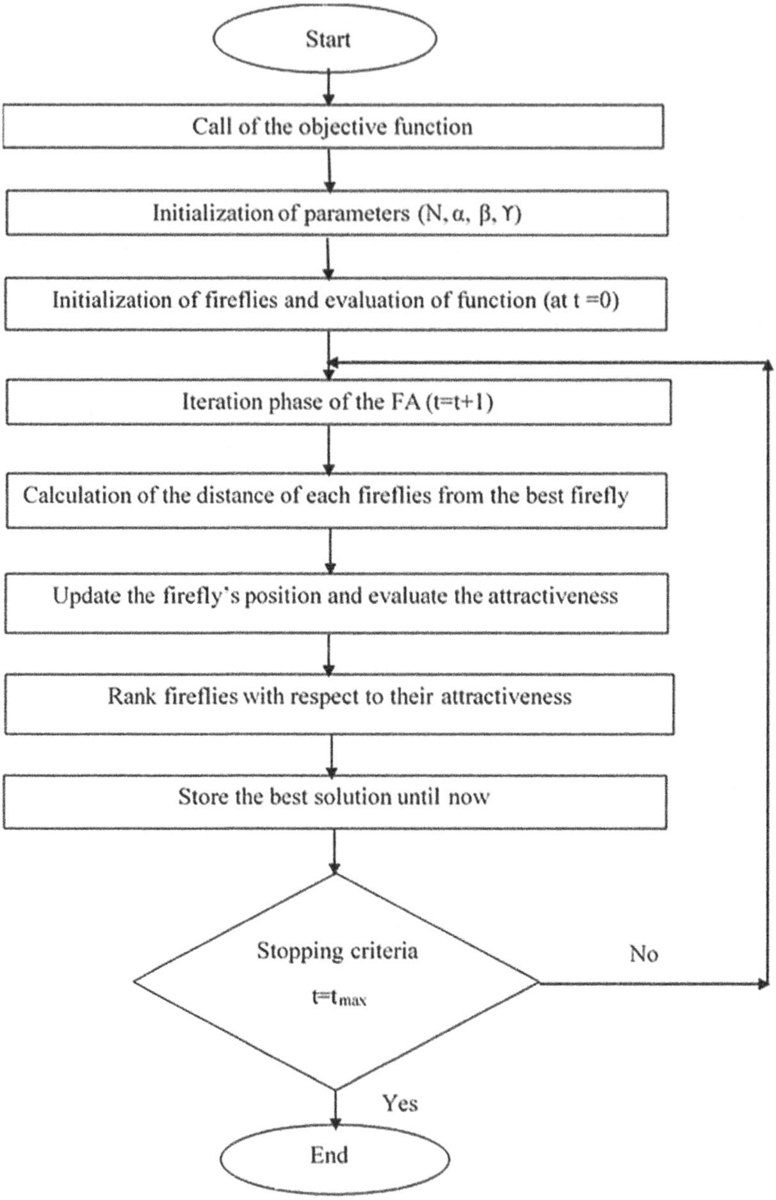


**Fig. 1** Flow chart of Firefly algorithm

AlSi10Mg powder (supply from EOS GmbH - Electro Optical Systems, München, Germany) is as given in Table 1. Selectively melting and fusion of metallic powder prepare the EDM tool electrode by the help of the laser beam in the SLS machine EOSINT M 280 (manufactured by: EOS GmbH - Electro Optical Systems,

**Table 1** Chemical composition (weight %) of AlSi10Mg [24]

| Material | Si | Fe | Cu | Mn | Mg | Ni | Zn | Pb | Sn | Ti | Al |
|---|---|---|---|---|---|---|---|---|---|---|---|
| wt. % | 9–11 | ≤0.55 | ≤0.05 | ≤0.45 | 0.2–0.45 | ≤0.05 | ≤0.1 | ≤0.05 | ≤0.05 | ≤0.15 | Balance |

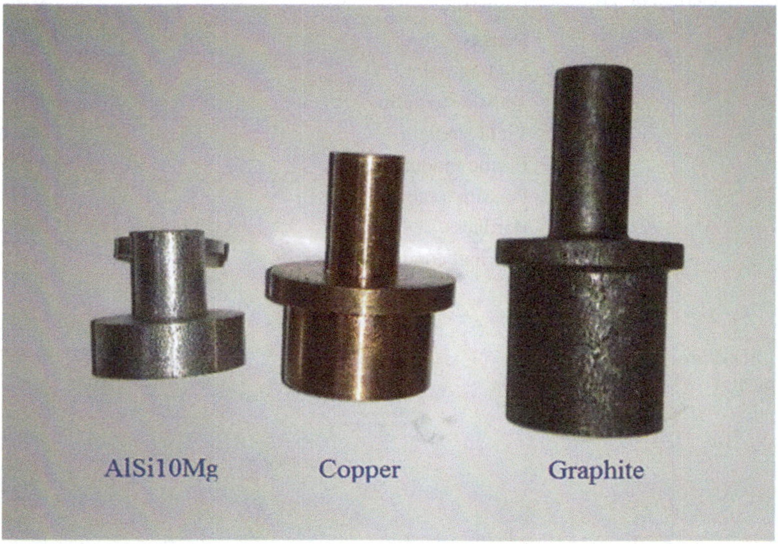


**Fig. 2** Tool electrodes

München, Germany). Here thin layer of metallic powder is spread by a roller on the surface of the build cylinder and heated below the melting point by infrared heating panels and the laser beam traced out the cross-section of the slice part. The laser beam fuse the affected powder collectively (sinter) to form the 3D compound i.e. the EDM tool from the metallic powders. The SLS parameters used to prepared the AlSi10Mg composite RP electrode are layer thickness: 30 μm, laser power: 400 W, scan speed: 32 mm/s, build platform temperature: 200 °C and building environment: Argon (at pressure 4.1 bar). The EDM tool electrode prepared is of diameter of 25 mm. The copper and graphite electrodes are prepared from solid circular rod by the turning process having diameter of 25 mm. The three tool electrodes used in this experiment are shown in Fig. 2.

To study the machining performance of these electrodes during EDM Ti-alloy (Ti6Al4V; size: 60 × 60 × 10 mm) is use as work piece materials. Ti-alloy is high corrosion resistance, high strength to weight ratio, highly strong, lightweight, durable and long lasting. Therefore, Ti-alloy is use in biomedical, automobile, aerospace, military and chemical industries. The chemical composition and properties of the Ti-alloy are presented in Tables 2 and 3 respectively.

The AlSi10Mg electrode prepared via SLS process is used for the EDM process along with conventional copper and graphite electrodes by taking Ti-alloy as work

**Table 2** Chemical composition (weight %) of Ti-alloy

| Material | C | Fe | Al | O | N | V | H | Ti |
|---|---|---|---|---|---|---|---|---|
| wt. % | 0.018 | 0.22 | 6.08 | 0.18 | 0.05 | 4.02 | 0.1 | Balance |

**Table 3** Properties of Ti-alloy (Ti6Al4V)

| Properties | Value |
|---|---|
| Density | 4.42 g/cc |
| Melting point | 1650 °C |
| Tensile strength | 1200 MPa |
| Yield strength | 880 MPa |
| Elastic modulus | 113.8 GPa |
| Poisson's ratio | 0.31–0.37 |
| Hardness | 334 HB |
| Thermal conductivity | 6.7 W/mK |

**Fig. 3** EDM process during experiment

piece in a die sinking EDM (ELECTRA EMS 5535, India) by taking commercially available EDM 30 oil (specific gravity: 0.763) as the dielectric fluid. In this work straight polarity (i.e., work piece: anode and tool: cathode) is used for EDM process. The EDM process during machining is shown in Fig. 3. The EDM process is carried out by taking five different process parameters like open circuit voltage (V), discharge current (Ip), duty cycle (τ), pulse-on-time (Ton) and type of tool (i.e.

**Table 4** Input parameters with different levels

| Parameters | Unit | Level 1 | Level 2 | Level 3 |
|---|---|---|---|---|
| A-Open circuit voltage (V) | V | 20 | 25 | 30 |
| B-Discharge current ($I_p$) | A | 10 | 15 | 20 |
| C-Duty cycle ($\tau$) | % | 67 | 75 | 83 |
| D-Pulse-on-time ($T_{on}$) | μs | 100 | 200 | 300 |
| E- Tool material | – | AlSiMg RP Tool | Copper | Graphite |

AlSi10Mg, copper and graphite). The different process parameters with their levels are listed in Table 4. The different variable parameters used in this EDM process are discussed as follows.

(a) Open circuit voltage: It is the potential difference applied between the tool electrodes and work piece.
(b) Discharge current: It is also called as peak current. It is the most dominant process parameter that governs the spark energy.
(c) Duty cycle: It is the percentage of the pulse-on-time with respect to total cycle time.

$$\tau = \frac{T_{on}}{T_{on} + T_{off}} \tag{9}$$

(d) Pulse-on-time: It is also known as spark-on-time or pulse duration. It is the duration (per cycle) in which the current is allowed to flow through the electrode gap.
(e) Tool material: The different type tool electrode used during the EDM process.

To reduce the total number of the experiment, the design of experiment (DOE) approach like Taguchi's $L_{27}$ orthogonal array is used to plan the experiment. In this work, 5-factor-3-level is used as in Table 6. Each experiment is run for 10 min. Different process parameters varied during the electrical discharge coating process are open circuit voltage (V), discharge current (Ip), duty cycle ($\tau$), pulse-on-time (Ton) and type of tool material. The corresponding output responses consider to study the EDM process are (1) material removal rate (MRR), (2) tool wear rate (TWR) and (3) average surface roughness ($R_a$) of the machined surface. The Taguchi's $L_{27}$ orthogonal array (OA) along with the output responses for EDM process are presented in Table 5.

## 4.1 *Material Removal Rate (MRR)*

Material removal rate (MRR) is defined as the rate at which material removal occurred from the surface of the work piece. The MRR can be determined by the weight loss criteria of the work piece as shown in the Eq. (10).

**Table 5** Densities if tool electrodes and titanium alloy work piece

| Material | Ti-alloy | AlSi10Mg | Copper | Graphite |
|---|---|---|---|---|
| Density (g/cm$^3$) | 4.42 | 2.664 | 8.96 | 2.267 |

$$\mathrm{MRR} = \frac{(\mathrm{W_i} - \mathrm{W_f})}{(\mathrm{t} \times \rho_\mathrm{w})} \tag{10}$$

Where

$\mathrm{W_i}$ = Initial weight of the work piece before EDM,
$\mathrm{W_f}$ = Final weight of the work piece after EDM,
t = Machining time,
$\rho_\mathrm{w}$ = Density of work piece Titanium-alloy.

## 4.2 *Tool Wear Rate (TWR)*

Tool wear rate (TWR) is defined as the rate at which material loss occurred from the tool electrode. The TWR calculated by the weight loss criteria of the tool electrode as given in the Eq. (11).

$$\mathrm{TWR} = \frac{(\mathrm{W_{ti}} - \mathrm{W_{tf}})}{(\mathrm{t} \times \rho_\mathrm{t})} \tag{11}$$

Where

$\mathrm{W_{ti}}$ = Initial weight of the tool electrode before EDM,
$\mathrm{W_{tf}}$ = Final weight of the tool electrode after EDM,
t = Machining time,
$\rho_\mathrm{t}$ = Density of tool electrodes.

The densities of the tool electrodes and work piece are given in Table 5. The initial and final weight of the work piece and tool electrodes before machining and after machining are measured by a weight measurement machine of least count 0.05 g.

## 4.3 *Average Surface Roughness ($R_a$)*

Average surface roughness is the arithmetic mean of the absolute height of the profiles (i.e. peak height and valley) over the sampling length. The average surface roughness of the EDM machined surface is measured by surface roughness measurement machine (Taylor-Hobson-PNEUNO-Suetronic 3+). The equation of average surface roughness is given in Eq. (12) as follows.

**Table 6** Taguchi's $L_{27}$ OA and output responses

| Sl. No. | A | B | C | D | E | MRR ($mm^3$/min) | TWR ($mm^3$/min) | Ra (μm) |
|---|---|---|---|---|---|---|---|---|
| 1 | 1 | 1 | 1 | 1 | 1 | 0.5454 | 0.6706 | 6.4 |
| 2 | 1 | 1 | 1 | 1 | 2 | 0.5820 | 0.6898 | 6.7 |
| 3 | 1 | 1 | 1 | 1 | 3 | 1.0643 | 0.4955 | 8.2 |
| 4 | 1 | 2 | 2 | 2 | 1 | 0.9226 | 1.5553 | 7.2 |
| 5 | 1 | 2 | 2 | 2 | 2 | 0.9654 | 1.0678 | 7 |
| 6 | 1 | 2 | 2 | 2 | 3 | 1.3029 | 0.5744 | 8.3 |
| 7 | 1 | 3 | 3 | 3 | 1 | 1.4221 | 1.5100 | 7.8 |
| 8 | 1 | 3 | 3 | 3 | 2 | 1.4516 | 1.5529 | 8.3 |
| 9 | 1 | 3 | 3 | 3 | 3 | 1.6093 | 0.8402 | 9.2 |
| 10 | 2 | 1 | 2 | 3 | 1 | 0.8291 | 1.6003 | 7.2 |
| 11 | 2 | 1 | 2 | 3 | 2 | 1.1820 | 1.3972 | 7.3 |
| 12 | 2 | 1 | 2 | 3 | 3 | 1.1784 | 0.5556 | 8.2 |
| 13 | 2 | 2 | 3 | 1 | 1 | 1.1204 | 1.7030 | 7.1 |
| 14 | 2 | 2 | 3 | 1 | 2 | 1.2286 | 1.6353 | 7.4 |
| 15 | 2 | 2 | 3 | 1 | 3 | 1.2890 | 0.5739 | 9.4 |
| 16 | 2 | 3 | 1 | 2 | 1 | 1.3368 | 1.5642 | 7.3 |
| 17 | 2 | 3 | 1 | 2 | 2 | 1.6033 | 1.5320 | 8.4 |
| 18 | 2 | 3 | 1 | 2 | 3 | 1.6753 | 0.2556 | 9 |
| 19 | 3 | 1 | 3 | 2 | 1 | 1.0109 | 1.6835 | 7.2 |
| 20 | 3 | 1 | 3 | 2 | 2 | 1.2899 | 1.4414 | 7.4 |
| 21 | 3 | 1 | 3 | 2 | 3 | 1.1223 | 0.5614 | 9.5 |
| 22 | 3 | 2 | 1 | 3 | 1 | 1.1701 | 1.5359 | 7.4 |
| 23 | 3 | 2 | 1 | 3 | 2 | 1.3438 | 1.7646 | 7.8 |
| 24 | 3 | 2 | 1 | 3 | 3 | 1.6224 | 0.3751 | 9.6 |
| 25 | 3 | 3 | 2 | 1 | 1 | 1.3436 | 1.7590 | 8.1 |
| 26 | 3 | 3 | 2 | 1 | 2 | 1.6286 | 1.5441 | 8.4 |
| 27 | 3 | 3 | 2 | 1 | 3 | 1.7113 | 0.7015 | 9.8 |

$$\text{Average roughness, } R_a = \frac{1}{L}\int_0^L y(x)dx \tag{12}$$

Where L is the sampling length, y is the profile curve, x is the profile direction. The surface roughness values are measured within a sampling length, L = 0.8 mm and having cut off length, $L_c = 0.4$ mm.

## 5 Result and Discussion

The experiment of EDM is performed by taking three different types of tool electrodes (AlSi10Mg SLS electrodes, conventional copper and graphite electrodes) and by varying different process parameters like open circuit voltage, discharge current, duty cycle and pulse-on-time and corresponding output responses like material removal rate, tool wear rate and average surface roughness are tabulated in Table 6. The effect different process parameters on the output responses are discuss as follows.

### *5.1 Effect of Parameters on MRR*

With increase in the process parameters like open circuit voltage, discharge current, duty cycle and pulse-on-time, the spark energy generated increase that lead to increase in material removal. Therefore, material removal rate increase with increase in the process parameters as stated above. The main effect plot and interaction plot of the MRR is given in Figs. 4 and 5. The MRR is maximum when graphite tool electrode is use followed by copper and AlSi10Mg SLS electrode. The analysis of variance (ANOVA) for MRR is presented in Table 7 with $R^2$ value of 96.2%. The

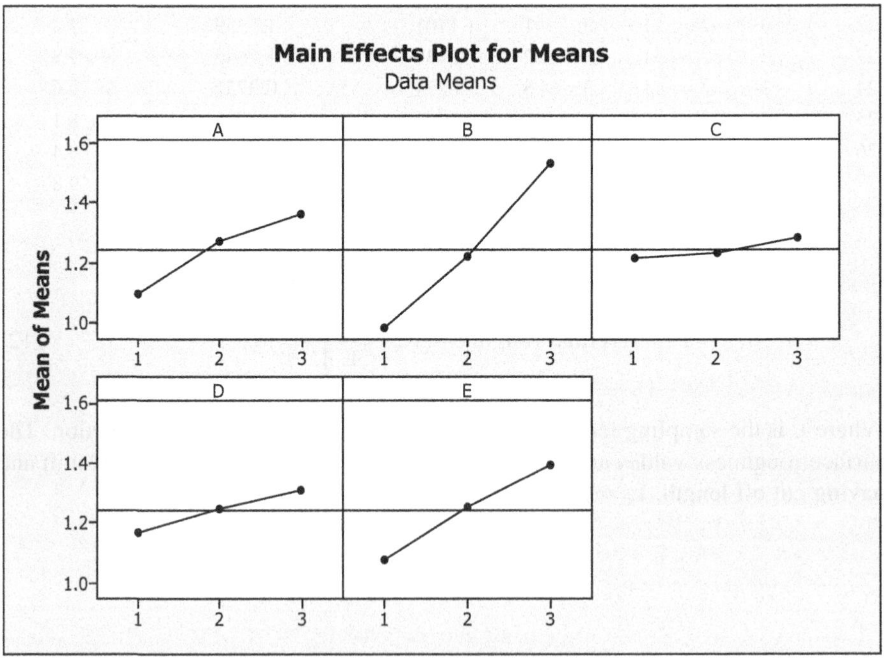


**Fig. 4** Main effects plot for Means of MRR

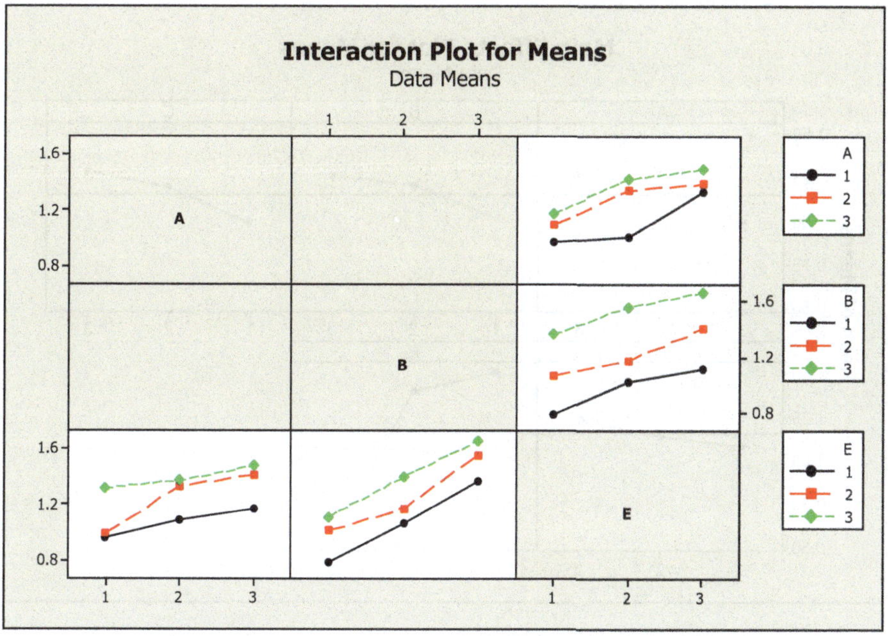


**Fig. 5** Interaction plot for Means of MRR

**Table 7** ANOVA for MRR

| Source | DF | Seq SS | Adj SS | Adj MS | F | P | % Contribution |
|---|---|---|---|---|---|---|---|
| A | 2 | 0.32516 | 0.32516 | 0.162582 | 13.88 | 0.003[a] | 13.12 |
| B | 2 | 1.38444 | 1.38444 | 0.692218 | 59.12 | 0.000[a] | 55.85 |
| C | 2 | 0.02244 | 0.02244 | 0.011221 | 0.96 | 0.424 | 0.91 |
| D | 2 | 0.09360 | 0.09360 | 0.046800 | 4.00 | 0.063 | 3.78 |
| E | 2 | 0.46034 | 0.46034 | 0.230169 | 19.66 | 0.001[a] | 18.57 |
| A*E | 4 | 0.08147 | 0.08147 | 0.020367 | 1.74 | 0.234 | 3.29 |
| B*E | 4 | 0.01767 | 0.01767 | 0.004418 | 0.38 | 0.819 | 0.71 |
| Residual Error | 8 | 0.09367 | 0.09367 | 0.011709 | | | 3.77 |
| Total | 26 | 2.47880 | | | | | 100 |

[a]Significant parameters at 95% confidence interval

ANOVA, main effect plot and interaction plot are generated by using statistical software MINITAB 16. From ANOVA, it is found that discharge current has highest effect followed by tool type and open circuit voltage with percentage contribution of 55.85%, 18.57% and 13.12% respectively.

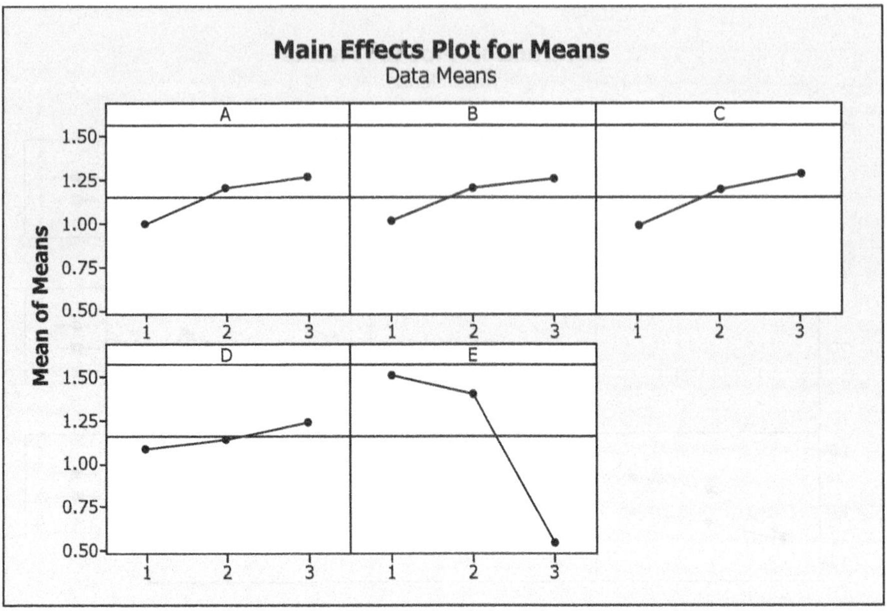


**Fig. 6** Main effects Plot for Means of TWR

## 5.2 *Effect of Parameters on TWR*

The tool wear is increase with increase in open circuit voltage, discharge current, duty cycle and pulse-on-time. With increase in this parameters the spark energy increase and mare tool wear rate occur. The graphite tool has less tool wear rate as compare to other two tool. The AlSi10Mg tool has more tool wear rate and copper has tool wear rate in between them. The main effect plot and interaction plot of the TWR is given in Figs. 6 and 7. The analysis of variance (ANOVA) for TWR is presented in Table 8 with $R^2$ value of 96.4%. From ANOVA, it is found that tool type has highest effect followed by duty cycle, open circuit voltage and discharge current with percentage contribution of 72%, 5.83%, 5.11% and 4.15% respectively.

## 5.3 *Effect of Parameters on Average Surface Roughness*

The average surface roughness is increase with increase in open circuit voltage, discharge current, duty cycle and pulse-on-time. With increase in these parameters, more material removal occur with increase in crater depth. Therefore, average surface roughness become more. The AlSi10Mg SLS tool is produce good surface finish on the machined specimen as compare to other two tool. The graphite tool exhibit worst surface finish specimens and copper tool exhibit surface on the machined surface in between the other two electrodes. The main effect plot and

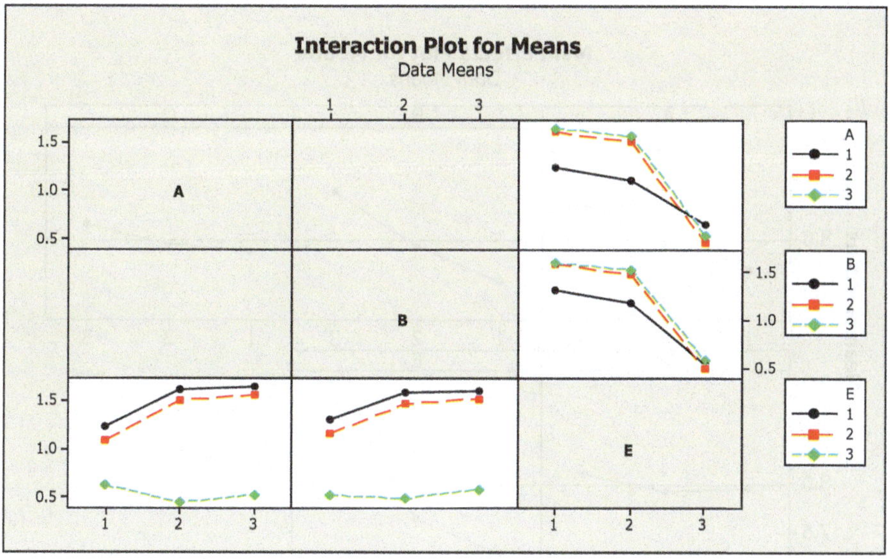


**Fig. 7** Interaction plot for Means of TWR

**Table 8** ANOVA for TWR

| Source | DF | Seq SS | Adj SS | Adj MS | F | P | % Contribution |
|---|---|---|---|---|---|---|---|
| A | 2 | 0.3545 | 0.3545 | 0.17726 | 5.68 | 0.029[a] | 5.11 |
| B | 2 | 0.2876 | 0.2876 | 0.14379 | 4.61 | 0.047[a] | 4.15 |
| C | 2 | 0.4043 | 0.4043 | 0.20216 | 6.48 | 0.021[a] | 5.83 |
| D | 2 | 0.1061 | 0.1061 | 0.05305 | 1.70 | 0.242 | 1.53 |
| E | 2 | 4.9955 | 4.9955 | 2.49776 | 80.10 | 0.000[a] | 72.00 |
| A*E | 4 | 0.4155 | 0.4155 | 0.10386 | 3.33 | 0.069 | 5.99 |
| B*E | 4 | 0.1253 | 0.1253 | 0.03132 | 1.00 | 0.459 | 1.81 |
| Residual Error | 8 | 0.2495 | 0.2495 | 0.03118 | | | 3.58 |
| Total | 26 | 6.9382 | | | | | 100 |

[a]Significant parameters at 95% confidence interval

interaction plot for the average surface finish is given in Figs. 8 and 9. The analysis of variance (ANOVA) for average surface finish is presented in Table 9 with $R^2$ value of 96.7%. From ANOVA, it is found that tool type has highest effect followed by discharge current and open circuit voltage with percentage contribution of 64.86%, 16.45% and 9.16% respectively.

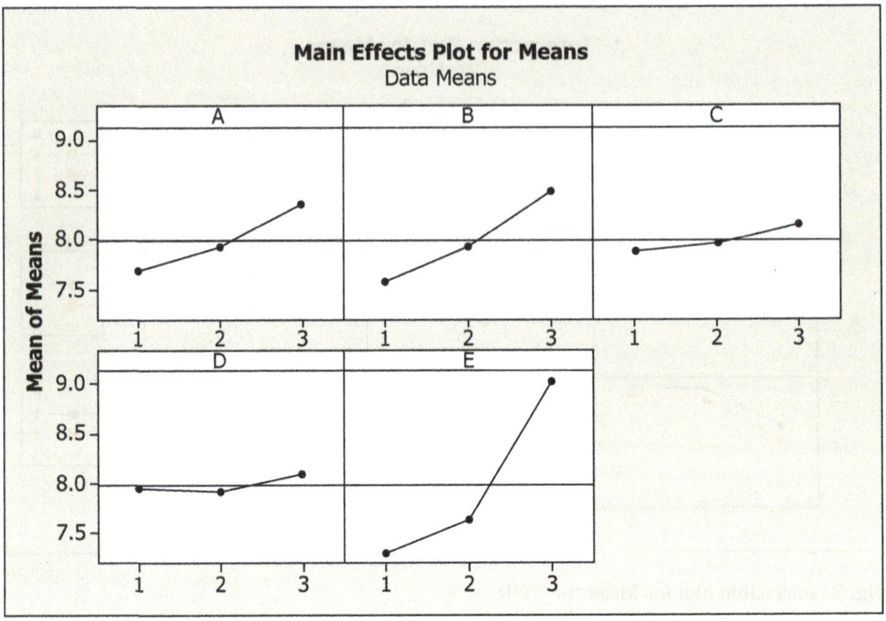


**Fig. 8** Main effects Plot for Means of $R_a$

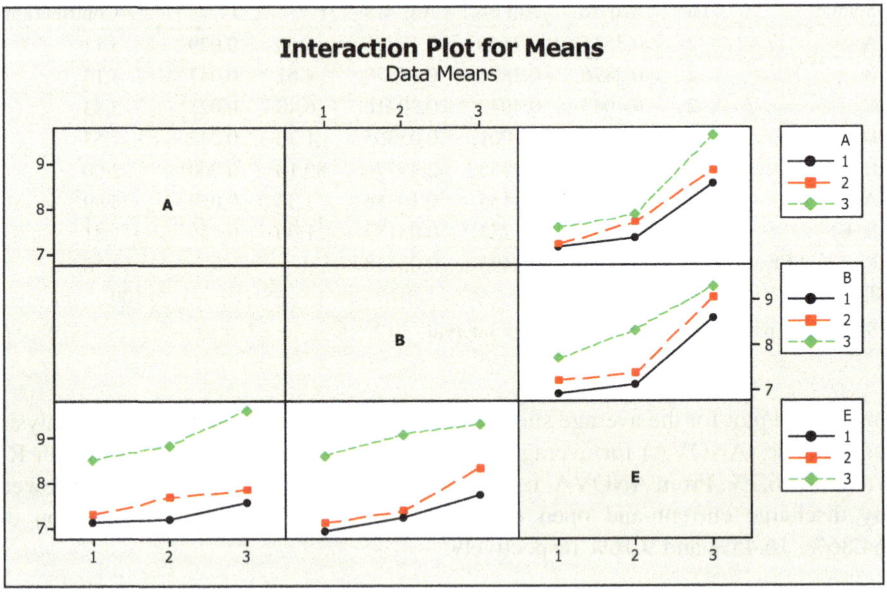


**Fig. 9** Interaction plot for Means of $R_a$

**Table 9** ANOVA for average surface roughness

| Source | DF | Seq SS | Adj SS | Adj MS | F | P | % Contribution |
|---|---|---|---|---|---|---|---|
| A | 2 | 2.1207 | 2.1207 | 1.06037 | 11.15 | 0.005[a] | 9.16 |
| B | 2 | 3.8096 | 3.8096 | 1.90481 | 20.03 | 0.001[a] | 16.45 |
| C | 2 | 0.3696 | 0.3696 | 0.18481 | 1.94 | 0.205 | 1.60 |
| D | 2 | 0.1474 | 0.1474 | 0.07370 | 0.78 | 0.492 | 0.64 |
| E | 2 | 15.0185 | 15.0185 | 7.50926 | 78.97 | 0.000[a] | 64.86 |
| A*E | 4 | 0.4681 | 0.4681 | 0.11704 | 1.23 | 0.371 | 2.02 |
| B*E | 4 | 0.4593 | 0.4593 | 0.11481 | 1.21 | 0.379 | 1.98 |
| Residual Error | 8 | 0.7607 | 0.7607 | 0.09509 | | | 3.29 |
| Total | 26 | 23.1541 | | | | | 100 |

[a]Significant parameters at 95% confidence interval

# 6 Optimization by Grey Relational Analysis Based Firefly Algorithm

To enhance the EDM performance material removal rate correspond to higher-is-better (HB) and tool wear rate and average surface roughness of the machined surface correspond to lower-is-better (LB). By following the procedure of GRA method as mentioned in Eqs. (5–8), the scaled value ($Y_{ij}$) of responses, grey relational coefficient ($\gamma_{ij}$) and grey relational grade (GRG) are calculated and tabulated in Table 10.

The analysis of variance (ANOVA) for the means of grey relational grade (GRG) is given in Table 11 with $R^2 = 87.6\%$ at 95% confidence interval. The main effect plots for the grey relational grade values are plotted in Fig. 10. The ANOVA (Table 11) and response table (Table 12) are found out by using MINITAB 16. From the ANOVA table it is found that type of tool electrodes has the highest effect towards the output responses followed by duty cycle with percentage contribution of 38.59% and 19.04% respectively. From Fig. 8, it is found that better machining performance can be achieved with lower parametric setting of open circuit voltage, duty cycle and pulse-on-time with higher value of discharge current and by the use of graphite electrode. But, to get better surface finish machined parts AlSi10Mg SLS tool electrode must be used.

By using linear regression analysis between the GRG and the EDM process parameters the objective function has been found out and expressed in Eq. (13) with the $R^2$ value of 99.8%. This equation is used as the objective function in the Firefly algorithm and the optimal parametric setting is obtained by using the Firefly algorithm with the help of software MATLAB R2014a. In the Firefly algorithm various parameters taken are number of firefly (n) = 40, number of iteration (N) = 500, randomization ($\alpha$) = 0.5, attractiveness ($\beta$) = 0.2 and absorption coefficient ($\gamma$) = 1, which leads to total number of function evaluations of 20,000.

**Table 10** Scaled value ($Y_{ij}$) of responses, grey relational coefficient ($\gamma_{ij}$) and grey relational grade (GRG)

| Sl. No. | $Y_{ij}$ (MRR) | $Y_{ij}$ (TWR) | $Y_{ij}$ (Ra) | $\Delta_{ij}$ (MRR) | $\Delta_{ij}$ (TWR) | $\Delta_{ij}$ (Ra) | $\gamma_{ij}$ (MRR) | $\gamma_{ij}$ (TWR) | $\gamma_{ij}$ (Ra) | GRG |
|---|---|---|---|---|---|---|---|---|---|---|
| 1 | 0 | 0.7250 | 1 | 1 | 0.2750 | 0 | 0.25 | 0.3922 | 0.5 | 0.3807 |
| 2 | 0.0314 | 0.7122 | 0.9118 | 0.9686 | 0.2877 | 0.0882 | 0.2540 | 0.3883 | 0.4595 | 0.3672 |
| 3 | 0.4451 | 0.8410 | 0.4706 | 0.5549 | 0.1590 | 0.5294 | 0.3216 | 0.4314 | 0.3269 | 0.3600 |
| 4 | 0.3235 | 0.1387 | 0.7647 | 0.6765 | 0.8613 | 0.2353 | 0.2982 | 0.2686 | 0.4048 | 0.3239 |
| 5 | 0.3602 | 0.4618 | 0.8235 | 0.6398 | 0.5382 | 0.1765 | 0.3049 | 0.3250 | 0.425 | 0.3517 |
| 6 | 0.6497 | 0.7887 | 0.4412 | 0.3503 | 0.2113 | 0.5588 | 0.3703 | 0.4128 | 0.3208 | 0.36797 |
| 7 | 0.7520 | 0.1687 | 0.5882 | 0.2480 | 0.8313 | 0.4118 | 0.4006 | 0.2730 | 0.3542 | 0.3426 |
| 8 | 0.7773 | 0.1403 | 0.4412 | 0.2227 | 0.8597 | 0.5588 | 0.4089 | 0.2689 | 0.3208 | 0.3328 |
| 9 | 0.9125 | 0.6126 | 0.1765 | 0.0875 | 0.3874 | 0.8235 | 0.4598 | 0.3604 | 0.2742 | 0.3648 |
| 10 | 0.2433 | 0.1089 | 0.7647 | 0.7567 | 0.8911 | 0.2353 | 0.2846 | 0.2644 | 0.4048 | 0.3179 |
| 11 | 0.5460 | 0.2435 | 0.7353 | 0.4540 | 0.7565 | 0.2647 | 0.3439 | 0.2847 | 0.3953 | 0.3413 |
| 12 | 0.5429 | 0.8012 | 0.4706 | 0.4571 | 0.1988 | 0.5294 | 0.3432 | 0.4171 | 0.3269 | 0.3624 |
| 13 | 0.4932 | 0.0408 | 0.7941 | 0.5068 | 0.9592 | 0.2059 | 0.3318 | 0.2552 | 0.4146 | 0.3339 |
| 14 | 0.5860 | 0.0857 | 0.7059 | 0.4140 | 0.9143 | 0.2941 | 0.3536 | 0.2612 | 0.3864 | 0.3337 |
| 15 | 0.6378 | 0.7891 | 0.1176 | 0.3622 | 0.2109 | 0.8824 | 0.3671 | 0.4129 | 0.2656 | 0.3485 |
| 16 | 0.6788 | 0.1328 | 0.7352 | 0.3212 | 0.8672 | 0.2647 | 0.3784 | 0.2678 | 0.3953 | 0.3472 |
| 17 | 0.9074 | 0.1541 | 0.4118 | 0.0926 | 0.8459 | 0.5882 | 0.4576 | 0.2709 | 0.3148 | 0.3478 |
| 18 | 0.9691 | 1 | 0.2353 | 0.0309 | 0 | 0.7647 | 0.4850 | 0.5 | 0.2833 | 0.4228 |
| 19 | 0.3993 | 0.0537 | 0.7647 | 0.6007 | 0.9463 | 0.2353 | 0.3124 | 0.2569 | 0.4048 | 0.3247 |
| 20 | 0.6386 | 0.2142 | 0.7059 | 0.3614 | 0.7858 | 0.2941 | 0.3673 | 0.2800 | 0.3864 | 0.3445 |
| 21 | 0.4948 | 0.7973 | 0.0882 | 0.5052 | 0.2027 | 0.9118 | 0.3322 | 0.4157 | 0.2615 | 0.3365 |
| 22 | 0.5358 | 0.1516 | 0.7059 | 0.4642 | 0.8484 | 0.2941 | 0.3415 | 0.2705 | 0.3864 | 0.3328 |
| 23 | 0.6848 | 0 | 0.5882 | 0.3152 | 1 | 0.4118 | 0.3802 | 0.25 | 0.3542 | 0.3281 |
| 24 | 0.9238 | 0.9208 | 0.0588 | 0.07635 | 0.0792 | 0.9412 | 0.4646 | 0.4633 | 0.2576 | 0.3952 |
| 25 | 0.6846 | 0.0037 | 0.5 | 0.3154 | 0.9963 | 0.5 | 0.3801 | 0.2505 | 0.3333 | 0.3213 |
| 26 | 0.9291 | 0.1461 | 0.4118 | 0.0709 | 0.8539 | 0.5882 | 0.4667 | 0.2697 | 0.3148 | 0.3505 |
| 27 | 1 | 0.7045 | 0 | 0 | 0.2955 | 1 | 0.5 | 0.3860 | 0.25 | 0.3787 |

**Table 11** ANOVA for grey relational grade (GRG)

| Source | DF | Seq SS | Adj SS | Adj MS | F | P | % Contribution |
|---|---|---|---|---|---|---|---|
| A | 2 | 0.000352 | 0.000352 | 0.000176 | 0.74 | 0.506 | 2.30 |
| B | 2 | 0.000531 | 0.000531 | 0.000266 | 1.12 | 0.373 | 3.47 |
| C | 2 | 0.002916 | 0.002916 | 0.001458 | 6.15 | 0.024[a] | 19.04 |
| D | 2 | 0.000210 | 0.000210 | 0.000105 | 0.44 | 0.657 | 1.37 |
| E | 2 | 0.005911 | 0.005911 | 0.002955 | 12.46 | 0.003[a] | 38.59 |
| B*E | 4 | 0.001836 | 0.001836 | 0.000459 | 1.94 | 0.198 | 11.99 |
| C*E | 4 | 0.001667 | 0.001667 | 0.000417 | 1.76 | 0.230 | 10.88 |
| Residual Error | 8 | 0.001897 | 0.001897 | 0.000237 | | | 12.36 |
| Total | 26 | 0.015320 | | | | | 100 |

[a]Significant parameters at 95% confidence interval

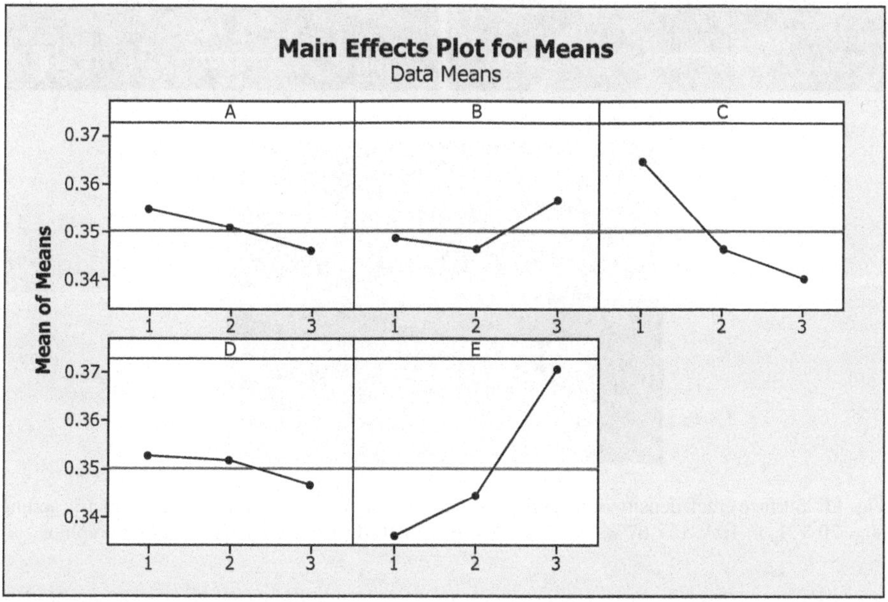


**Fig. 10** Main effects plot for means of GRG

**Table 12** Response table grey relational grade (GRG)

| Level | A | B | C | D | E |
|---|---|---|---|---|---|
| 1 | 0.3546 | 0.3484 | 0.3646 | 0.3527 | 0.3361 |
| 2 | 0.3506 | 0.3462 | 0.3462 | 0.3519 | 0.3442 |
| 3 | 0.3458 | 0.3565 | 0.3402 | 0.3464 | 0.3707 |
| Delta | 0.0088 | 0.0103 | 0.0244 | 0.0063 | 0.0346 |
| Rank | 4 | 3 | 2 | 5 | 1 |

**Table 13** Optimal parametric setting

| Parameters | V (A) | Ip (B) | τ (C) | Ton (D) | Tool type (E) | Fitness value |
|---|---|---|---|---|---|---|
| Optimal setting | 20 V | 20A | 67% | 100 μs | Graphite | 0.27281105 |

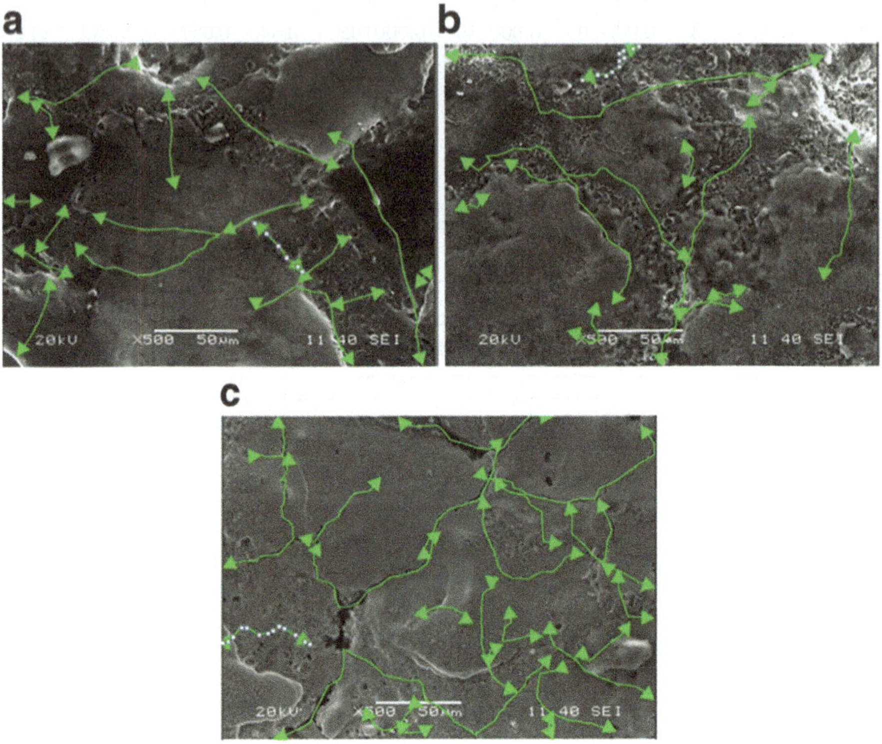


**Fig. 11** Surface crack density of machined surface by different tool electrodes at parametric setting $V = 20$ V, $I_p = 10$A, $\tau = 67\%$, $T_{on} = 100$ μs (**a**) AlSi10Mg RP, (**b**) Copper, and (**c**) Graphite

The firefly algorithm provided the best parametric setting to get optimal EDM performance that is given in Table 13.

$$f(x) = 0.349 \times A^{-0.021} \times B^{0.020} \times C^{-0.065} \times D^{-0.014} \times E^{0.086} \tag{13}$$

## 7 Surface Crack Density of Machined Surface

The scanning electron micrograph of the machined surface by three different types of tool electrodes are taken by SEM (Model: Jeol JSM-6480LV, Japan) and the surface crack density (SCD) of the machined surface are measured by pdf xchange viewer software (as shown in Fig. 11). The SCD of work piece machined by AlSi10Mg, cooper and graphite tool electrodes are 0.01926 μm/μm$^2$, 0.02074 μm/μm$^2$ and 0.03407 μm/μm$^2$ respectively. The SCD of the machined surface machined by AlSi10Mg electrode is less as compare to copper and graphite electrodes. The SCD of the machined surface machined by graphite tool electrode is highest. Here, by the use of graphite electrode, more spark energy generated with increase in the thermal stress. Higher thermal stress is generated with increase in spark energy during the sparking. Therefore, the SCD on the machined surface by the use of graphite electrode is more.

## 8 Conclusion

In the present work, machining performance of AlSi10Mg composite tool electrode prepared via selective laser sintering process along with conventional copper and graphite tool electrodes have been investigated by taking Titanium alloy (Ti6Al4V) as work piece material and commercially available EDM 30 oil as dielectric fluid during EDM. The design of experiment approach like Taguchi's L27 orthogonal array has been used to conduct the experiment by varying the process parameters like oven circuit voltage (V), discharge current (Ip), duty cycle ($\tau$), pulse-on-time (Ton) and type of tool electrodes. The EDM performance like material deposition rate (MRR), tool wear rate (TWR) and average surface roughness (Ra) have been evaluated and the effect of different process parameters on these output responses have been studied. To obtain the best EDM performance, grey relational analysis based Firefly algorithm has been used. From this work following conclusion can be drawn.

1. Better MRR can be achieved with increase in the process parameters like open circuit voltage, discharge current, duty cycle and pulse-on-time and by the use of graphite tool electrode. Graphite tool electrode have higher material removal capacity followed by cooper and AlSi10Mg SLS tool electrode.
2. Lower value of TWR can be achieve with lower parametric setting of open circuit voltage, discharge current, duty cycle and pulse-on-time and by the use of graphite tool electrode. Graphite tool electrode has less tool wear as followed by cooper and AlSi10Mg SLS tool electrodes.
3. The average surface roughness value of the machined surface is increased with increase in the process parameters like open circuit voltage, discharge current, duty cycle and pulse-on-time and better surface finish achieved by using AlSi10Mg SLS tool electrode followed by copper and graphite tool electrodes.
4. The SCD of the machined surface machined by AlSAi10Mg RP electrode is less as compare to the machined surfaces by the use of copper and graphite electrodes. Therefore, AlSi10Mg RP tool electrodes must be use to achieve less surface crack density on the machined surfaces.

5. By using GRA based Firefly algorithm best optimal setting of the machining parameters has been obtained to increase MRR and to decrease TWR and average surface roughness of the machined surface. The best parametric setting obtained is $V = 20$ V, $I_p = 20$A, $\tau = 67\%$, $T_{on} = 100$ μs and graphite electrode.

## References

1. Arthur, A., Dickens, P.M., Cobb, R.C.: Using rapid prototyping to produce electrical discharge machining electrodes. Rapid Prototyping Journal, 2, 1, 4–12(1996)
2. Czelusniak, T., Amorim, F.L., Lohrengel, A., Higa, C.F.: Development and application of copper–nickel zirconium diboride as EDM electrodes manufactured by selective laser sintering. Int J Adv Manuf Technol, 72, 905–917 (2014)
3. Sahu, A.K., Mohanty, P.P., Sahoo S.K.: Electro discharge machining of Ti-alloy (Ti6Al4V) and 316L Stainless Steel and Optimization of process parameters by Grey relational analysis (GRA) method, In: Wimpenny D.I.et al (eds.) Advances in 3D Printing & Additive Manufacturing Technologies, pp. 65–78. Springer Science+Business Media Singapore (2017)
4. Nourbakhsh, F., Rajurkar, K.P., Malshe, A.P., Cao, J.: Wire electro-discharge machining of titanium alloy, Procedia CIRP 5,13–18 (2013)
5. Rahul, Datta, S., Biswal, B.B., and Mahapatra, S.S.: A Novel Satisfaction Function and Distance-Based Approach for Machining Performance Optimization During Electro-Discharge Machining on Super Alloy Inconel 718, Arabian Journal for Science and Engineering, 42, 1999–2020 (2017)
6. Mohanty, C.P., Mahapatra, S.S., and Singh, M.R.: An intelligent approach to optimize the EDM process parameters using utility concept and QPSO algorithm, Engineering Science and Technology an International Journal, 20, 552–562 (2017)
7. Sahu, A.K., Mahapatra, S.S., Chatterjee, S.: Optimization of electro-discharge coating process using harmony search. Materials today: Proceedings, 5, 12673–12680 (2018)
8. Dewangan, S., Gangopadhyay, S., Biswas, C.K.: Study of surface integrity and dimensional accuracy in EDM using Fuzzy TOPSIS and sensitivity analysis, Measurement, 63 364–376 (2015)
9. Czelusniak, T., Amorim, F.L., Higa, C.F., Lohrengel, A.: Development and application of new composite materials as EDM electrodes manufactured via selective laser sintering, Int J Adv Manuf Technol 72, 1503–1512 (2014)
10. Durr, H., Pilz, R., Eleser, N. S.: Rapid tooling of EDM electrodes by means of selective laser sintering. Computers in Industry, 39, 35–45 (1999)
11. Zhao, J., Li, Y., Zhang, J., Yu, C., Zhang, Y.: Analysis of the wear characteristics of an EDM electrode made by selective laser sintering, Journal of Materials Processing Technology, 138, 475–478 (2003)
12. Tang, Y., Hong, J., Zhou, H., Lu, B.: A new technique for the fabrication of graphite EDM electrodes, Journal of Materials Processing Technology, 166, 199–204 (2005)
13. Ding, Y., Lan, H., Hong, J., Wu, D.: An integrated manufacturing system for rapid tooling based on rapid prototyping, Robotics and Computer-Integrated Manufacturing, 20, 281–288 (2004)
14. Datta, S., Mahapatra, S.S.: Modeling , simulation and parametric optimization of wire EDM process using response surface methodology coupled with grey-Taguchi technique, International Journal of Engineering, Science and Technology, 2, 5, 162–183 (2010)
15. Raja, S.B., Pramod, C.V.S., Krishna, K.V., Ragunathan, A., Vinesh, S.: Optimization of electrical discharge machining parameters on hardened die steel using Firefly Algorithm, Engineering with Computers, 31, 1–9 (2015)

16. Shukla, R., Singh, D.: Selection of parameters for advanced machining processes using firefly algorithm, , Engineering Science and Technology an International Journal, 20, 212–221 (2017)
17. Mishra, S.B., Malik, R., Mahapatra, S.S.: Effect of external perimeter on flexural strength of FDM build parts, Arabian Journal of Science and Engineering, 42, 4587–4595 (2017)
18. Julong, D.: Control problems of grey systems, Systems & control letters, 1, 5, 288–294 (1982)
19. Julong, D.: Introduction to Grey System Theory, The Journal of Grey Systems, 1, 1–24 (1988)
20. Yang, X. S.: Firefly Algorithms for Multimodal Optimization, In: Watanabe, O., Zeugmann, T. (eds.): SAGA 2009, LNCS 5792, pp. 169–178, Springer-Verlag Berlin Heidelberg (2009)
21. Yang, X. S.: Multiobjective firefly algorithm for continuous optimization, Engineering with Computers, 29, 175–184 (2013)
22. Yang, X. S.: Nature-inspired metaheuristic algorithms, Luniver press, United Kingdom (2010)
23. Sahu, A.K., Mahapatra, S.S., Chatterjee, S.: Optimization of electrical discharge coating process using MOORA based firefly algorithm, Proceeding of the ASME Gas Turbine Conference GTINDIA2017, December 7–8, Bangalore, India (2017)
24. ASM Material Data Sheet, http://asm.matweb.com/search/SpecificMaterial.asp?bassnum=mtp641

# Laser-Based Additive Manufacturing of Lightweight Metal Matrix Composites

Eskandar Fereiduni, Mostafa Yakout, and Mohamed Elbestawi

## 1 Introduction

In recent decades, there has been an increasing demand in modern industries for the development of high-performance lightweight materials having outstanding mechanical properties. As well-known lightweight metallic materials, the poor wear resistance and low hardness of aluminum (Al) and titanium (Ti) are serious impediments limiting their further development especially in automotive and aerospace industries. If reinforced with ceramic reinforcing particles, the obtained materials are capable of meeting most of the industrial requirements. These materials are known as light-weight metal matrix composites (MMCs) and offer outstanding combination of properties including low density, low coefficient of thermal expansion (CTE), high hardness, improved wear resistance and comprehensive mechanical properties.

Lightweight MMCs have been fabricated for many years using a wide variety of manufacturing processes. Based on the temperature of the metallic matrix during processing, the conventional fabrication methods used for producing MMCs can be categorized into five groups as follows [1–4]:

(i) Liquid phase processes (liquid metal ceramic particulate mixing; melt infiltration; squeeze casting or pressure infiltration; reaction infiltration; melt oxidation processing).
(ii) Solid phase processes (powder metallurgy processes such as pressing and sintering, forging and extrusion; high energy and high rate processes; diffusion bonding; plastic deformation processes such as friction stir welding and superplastic forming).

E. Fereiduni · M. Yakout · M. Elbestawi (✉)
Department of Mechanical Engineering, McMaster University, Hamilton, ON, Canada
e-mail: elbestaw@mcmaster.ca



B. AlMangour (ed.), *Additive Manufacturing of Emerging Materials*,
https://doi.org/10.1007/978-3-319-91713-9_3

(iii) Solid/liquid two-phase processes (rheo-casting/compo-casting; variable co-deposition of multiphase materials)
(iv) Deposition techniques (spray deposition; chemical vapour deposition (CVD); physical vapour deposition (CVD); spray forming techniques)
(v) In-situ processes in which the reinforcements are formed in-situ during the fabrication process.

While the conventional methods of MMCs fabrication could be employed to fabricate parts with non-complex geometries, they are not suitable candidates when it comes to the manufacturing of parts with complex shapes. The design freedom associated with AM technology demonstrates an outstanding capability in fabrication of novel structures with unique geometrical designs. In addition, this technology provides manufacturers with the ability to incorporate tailored materials (e.g. reinforcements) into the structure to produce parts with desired properties. By combining the design freedom with the ability to produce lightweight MMCs, AM technology provides an exceptional opportunity to fabricate parts with noticeable weight saving while maintaining the strength. The obtained MMCs parts are good candidates to meet the requirements of various industries such as automotive and aerospace among others. Considering above, this chapter focuses on the use of AM technology in fabricating lightweight MMCs, metal matrix nano-composites (MMnCs) and functionally graded metal matrix composites (FGMMCs). The AM processes and processing techniques employed for producing various types of lightweight MMCs are discussed, and the effect of reinforcements features on the mechanical properties are explored.

## 2 Additive Manufacturing Processes

Additive manufacturing (AM) is a relatively new manufacturing technique, having only been created in 1986. *The father of AM* title is granted to Charles W. Hull, who first patented the technique of Stereolithography (SLA) [5]. In addition to the first technique for AM, he created an intermediary file system, STL, which breaks down the CAD drawing into a series of layers to be interpreted by the AM machine. This file format is currently the de facto standard of file preparation software for all AM machines. After this, Fused Deposition Modeling (FDM) was introduced by Stratasys Inc. in 1991 [6], followed shortly by the introduction of Laminated Object Manufacturing (LOM), according to the Wohlers Report of 2014. Selective Laser Sintering (SLS), the next largest technique, which is capable of additively producing metal and ceramic parts, was created in 1992. Continuing on the rapid growth of AM technology, the following year Inkjet Binder Printing was invented and patented by MIT. It wasn't until 1999 when a technique to produce fully dense metal parts, Selective Laser Melting (SLM), was invented. A notable AM technique invented shortly after this in 2002 was the Direct Metal Deposition (DMD) [7].

The leading companies for AM systems around the world are 3D Systems in USA, Electro Optical Systems (EOS) in Germany, Stratasys in USA, Renishaw in England, and Arburg in Germany. Table 1 shows a list of leading companies for AM systems arranged according to the year of foundation or when they launched their AM technology. For each company, the table summarizes the materials used, the technologies offered, and the applications served. In addition, the table shows the industrial terms used by each company for the technologies offered and their corresponding scientific terms.

**Table 1** List of leading companies in AM systems

| Company | Technology acronym(s) | | Material classes | Applications | Developed |
|---|---|---|---|---|---|
| | Academic | Commercial | | | |
| 3D Systems | SLA | SLA | Polymers | Aerospace, Defense, Architecture, Geo, Arts, Jewelry, Entertainment, Automotive, and Healthcare | 1987 in USA |
| | SLS | PLS | | | |
| | 3DP | CJP | | | |
| | MJM | MJP | | | |
| | SLA | Cast Urethane | | | |
| | 3DP | PJP | | | |
| | SLM | DMP | Metals | | |
| CMET | SLA | SLA | Polymers | Rubber-Like Prototypes and Super Heat Resistant Models | 1988 in Japan |
| EOS | SLM | DMLS | Metals | Aerospace, Automotive, Lifestyle Products, Medical, and Tooling | 1989 in Germany |
| | SLS | PLS | Polymers | | |
| Stratasys | 3DP | PolyJet 3DP | Polymers | Aerospace, Automotive, Industrial, Defense, Medical, and Dental | 1989 in USA |
| | FDM | FDM | | | |
| Sciaky | EBM | EBAM | Metals | Large Parts, Aerospace, Automotive, Nuclear, Medical, Power Generation, Marine | 1990 in USA |
| Arcam | EBM | EBM | Metals | Aerospace and Orthopedic Implants | 1997 in Sweden |
| Optomec | DED | LENS | Metals | Components Repair, Hybrid Manufacturing, Medical Implants, and Electronics | 1997 in USA |
| SLM GmbH | SLM | SLM | Metals | Automotive, Medical, Dental, Aeronautics, and Construction | 1998 in Germany |

(continued)

**Table 1** (continued)

| Company | Technology acronym(s) | | Material classes | Applications | Developed |
|---|---|---|---|---|---|
| | Academic | Commercial | | | |
| Voxeljet | 3DP | Continuous 3D Printing | Sands | Molds and Tools for Sand Casting | 1999 in Germany |
| | | | Plastics | Aerospace, Automotive, Medical, Art, and Architecture | |
| Carima | SLA | DLP | Polymers | Architecting, Machine Elements, Jewellery, Plaything, and Medical | 2000 in Korea |
| Concept Laser | SLM | Laser CUSING | Metals | Aerospace, Automotive, Dental, Medical, Jewellery, and Molds | 2000 in Germany |
| Envision TEC | LTP | DLP | Polymers | Aerospace, Biomedical, Electronics Architecture, and Automotive. | 2002 in Germany |
| | | 3SP | | | |
| ReaLizer | SLM | SLM | Metals and Ceramics | Dental, Automotive, Tools, Medical, and Jewellery | 2004 in Germany |
| Mcor Tec. | LOM | SDL | Papers | Geospatial, Arts, Architecture, Medical, and Manufacturing | 2004 in Ireland |
| ExOne | 3DP | Sand 3DP | Sands | Molds and Tools for Sand Casting | 2005 in USA |
| | | Metal 3DP | Metals | Aerospace, Automotive, Art, and Heavy Equipment | |
| DWS | LTP | X-Y Scanning | Polymers | Jewellery, Fashion, Dental, Biomedical, and Industrial | 2007 in Italy |
| | | DLP | | | |
| | | Galvanometer | | | |
| Blueprinter | SLS | SHS | Polymers | Architecting, Design, Engineering, Construction, and Manufacturing | 2008 in Denmark |
| ASIGA | SLA | SAS | Polymers | Dental, Jewellery, Audiology, and Medical | 2009 in USA |
| BeAM | DED | CLAD | Metals | Aeronautics, Aerospace, Defence, Nuclear, and Medical | 2009 in France |
| Farsoon | SLS | SLS | Polymers | Automotive, Architecture, Medical Aerospace, and Manufacturing | 2009 in China |
| Renishaw | SLM | LM | Metals | Dental, Aerospace, Medical, Automotive, and Tooling | 2010 in UK |

(continued)

**Table 1** (continued)

| Company | Technology acronym(s) | | Material classes | Applications | Developed |
|---|---|---|---|---|---|
| | Academic | Commercial | | | |
| LUXeXcel | 3DP | Printoptical | Transparent Polymers | LED Lighting Industry | 2012 in Netherlands |
| Arburg | FDM | AKF | Polymers | Manufacturing, Medical, and Dental | 2013 in Germany |
| Carbon 3D | SLA | CLIP | Polymers | Automotive, Architecture, Fashion, and Medical | 2013 in USA |
| XJET | 3DP | NPJ | Metals | Nano-Manufacturing | 2016 in Israel |

# 3 Additive Manufacturing Versus Conventional Manufacturing

AM technology opens the door to produce complex, customized, and cost-effective components with acceptable mechanical properties. It allows a vast innovation in material development and metallurgical engineering. In addition to polymers, metals, and ceramics, AM technology provides an outstanding opportunity for fabricating composites and complex materials.

AM technology has a lot of advantages over conventional manufacturing techniques. One of these advantages is the ability to produce complex designs from either traditional or newly developed materials. This will allow high flexibility in producing parts containing various elements within the same design. Figure 1 shows an example of a multi-directional preform having large numbers of yarn groups and a complex architecture. As shown, the unit-cell consists of three different yarn orientations. If not impossible, this design could be difficulty manufactured using conventional manufacturing methods. AM technology allows the manufacturing of multi-directional composite structures as well as the optimization of parts design [8].

New properties, forms, and functionalities can be also created by processing new materials to produce complex structures using AM technology. Despite the conventional manufacturing techniques, the design freedom in AM technology allows the designer to design parts with complex geometries or even to apply topology optimization to produce parts with the desired dimensions, complexities and properties [9]. For instance, while being difficult or even impossible for some of the cellular structures in Fig. 2 to be produced with conventional manufacturing processes, these complex geometries could be made using AM processes.

Parts with different materials or functionally graded objects could be also manufactured using AM technology. The possibility to use different feedstocks in some of the AM processes enables the manufacturers to produce multi-material structures and functionally graded materials (FGMs). Figure 3 shows a dual alloy blisk (as an example of multi-material component) fabricated by direct laser deposition (DLD) process using powder feedstock. FGMs are those structures having

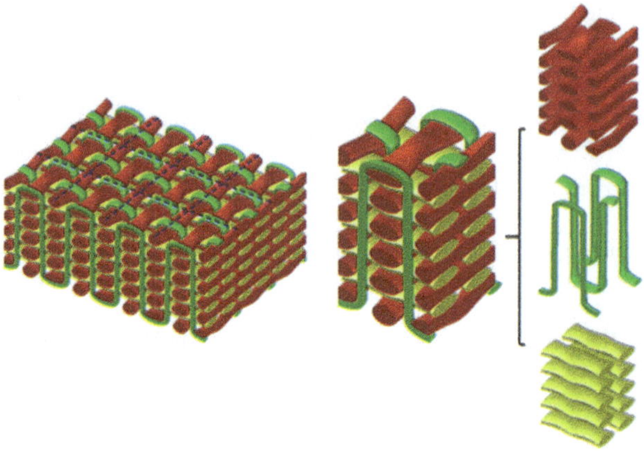

Fig. 1 An example of multi-directional structure that could be manufactured using AM processes [8]

Fig. 2 Various cell structures that could be manufactured using AM technology [9]

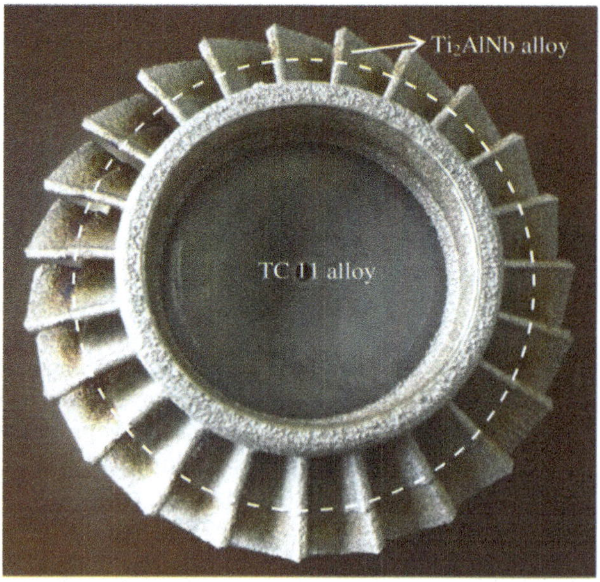


Fig. 3 DLD processed TC11/$Ti_2AlNb$ dual alloy blisk [10]

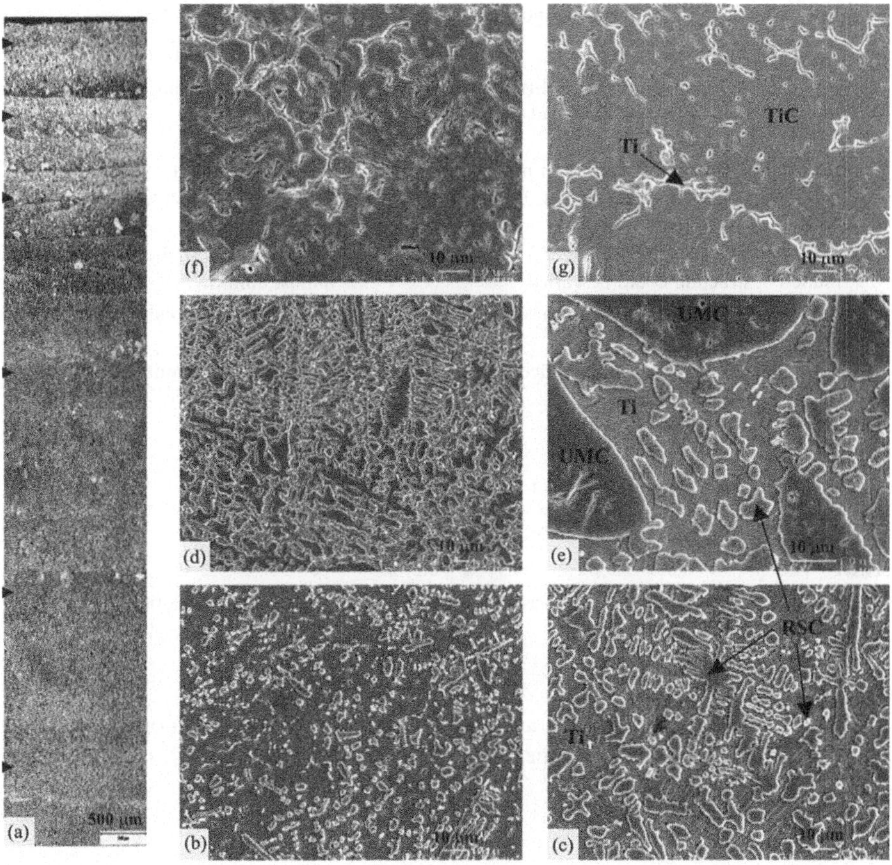


**Fig. 4** (**a**) Optical cross-section and (**b–g**) SEM micrographs of functionally graded TiC reinforced Ti matrix composite with increasing TiC content along the building direction (bottom to top). UMC and RSC refer to un-melted and re-solidified TiC reinforcement, respectively [11]

varying material properties along one direction of the structure. Figure 4 provides an example of functionally graded MMC benefiting from gradient change of reinforcement content along the building direction.

AM technology can be also employed to create customized alloys and composite materials. The possibility of using customized mixtures of powders in AM will provide the opportunity to fabricate composites with improved mechanical properties compared to the conventionally processed parts. The size, volume fraction, distribution pattern and orientation of reinforcement incorporated into the matrix affect the characteristics of fabricated MMCs. The control and optimization of these parameters are the main challenges in the manufacturing of MMCs. The high degree of complexity and customization as well as the flexibility in optimizing the characteristics of reinforcements are the main advantages of AM processes over conventional manufacturing methods used for fabricating MMCs.

**Table 2** AM versus conventional manufacturing [1, 12–15]

| Factor | AM processes | Conventional manufacturing processes |
|---|---|---|
| Part complexity | AM offers high flexibility in producing complex geometries and customized products | They require assembly for complex designs |
| Part design | AM offers flexibility in manufacturing lightweight components (i.e. hollow or lattice structures) | Conventional manufacturing methods are not applicable for hollow and/or lattice structures |
| Material distribution | The material distribution within a part structure could be varied | The material distribution within a part is almost uniform |
| Production volume | Low production volumes | Large production volumes |
| Production size | Available for the production of low and medium size parts | They provide the production of large-size components |
| Post-processing | Mandatory | Optional |
| Assembly | It reduces the number of part and assembly stages | It requires assembly |
| Material cost | High | Low |
| Material waste | Reduces the material waste | High amount of material waste |

Table 2 lists the advantages and disadvantages of AM technology over other conventional manufacturing processes.

# 4 Additive Manufacturing Processes for Fabricating MMCs

According to ISO/ASTM 52900 standard [16], the AM processes are classified into: VAT polymerization, material jetting, powder bed fusion (PBF), binder jetting, direct energy deposition (DED), and sheet lamination. These processes can be categorized according to the state of the raw material into: liquid, molten, powder, and solid-layer processes as shown in Fig. 5. While some of these processes are available for polymers, the others are available for metals and ceramics.

AM processes could allow the fabrication of MMCs reinforced with either particulates or fibres. While few attempts have been performed to produce fibre reinforced MMCs using solid layer AM processes, most of the literature existing on AM processing of MMCs refers to powder-based processes used to fabricate particulate reinforced MMCs. In fact, solid-layer AM processes (e.g. sheet lamination) might not be applicable for fabricating MMCs parts with complex geometries [17, 18]. This chapter focuses only on particulate reinforced MMCs made through powder-based AM processes.

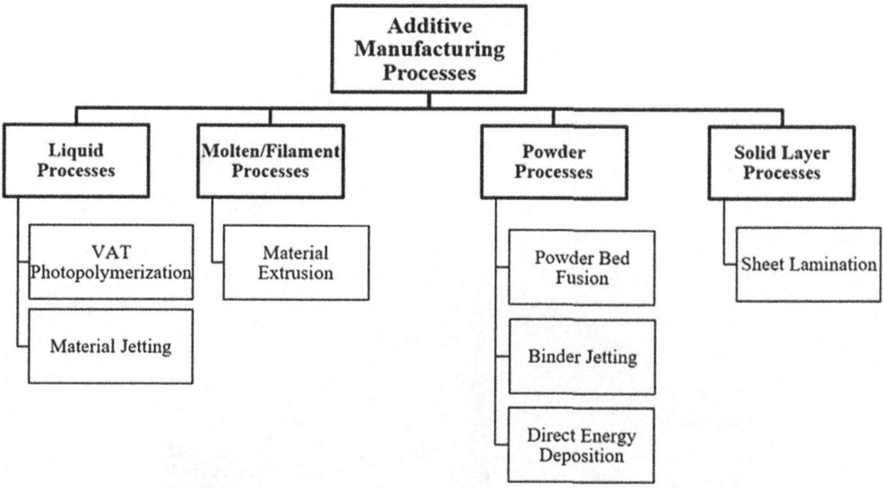


**Fig. 5** Classification of ISO/ASTM 52900 standard AM processes according to the state of material

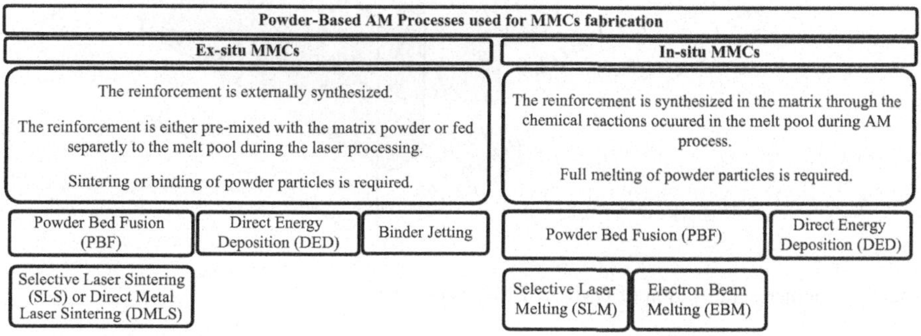


**Fig. 6** Powder-based AM processes used for fabricating two different categories of MMCs

In general, powder-based AM processes can be used to fabricate two major categories of MMCs, i.e. ex-situ and in-situ reinforced MMCs [181]. Figure 6 presents the AM processes employed for manufacturing each category of these MMCs.

## *4.1 Powder Bed Fusion (PBF) Processes*

The PBF processes such as DMLS, SLM and EBM are available for fabricating both ex-situ and in-situ reinforced MMCs. In DMLS or SLS processing of MMCs, a single powder layer (mixture of reinforcements and the matrix powders) is spread on a bed, then a focused laser beam is directed at the layer to sinter the mixed powder as

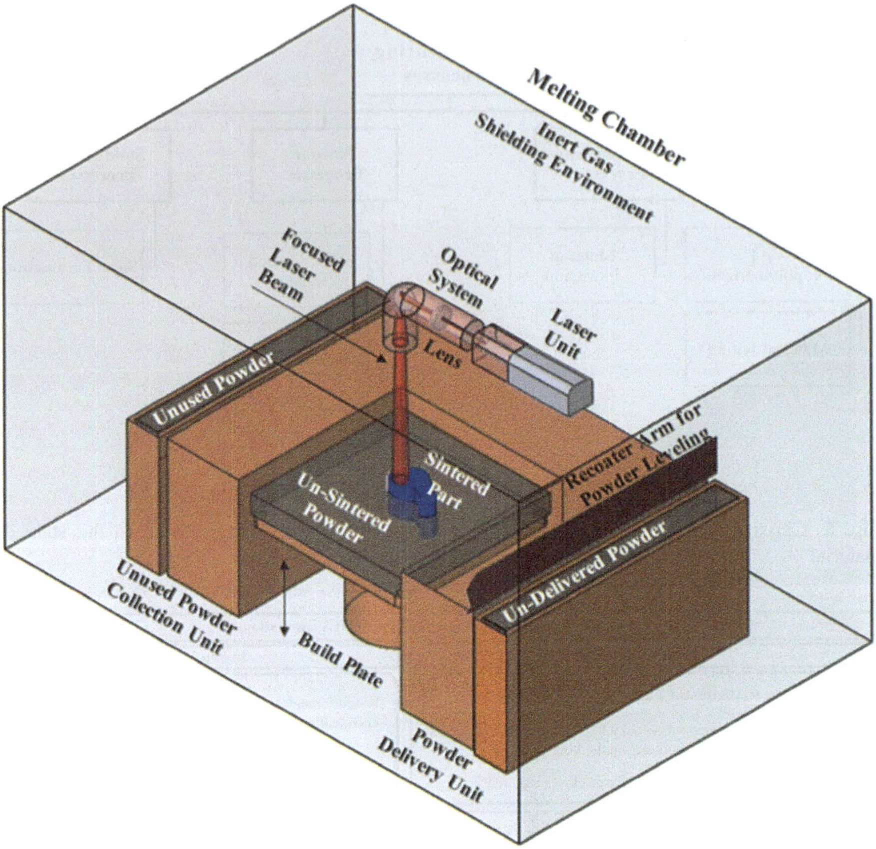


**Fig. 7** Schematic illustration of DMLS/SLM process [19]

shown in Fig. 7. The SLM process reflects the SLS and DMLS processes with full melting of the powder particles. The EBM process is similar to SLM process except that the powder is melted using an electron beam instead of a laser source.

The sintering phenomenon associated with AM processes with partial melting nature (e.g. SLS or DMLS) leads to the production of ex-situ reinforced MMCs. However, the AM processes with full-melting nature (e.g. SLM and EBM) could be used for fabricating in-situ reinforced MMCs through activating chemical reactions between the constituents existing in the system.

## 4.2 Direct Energy Deposition (DED) Processes

DED is a powder-fed AM process in which the powder is fed through a nozzle and then melted through a laser/electron beam as shown in Fig. 8 [20, 21]. DED

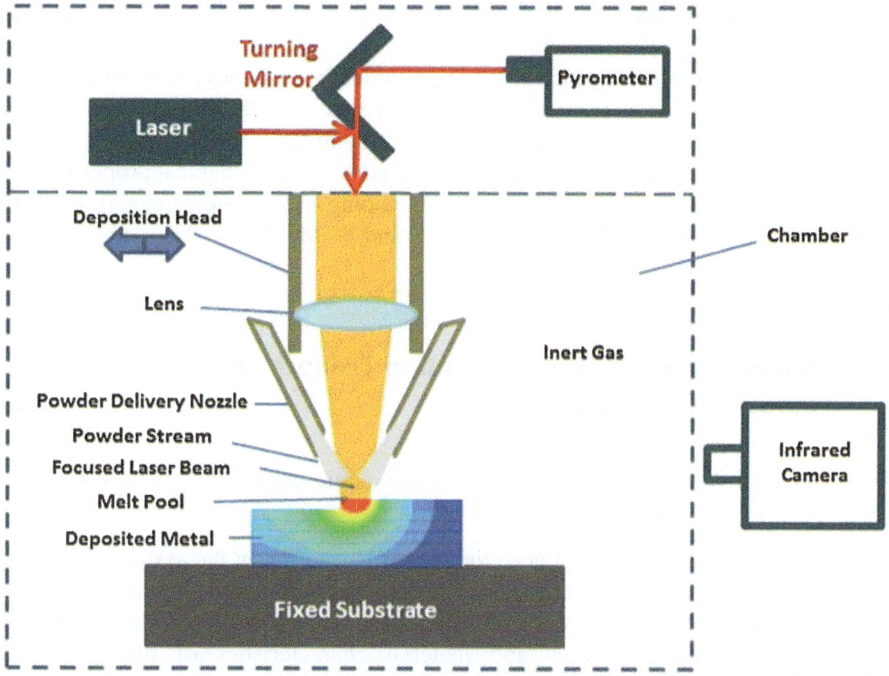


**Fig. 8** Schematic illustration of DLD process [25]

processes with a laser source are called Direct Laser Deposition (DLD), Laser Metal Deposition (LMD) [13], Laser Engineered Net Shaping (LENS) [22], Laser Consolidation (LC) [23] or Laser Cladding [24]. It has the ability to reface or add layers onto objects (cladding or hardfacing) [21] as well as to deposit multiple materials in one process loop. These multi-materials are FGMs and allow designing of material with varying compositions and microstructures depending on the position [20]. In terms of metallurgy, DED creates a mobile melt pool on the substrate with a cooling rate which is significantly dependent on the input of powders [13].

The DED processes utilize knowledge of thermal and kinetic parameters such as temperature, velocity and composition, which are essential to understand if the technique is successfully used [13]. The processes with DED nature such as LENS, LMD and Cladding are all available for fabricating both ex-situ and in-situ reinforced MMCs. The DED systems benefiting from several nozzles are outstanding systems enabling the fabrication of MMCs and FGMs.

Plasma Deposition Manufacturing (PDM) also belongs to the DED category of AM processes. This process starts with supplying a continuous powder feed to a plasma-melted zone where the powder then gets melted and solidified as the plasma beam scans across the deposited layer. In order to avoid or reduce oxidation, processing occurs in an environment with inert gas such as argon, helium or nitrogen. PDM is a feasible technique for fabricating 3D parts as well as FGMs.

The parts are metallurgically acceptable but further experiments are necessary to optimize this process [26].

In addition to the DED processes using powder as feedstock, some of the DED processes employ wire as feed-stock. During the wire-fed DED processes, a wire of small diameter is fed, melted, and deposited on the substrate. This repetitious process leads to the fabrication of very dense and large parts with mechanical properties that are competitive to cast material [27]. By developing DED systems benefiting from both wire and powder feeding, novel MMCs and FGMs can be also manufactured.

## 5 Challenges of MMCs Fabrication Using Additive Manufacturing Processes

Fabrication of MMCs with AM processes has some challenges which have been summarized as follows [19, 28–33, 180]:

1. The chemical reactions between the elements existing in the system. Although the chemical reactions between the powder particles are essential to develop in-situ reinforced MMCs, the occurrence of unwanted reactions may deteriorate the mechanical properties and material characteristics of the developed composites.
2. Irregular dissolution of elements that may take place during the AM process.
3. Extensive cracking between tracks or layers due to the high thermal gradients induced during the process.
4. Loss of specific material properties (e.g. the decrease in the content of alloying elements due to over-heating and excessive energy density subjected to the system).
5. Micro-segregation of reinforcements to the bottom of each melt pool, especially in the case of ex-situ reinforced MMCs.
6. Micro-cracking at the reinforcement/matrix interface which is mainly attributed to the significant difference in physical and mechanical properties of reinforcements and the matrix. These micro-cracks affect the functionality of fabricated parts and may cause failure.
7. The residual stresses induced in inter-track/inter-layer regions and consequently the whole part might affect the mechanical properties of fabricated components.
8. The balling effect that may happen due to the improper selection of process parameters. This phenomenon might inversely affect the surface integrity and densification level of parts.
9. The thermal history experienced by a single layer during the deposition of subsequent layers may cause microstructural changes in it.
10. The determination of optimum process parameters for each specific composite material.

These challenges could be addressed and solved using the following suggested considerations:

1. Optimizing the process parameters for various MMCs and finding their influence on the surface integrity, mechanical properties and material characteristics of the parts produced [34, 35].
2. Modifying the AM machines to produce new in-situ MMCs and developing new multi-material AM systems [36–38].
3. Design and topology optimization for AM [39, 40].
4. Creating a real-time process control for the AM systems [36].
5. Building a material database for AM of each MMC system [36, 41, 42].

# 6 Various Types of MMCs

## *6.1 Ex-Situ Reinforced MMCs*

Particle reinforced MMCs are fabricated by various conventional methods. In most of these processes, the ceramic reinforcing particles added to the matrix have been made separately prior to the composite fabrication process. As a result, such composites are called ex-situ reinforced MMCs.

Laser-based AM processes have been recently used to fabricate 3D MMC structures as well as coatings to improve the properties of materials [43]. One of the unique features of AM processes is their capability to incorporate reinforcements either ex-situ or in-situ into the structure. While the size and shape of reinforcements in ex-situ reinforced MMCs are limited by the initial powder condition [44], the in-situ reinforced MMCs benefit from desired reinforcing particles synthesized directly from the chemical reaction between the reactants in the melt pool.

The SiC particle reinforced Al matrix composites (AMCs) are one of the most common ex-situ reinforced MMCs fabricated by AM processes. Figures 9a, b show the microstructures of DMLS processed SiC reinforced $Al_{4.5}Cu_3Mg$ matrix composites having two different particle sizes of SiC (i.e. coarse and fine) [45]. Although the SiC ceramic reinforcing particles may partially melt and dissolve into the Al matrix and probably form a few in-situ phases, the remaining of primarily added SiC mostly as un-melted particles in the matrix can introduce the obtained parts as ex-situ reinforced MMCs. Besides the applied AM process (e.g. laser sintering (LS) or laser melting (LM) nature of the process) and the selected process parameters, the degree of melting experienced by the ceramic reinforcing particles is also affected by their size. While the large size reinforcing particles may remain relatively un-melted in the solidified microstructure, the finer ones could be partially melted during the laser processing.

The main drawbacks associated with the ex-situ reinforced MMCs are the relatively weak interfacial bonding between the ceramic reinforcing particles and the matrix as well as the poor wettability between them due to the surface contaminations existing on pre-processed reinforcing particles. Combination of these two

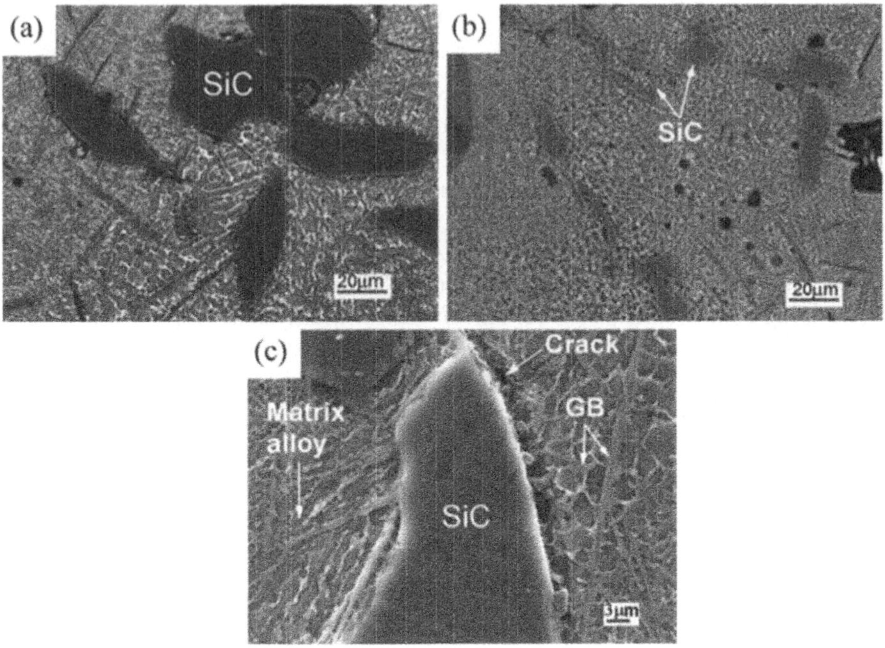


**Fig. 9** SEM micrographs of laser sintered $SiC/Al_{4.5}Cu_3Mg$ powder mixtures having: (**a**) coarse and (**b**) fine SiC ceramic particles. (**c**) Cracking at the SiC/Al matrix interface of laser processed parts [45]

**Table 3** AM processed lightweight MMCs having hybrid ex-situ/in-situ reinforcements

| Mixed powder system | Applied AM process | Reinforcement | | Morphology of In-Situ reinforcements | Reference |
|---|---|---|---|---|---|
| | | Ex-Situ | In-Situ | | |
| SiC/ AlSi10Mg | SLM | Un-melted SiC | $Al_4SiC_4$ | Combined Plate-like & Particle-shape | [46] |
| SiC/A356 | DMLS | | $Al_4SiC_4$ | Plate-like | [47, 48] |
| | | | Si | Particle-shape | |
| $Al_2O_3$/Al | SLM | Un-melted $Al_2O_3$ | $Al_2O_3$ | Long-strip | [49] |
| TiN/Ti6Al4V | LENS | Un-melted TiN | TiN | Elongated rods | [50] |
| SiC/Ti | LENS | Un-melted SiC | TiC | Dendritic | [51, 52] |
| | | | $Ti_5Si_3$ | Faceted particles | |

phenomena and the noticeable difference between the CTEs of ceramic particles and the metallic matrix introduces the particle/matrix interfaces as potential locations for crack nucleation and propagation (Fig. 9c). These cracks lower the load transfer efficiency from the matrix to the reinforcements and inversely affect the mechanical properties of MMCs. Moreover, when subjected to external loading, these cracks may easily propagate into the matrix and lead to premature failure of MMCs parts.

## 6.2 Hybrid Ex-Situ/In-Situ Reinforced MMCs

In recent years, several AM processes including LENS, DMLS and SLM have been used for fabricating lightweight AMCs and Ti matrix composites (TMCs) containing hybrid ex-situ/in-situ reinforcements. Table 3 summarizes various material systems used to produce these types of composites via AM processes. It should be noted that the ex-situ part of the reinforcements refer to the primarily added reinforcing particles remained un-melted in the microstructure. However, the in-situ part represents those reinforcements synthesized in an in-situ manner during the fabrication process. The in-situ formed reinforcements may be either as the same or as a different phase from the starting reinforcing particle.

The microstructures shown in Fig. 10 are hybrid ex-situ/in-situ reinforced AMCs and TMCs fabricated by AM processing of $Al_2O_3$/Al, TiC/Ti6Al4V and $TiB_2$/Ti6Al4V powder mixtures. The obtained MMCs contain un-melted or partially melted $Al_2O_3$, TiC and $TiB_2$ particles remained even after the solidification process. The remaining of un-melted particles in the microstructure can be ascribed to several reasons including relatively low laser energy densities applied to the system, and large-size and/or high contents of high melting point ceramic reinforcing particles present in the powder mixture. Under these conditions, sufficient energy and time may not be provided for complete melting of reinforcing particles, leading to the

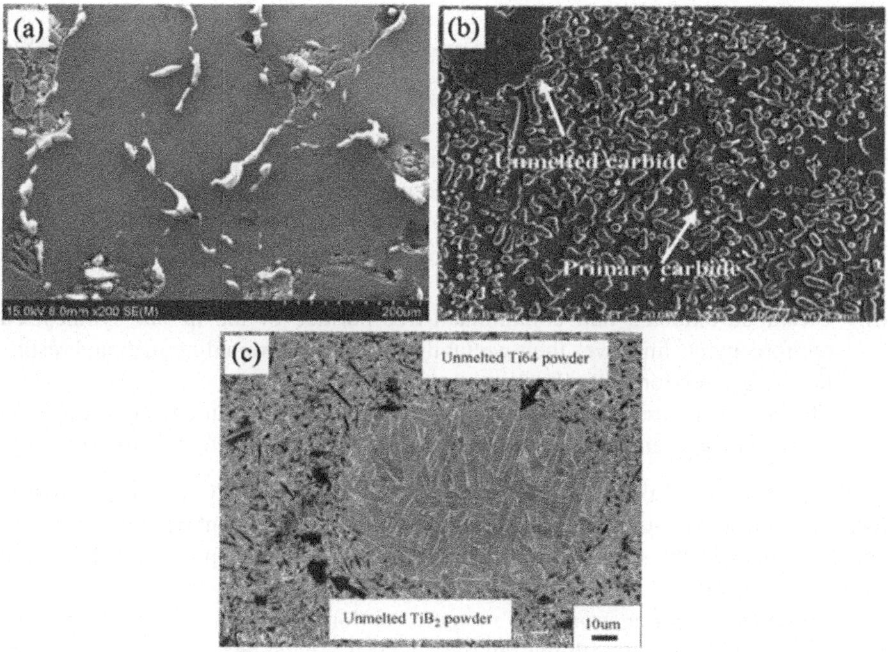


**Fig. 10** SEM micrographs of: (**a**) SLM processed $Al_2O_3$/Al [49], (**b**) Direct laser fabrication (DLF) processed TiC/Ti6Al4V [53], and (**c**) DLF processed $TiB_2$/Ti6Al4V [54] systems

remaining of un-melted or partially melted particles in the microstructure along with the in-situ formed precipitates. Figure 10a shows the un-melted $Al_2O_3$ particles gathered at the grain boundaries to form colonies. This is while the un-melted TiC and $TiB_2$ particles in Figs. 10b, c are found all over the matrix. The part of $Al_2O_3$ particles melted and dissolved into the matrix tends to reprecipitate as $Al_2O_3$ phase in a long-strip morphology along the grain boundaries of the Al matrix [49]. To the same, TiC powder particles melted and dissolved into the matrix also precipitate again as TiC in the microstructure. However, the in-situ reaction between the $TiB_2$ reinforcing particles and the Ti matrix results in the synthesis of a new phase (TiB).

## 6.3 *In-Situ Reinforced MMCs*

Depending on the applied processing parameters (e.g. laser power, scan speed, layer thickness and hatch spacing) and the properties of mixed powder system (e.g. melting points of constituents, size and volume fraction of powder particles), varying amounts of reaction may proceed in the system to form in-situ products. Compared with the ex-situ and hybrid ex-situ/in-situ reinforced MMCs, the primarily added reinforcing particles are fully consumed in the case of in-situ reinforced MMCs. This results in the formation of microstructures free of primarily added reinforcing particles, containing desired reinforcements synthesized directly from the induced chemical reaction among the starting materials. The in-situ reinforced MMCs have some advantages over the ex-situ reinforced ones as follows [55–58]:

(i) While the size of reinforcements in ex-situ reinforced MMCs is dictated by their size in the staring powder mixture and is typically in the order of microns, the fine in-situ synthesized reinforcements with improved distribution state in the matrix offer significantly enhanced mechanical properties.
(ii) The higher thermodynamic stability of in-situ synthesized reinforcements than the ex-situ ones introduces them as better candidates for MMCs having high temperature applications.
(iii) Clean, un-oxidized and contaminant-free surface of the in-situ synthesized reinforcements improves their wettability with the surrounding melt and results in stronger reinforcement/matrix interfacial bonding.
(iv) The in-situ processes are potentially cost-effective and require fewer processing steps since the reinforcements are synthesized during the fabrication process.

Because of benefitting from these unique advantages, MMCs with in-situ synthesized reinforcements have attracted a great attention in recent decades. Tables 4 and 5 summarize the works performed to fabricate in-situ reinforced AMCs and TMCs employing AM processes, respectively.

Because of benefiting from two or more reinforcements in a single composite structure, the hybrid and multiple reinforced MMCs have been recently developed and have attracted a great attention. The in-situ reactions in these MMCs are controlled by several parameters including the type and ratio of powders in the

**Table 4** Additively manufactured AMCs containing in-situ synthesized reinforcements

| Mixed powder system | Applied AM process | In-Situ reinforcement(s) | Morphology of reinforcement(s) | Reference |
|---|---|---|---|---|
| TiC/AlSi10Mg | SLM | TiC | Particulate | [59–64] |
| SiC/AlSi10Mg | SLM | $Al_4SiC_4$ | Plate-like & Particle-shaped both together | [65] |
| $Al_2O_3$/AlSi10Mg | SLM | $Al_2Si_4O_{10}$ | Particulate | [66] |
| AlN/AlSi10Mg | SLM | AlN | Particulate | [67] |
| Al-20Si-5Fe-3Cu-1Mg | SLM | $Al_4FeSi_2$ | Needle-like | [68] |
| TNM/Al-12Si | SLM | $Al_6MoTi$ | Irregularly-shaped dendrites and particles | [69] |
| Al85Nd8Ni5Co2 | SLM | $AlNd_3$, $Al_4CoNi_2$, $AlNdNi_4$ | Platelet | [70] |
| $Fe_2O_3$/(Al) | SLM | Al-Fe IMCs[a], α-$Al_2O_3$ | Particulate | [71] |
| $Fe_2O_3$/(AlMg1SiCu) | | Al-Fe IMCs, α-$Al_2O_3$ | | |
| $Fe_2O_3$/(AlSi10Mg) | | Al-Fe IMCs, $Al_{0.5}Fe_3Si_{0.5}$ | | |
| SiC/AlSi10Mg | DMLS | $Al_4C_4$ | Elongated acicular | [72] |
| $TiB_2$/AlSi10Mg | SLM | $TiB_2$ | Rod-like | [73] |

[a]Intermetallic compounds

mixed powder system as well as the applied processing parameters affecting the temperature and the duration of reaction in the melt pool.

Typical microstructures of hybrid (TiB + TiC) and (TiB + TiN) reinforced TMCs fabricated by LENS process are presented in Figs. 11a, b, respectively. The hybrid (TiB + TiC) reinforced TMC is produced by in-situ reaction between $B_4C$ and Ti powders while the (TiB + TiN) one is made as a result of the induced reaction between BN and Ti part of Ti6Al4V powder.

Although having different size and distribution pattern, TiB reinforcing phase shows a needle-like morphology in both of the hybrid reinforced microstructures. The differences between the size and distribution pattern of TiB reinforcements are related to the differences in size of primarily added ceramic powder particles (BN, $B_4C$) as well as the B content of the systems. The partial melting of relatively large size (30–45 μm) $B_4C$ powder particles as well as the hypereutectic composition of the Ti-B system result in the formation of coarse primary TiB precipitates distributed in the matrix. However, very small size of BN powder particles (0.5–2 μm) combined with the exothermic nature of reaction between BN and liquid Ti leads to the complete melting of BN powder particles. Although B concentration is as high as 2.5 and 7.5 wt.% in the system, TiB precipitates form as very fine eutectic wisker/nano-rod morphology distributed as a quasi-continuous network in the microstructure (Fig. 11b). To the same feature as the TMCs having either TiC or TiN as

**Table 5** Additively manufactured TMCs containing in-situ synthesized reinforcements

| Mixed powder system | Applied AM process | Reinforcement (s) | Morphology | Parameter (s) affecting morphological change | Ref. |
|---|---|---|---|---|---|
| TiC/Ti | SLM | TiC | Rod-like<br>Lamellar | Increase in laser energy density | [74] |
| | | | Lamellar<br>Accumulated whisker<br>Coarsened dendrite | Increase in laser energy density | [75] |
| | | | Lamellar<br>Feather-structured<br>Dendritic | Enhancement of TiC content | [76] |
| $B_4C$/Ti | DLD | TiB | Needle-like & Prismatic | – | [77] |
| | | TiC | Granular | | |
| $TiB_2$/Ti-Nb-Zr-Ta | LENS | TiB | Needle like & Prismatic | – | [78] |
| $Si_3N_4$/Ti | SLM | TiN | Irregular polyangular<br>Refined near round shape<br>Coarsened dendrite | Increase in laser energy density | [79] |
| Elemental B/Ti6Al4V | DLD or LENS | TiB | Needle-like | – | [54, 80–82] |
| Elemental B/Ti | | | | | |
| $TiB_2$/Ti6Al4V | | | | | |
| $TiB_2$/Ti | SLM | TiB | Needle-like | – | [83–85] |
| SiC/Ti | SLM | TiC | Dendritic | – | [86] |
| | | $Ti_5Si_3$ | Faceted particles | | |
| BN/Ti6Al4V | LENS | TiB | Very fine whiskers and nano-rods | Increased laser energy density and/or Enhanced BN content | [87] |
| | | TiN | Plate like or Flowery-dendritic | | |
| Ti, Al and Graphite elemental powders | SLM | TiC | Round-shaped | – | [88] |
| BN/Ti6Al4V | LENS | TiB | Fine needle like | – | [89] |
| | | TiN | Flowery or Dendritic | | |

reinforcement, the TiC and TiN precipitates show also granular and dendritic/plate-like morphology, respectively in the hybrid reinforced TMCs.

Recently, DMLS process was combined with self-propagating high-temperature synthesis (SHS) to fabricate in-situ reinforced AMCs with multiple reinforcements.

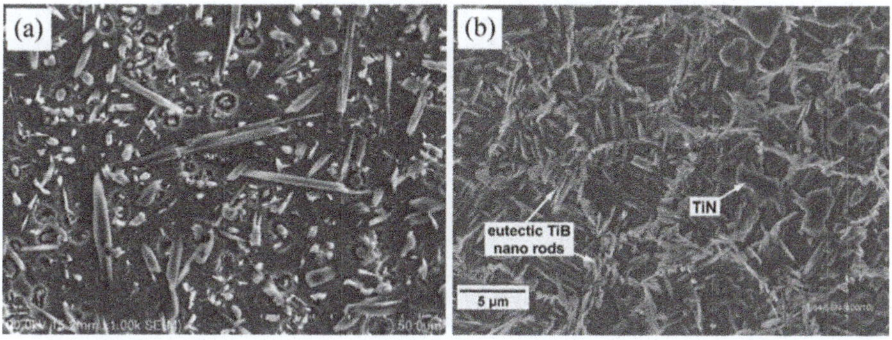


**Fig. 11** Microstructures of hybrid: (**a**) (TiB + TiC) reinforced TMC fabricated by LENS processing of $B_4C$/Ti mixed powder system [77] and (**b**) (TiB + TiN) reinforced TMC made by LENS processing of BN/Ti6Al4V system [87]

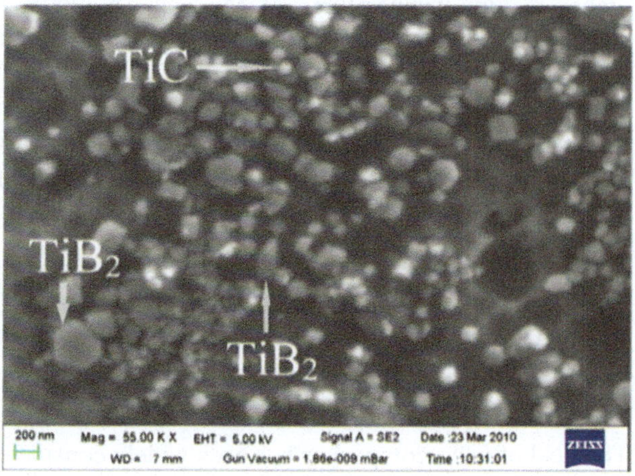


**Fig. 12** Microstructure of DMLS processed $B_4C/TiO_2$/Al mixed powder system [90]

For this purpose, mixture of Al, $TiO_2$ and $B_4C$ powders were used, resulting in the formation of in-situ $Al_2O_3$, TiC and $TiB_2$ reinforcements [90]. Figure 12 shows the particulate shape TiC and $TiB_2$ reinforcements distributed in the Al matrix of obtained AMCs.

## 6.4 TMCs with Metallic Reinforcements

By SLM processing of 10 wt.% Mo/Ti6Al4V powder mixture, Vrancken et al. [91] recently developed TMCs having Mo metallic particles as reinforcements randomly dispersed in a β-Ti matrix (Fig. 13a). TiTa solid solution matrix composites with few un-melted Ta particles as reinforcements have been also developed by SLM processing of 50 wt.% Ta/Ti mixed powder system (Fig. 13b) [92]. Due to the

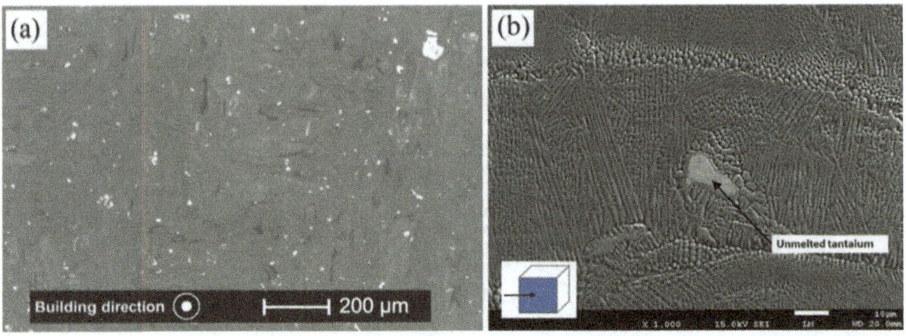


**Fig. 13** Microstructures of SLM processed: (**a**) 10 wt.% Mo/Ti6Al4V system showing un-melted Mo particles randomly dispersed in the Ti matrix [91] and (**b**) 50 wt.% Ta/Ti system with un-melted Ta particles remained in the solidified structure [92]

higher melting temperatures of Mo and Ta (2625 and 3020 °C, respectively) compared with that of Ti6Al4V (1668 °C), a part of them may not be melted when subjected to relatively low laser energy densities. Moreover, the higher reflectivity of Mo and Ta than Ti6Al4V to the laser beam could be also responsible for their lower laser absorptivity and the subsequent partial melting. Therefore, a part of Mo and Ta powder particles may remain as unmelted particles and play the role of metallic reinforcement for the matrix.

## 6.5 *Additively Manufactured Lightweight Metal Matrix Nano-Composites (MMnCs)*

The field of nano-composites has been widely considered in recent decades because of the significant differences in mechanical, electrical, thermal, electrochemical and optical properties of nano-composites compared with the component materials. The main concept behind the development of lightweight MMnCs is benefiting from the great potential of nano-scale reinforcements for further improvement of mechanical properties beyond those of conventional MMCs.

The desired properties of MMnCs are obtained when the nano-scale reinforcements are uniformly distributed throughout the matrix. However, the low wettability of ceramic nanoparticles with the molten metal matrix and the extremely large van der Waals attractive force between the reinforcing nanoparticles are the main issues limiting the production of MMnCs by conventional fabrication processes. The main problems with these processes are agglomeration of nanoparticles to the coarsened clusters or even the disappearance of nanoscale nature of reinforcements. This phenomenon paves the way for microstructural inhomogeneity and makes it rather difficult to achieve the optimal exploitation of the strengthening potentials of nano-scale reinforcements. Therefore, it would be necessary to develop novel fabrication methods avoiding the agglomeration of reinforcing nanoparticles and guaranteeing

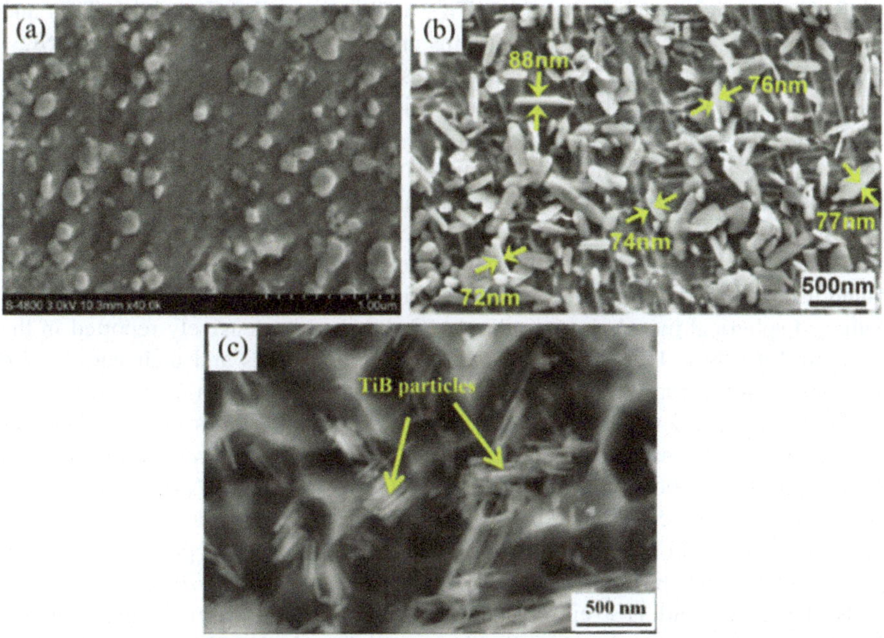


**Fig. 14** SEM images of SLM processed: (**a**) TiC reinforced Al matrix [95], (**b**) TiC reinforced Ti matrix [76] and (**c**) TiB reinforced Ti matrix [85] nano-composites

their homogenous distribution throughout the matrix. In addition to the problems associated with the distribution of nano-scale reinforcements, fabrication of nano-composites still have some challenges including processing, cost, consistency in volume production as well as oxidation and thermal instability of nano-materials [58].

AM technology is believed to be one of the promising approaches able to mitigate some of these limitations [93]. Due to the ease of incorporating nano-materials as well as the significant progress in polymer-based AM techniques, the recent developments in AM processing of nano-composites have been mostly focused on polymer-matrix composites. However, limited numbers of works have been performed in the field of MMnCs. This could be due to the difficulties accompanied by incorporation of nanoparticles in metal powder-based AM techniques as well as the limitation with the variety of existing materials that can be used. The development of new processes and innovative materials may open new windows for wider fabrication of MMnCs adopting AM technology.

Recently, novel MMnCs with nanometer-scale reinforcements have been fabricated using AM processes, especially SLM. TiC is the most widely used reinforcement for fabricating Al matrix nano-composites (AMnCs). This is while TiC and TiB are both used in the literature as reinforcement to the Ti matrix nano-composites (TMnCs). Figure 14 shows typical microstructures of TiC reinforced AMnCs as well as TiC and TiB reinforced TMnCs. Due to the extremely small size of ceramic reinforcing particles (nanometer-scale TiC, and very small 2–5 μm size $TiB_2$ and elemental B) in the mixed powder systems and the full melting nature of the SLM process, the in-situ TiC and TiB precipitates are most likely synthesized through solution precipitation mechanism, during which the reinforcement nuclei is formed by heterogeneous nucleation and then grows.

The nanoscale TiC reinforcements in AMnCs are also formed through solution precipitation mechanism by multiple nucleation and growth of TiC nuclei from carbon saturated Al-Ti melt [94, 95]. Microstructural observation of the TiC reinforced AMnCs reveals that the nanometer scale TiC precipitates are not only round in shape, but also uniform in size. The in-situ formation of TiC precipitates as equiaxed spherical morphology in the AMCs has been extensively reported in the literature [96–100]. The nanometer-size of TiC precipitates and their narrow size distribution in the matrix both give evidence of easier nucleation than growth.

When in nanometer scale, the TiC precipitates in TMnCs show a lamellar morphology with a mean thickness less than 100 nm. It is interesting to point out that despite the particulate morphology of TiC precipitates in AMnCs for a wide range of processing conditions, the TiC precipitates in TMnCs are associated with extensive changes in size and morphology based on the applied processing parameters as well as the characteristics of mixed powder system [74–76].

To the same as all TiB reinforced TMCs made by various processing methods, TiB reinforcements also show needle-like morphology in the AM processed TMCs. However, by using appropriate processing parameters and engineered mixed powder systems, TiB reinforced TMnCs containing needle-like TiB precipitates with diameters in nanoscale can be fabricated through AM technology [82, 84, 85].

## 7 Temperature-Driven Forces and Flows and Viscosity in Laser-Induced Melt Pools

### *7.1 Surface Tension and Marangoni Flow*

The temperature variation in the melt pool during laser-based AM is in a way that compared to the edges, the central regions are associated with higher temperatures. Moreover, there exist chemical potential gradients of solute elements in different regions of the melt pool. These thermal and chemical gradients can induce a surface tension gradient which is the main driving force for the movement of molten material in the melt pool usually from the center toward the edges. The surface tension gradient induced in the melt pool results in the formation of a so called Marangoni flow which is significantly dependent on temperature sensitivity of the liquid-to-solid surface tension of the material. When Marangoni flow is active, the induced radial temperature gradients exceed the depth-wise temperature gradients [101].

As shown in Fig. 15, the surface tension is inversely proportional to the melt pool temperature for most pure metals with low oxygen content and follows an increasing

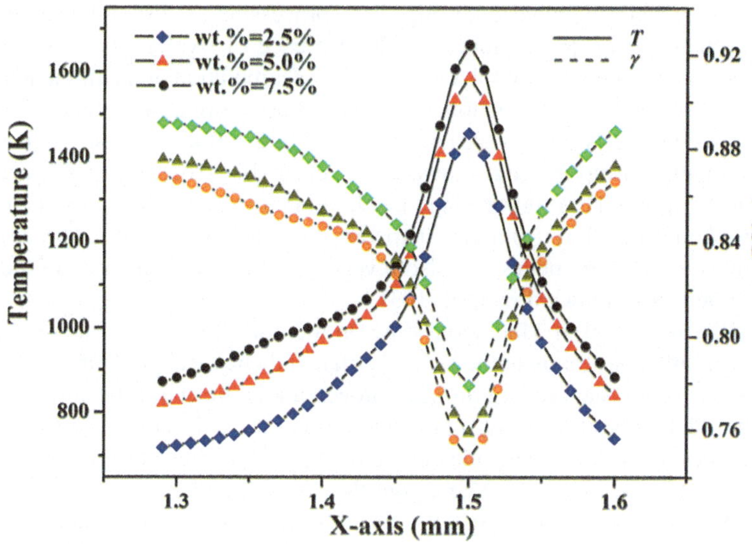


**Fig. 15** Variation of surface temperature and surface tension in the melt pool of SLM processed TiC/AlSi10Mg mixed powder system containing different amounts of TiC [64]

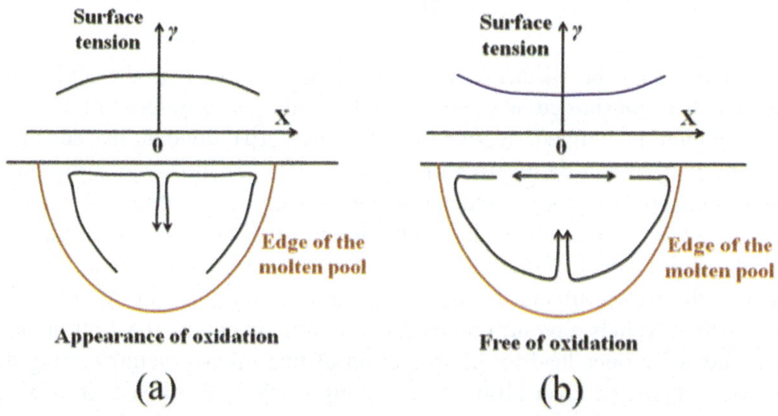


**Fig. 16** Schematic illustration of material flow in the melt pool without and with oxidation [67]

trend by moving from center toward the periphery of melt pool [64, 102]. In addition, higher temperatures and consequently smaller surface tensions are induced in the melt pool when processing powder mixtures with higher contents of reinforcing particles. In fact, due the higher laser absorptivity of ceramic reinforcing particles than the metallic powder, the increase in volume fraction of reinforcing particles enhances the effective laser absorbance of the whole powder system and consequently elevates the melt pool temperature.

The surface tension induced during laser processing is also affected by the oxidation phenomenon [67]. Moreover, the presence of oxides can influence the convection pattern in the melt pool. Owing to the accumulation of insufficient heat at initial stages of melt pool formation, the oxide films existing at the edges of melt pool are not completely destroyed, leading to reverted surface tension gradient and convection pattern from the edges toward the center (Fig. 16a). When subjected to enough heat during laser processing, the braking of oxide films results in the reversion of material flow from center toward the edges (Fig. 16b). The addition of impurities to the pure metals can also reverse the fluid flow direction in the melt pool from outward to inward pattern [103].

Generally, the effect of Marangoni flow on the bulk flow of a melt is normally weaker than bulk convection at the same thermal and chemical gradients. However, under special circumstances, e.g. in laser processing and electron-beam melting of metals, the Marangoni flow may significantly affect the bulk flow. The Marangoni flow generates higher peak velocities in the melt pool and elevates the heat transfer coefficient. However, when the Marangoni flow is negligible, free convection flow becomes dominant, and smaller peak velocities are attained. The heat transfer for this type of melt pool is evaluated via conduction only [101]. The intensity of Marangoni flow can be evaluated using the dimensionless Marangoni number ($M_a$) as [104]:

$$\mathbf{M_a} = \frac{\Delta\sigma L}{\mu \upsilon_k} \quad (1)$$

here, $\Delta\sigma$ represents the surface tension difference (N/m), $L$ is the length of free surface which is considered as the melt pool length, and $\upsilon_k$ refers to the kinematic viscosity which is defined as dynamic viscosity (μ) divided by density. The Marangoni number can be positive or negative. The negative Marangoni numbers represent melt pools having a wide form factor with a flow direction from center toward the edges. However, positive Marangoni numbers describe a narrow melt pool with edge to center directional flow [105]. The positive/negative nature of Marangoni flow also affects the microstructural evolutions during solidification. While positive values favour microstructural refinement at the bottom of melt pool, the negative ones lead to the formation of fine microstructures along the top regions of a deposited layer [106]. The Marangoni flow also assists in mixing and stirring of the melt pool to avoid macro-segregation phenomenon [101].

## 7.2 Recoil Pressure

One of the interfacial forces induced during laser processing of metals which affects the temperature and morphology of melt pool is recoil force. The recoil force and the Marangoni flow are considered as the main driving forces for the instability of melt flow during laser processing [107]. Since subjected to the highest superheat during laser processing, the surface temperatures directly below the laser spot can easily reach the boiling temperature of the material. The extremely high superheat generated directly at and beneath the center of melt pool can pave the way for vapour recoil which is characterized by the nucleation and growth of metal vapour bubbles and trapped inert gas [108, 109]. When the vapour recoil pressure goes beyond the melt/gas surface tension, small droplets eject from the melt pool to the surrounding, resulting in a phenomenon known as balling effect in AM processes [110, 111]. The recoil vapour pressure varies exponentially with temperature as [107, 112]:

$$P(T) = 0.54 P_a e^{\left(\lambda/K_B \left(1/T - 1/T_b\right)\right)} \tag{2}$$

where $P_a$ is the ambient pressure, λ represents the evaporation energy per particle, $K_B$ is Boltzmann constant, $T$ is the surface temperature, and $T_b$ refers to the boiling temperature of the material.

### 7.3 Rayleigh-Benard Convection

The thermal variations during laser processing cause temperature difference (ΔT) between the top and bottom of the melt pool. The difference in density induced by these thermal variations generates a buoyancy force which acts to rearrange the lighter molten material at the top and the heavier one at the bottom of the melt pool. When ΔT is not so significant, the molten material remains stationery, and heat transfer occurs through conduction. However, as ΔT exceeds a critical value ($\Delta T_c$), energy transfer through rearrangement of molten material (convection) becomes dominant and leads to the motions in the melt pool known as Rayleigh-Benard convection. The magnitude of Rayleigh-Benard convection can be addressed by Rayleigh number ($R_a$) as [113]:

$$R_a = \frac{\beta g \Delta T d^3}{\alpha \upsilon_k} \tag{3}$$

where $\beta$ is the CTE, $g$ is gravitational acceleration, $d$ represents the melt pool thickness, and $\alpha$ signifies the thermal diffusivity.

### 7.4 Dynamic Viscosity of Solid-Liquid Mixed Melt Pools

The dynamic viscosity of the melt pool containing solid reinforcing particles distributed in the liquid metal is defined mathematically as [114–116]:

$$\mu = \mu_0 \left[ 1 - \frac{\Phi}{\Phi_m} \right]^{-2} \quad (4)$$

where $\mu_0$ refers to the viscosity of pure liquid metal (excluding the solid reinforcing particles), $\Phi$ is the volume fraction of solid phase in the liquid, and $\Phi_m$ corresponds to the critical volume fraction of solid phase above which the mixture has infinite viscosity. The dynamic viscosity of the melt pool is substantially increased by the retention of solid particles in it. Therefore, the viscosity of pure liquid metal ($\mu_0$) should be low enough to ensure the successful surrounding of solid particles. However, when the viscosity of pure liquid metal is too low, covering of solid particles with the liquid is limited.

Assuming that there exists a suitable wetting between the liquid and solid particles, the quality of laser processed part is significantly driven by the dynamic viscosity of pure liquid ($\mu_0$), assessed by [117]:

$$\mu_0 = \frac{16\sigma}{15} \sqrt{\frac{m}{K_B T}} \quad (5)$$

in which $\sigma$ indicates the surface tension of liquid metal, $m$ represents the atomic mass, and $T$ refers to the melt pool temperature. Based on Eq. 5, higher temperatures or lower surface tensions result in the formation of liquids with lower viscosities. Referring to Eqs. 4 and 5, the elevated temperature has two major effects on dynamic viscosity of solid-liquid mixed melt pool: (i) reducing the dynamic viscosity of the liquid metal, and (ii) facilitating the melting or at least partial melting of solid reinforcing particles which leads to the formation of higher liquid volumes. Both these factors decrease the melt pool viscosity and consequently improve the wetting of solid reinforcing particles by the surrounding molten metal. It should be noted that at extremely low melt viscosities, there may not be enough liquid to wet the solid particles. This may lead to balling effect. Therefore, the appropriate selection of solid-liquid ratio as well as the controlled superheating degree of the powder with lower melting point could be suggested to control both the $\mu_0$ and $\mu$ for fabricating high quality MMCs parts.

# 8 Parameters Affecting Microstructural Features of Reinforcements in Hybrid Ex-Situ/In-Situ Reinforced and In-Situ Reinforced MMCs

## 8.1 Laser Energy Density

The amount of laser energy density applied during AM processing has a significant influence on microstructural evolution of fabricated MMCs. The elevated heat

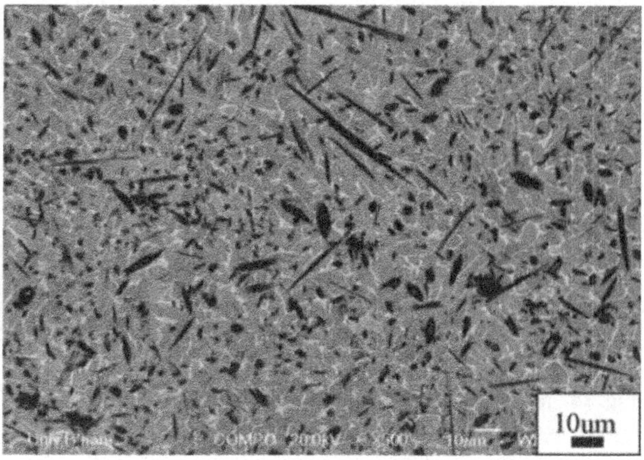


**Fig. 17** SEM micrograph of DLF processed $TiB_2$/Ti6Al4V system as in Fig. 10c but subjected to higher laser power [54]

accumulation and reduced cooling rates induced at relatively high energy densities may have three major effects on the system:

(i) The mixed powder system may experience higher levels of melting,
(ii) Higher temperatures and longer times are provided for the elements to diffuse into the system, and
(iii) The oscillation of reinforcing particles and dissolved elements in the melt are increased.

The effects of laser energy density on the amount, size and morphology of in-situ synthesized reinforcements are discussed in the following.

### 8.1.1 Amount of In-Situ Reaction

When using relatively low energy densities, the energy delivered to the mixed powder system may not be high enough for full melting of powder particles, especially those with high melting points (e.g. $Al_2O_3$, TiC and $TiB_2$ particles in Fig. 10), leading to the melting of only fine reinforcing particles or the edges of larger irregular-shape ones [65]. Accordingly, the microstructure may contain un-melted reinforcing particles along with the in-situ synthesized reinforcements. The enhanced melt pool temperature and larger degrees of melting caused by increased energy density can be an effective approach for decreasing the size and volume fraction of un-melted reinforcing particles and consequently increasing the amount of in-situ synthesized reinforcements [51, 52]. Figure 17 shows the microstructure for the same specimen in Fig. 10c but processed with higher laser power. As is indicative, the starting $TiB_2$ powder particles completely reacted with the fully melted Ti to form in-situ TiB precipitates with needle-shape morphology.

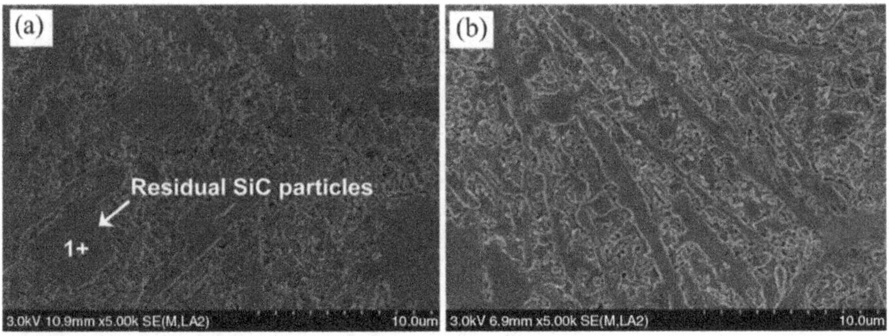


**Fig. 18** Microstructures of SLM processed SiC/AlSi10Mg mixed powder system subjected to laser energy densities of: (**a**) 800 and (**b**) 1000 J/m [118]

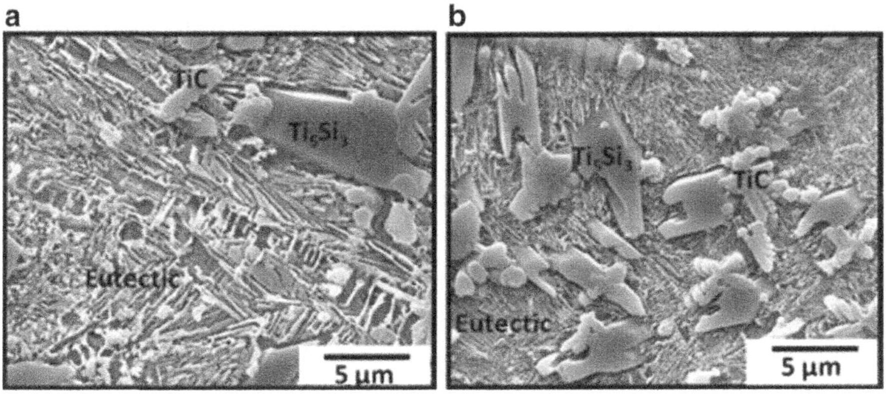


**Fig. 19** Microstructures of LENS processed SiC/Ti system subjected to energy densities of: (**a**) 33.95 and (**b**) 12.73 J/mm$^2$ [52]

### 8.1.2 Size of In-Situ Reaction Products

The greater heat accumulation and reduced cooling rates associated with higher laser energy densities provides the system with higher temperatures and longer times, leading to the growth of in-situ synthesized reaction products. When occurred, the obtained MMCs structures may contain fewer numbers of larger-size in-situ synthesized reinforcements. As shown in Fig. 18, the elevated melt pool temperature caused by application of higher energy densities increases the length and width of plate-like $Al_4SiC_4$ phase and favors the growth of particle-shape $Al_4SiC_4$ precipitates. However, the reduced melt pool temperatures combined with the increased cooling rates lead to extremely finer reaction products. This can be observed by comparing the size of $Ti_5Si_3$ precipitates as well as the microstructural features of lamellar eutectic structures in SiC/Ti powder mixtures processed with two different laser energy densities (Fig. 19).

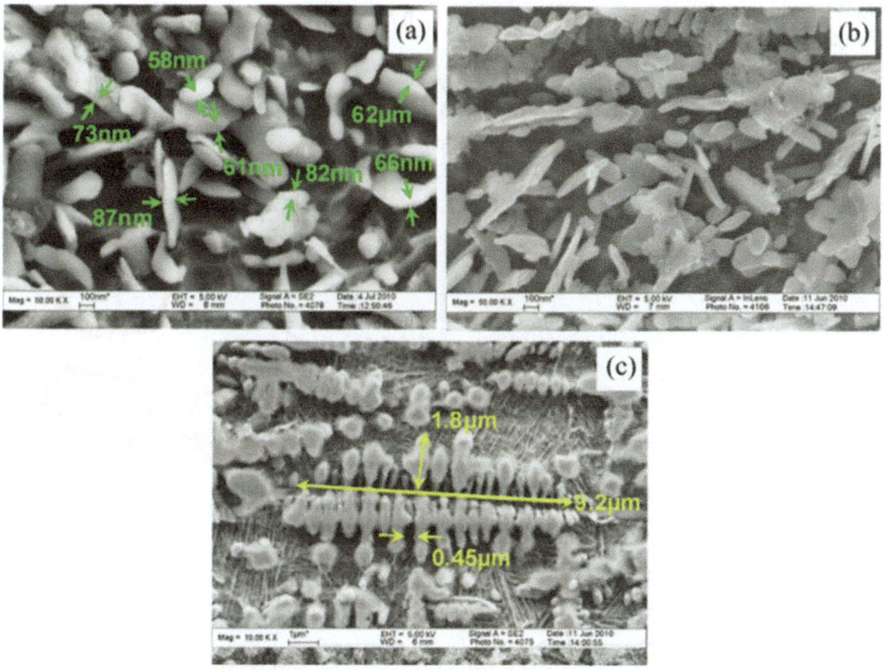


**Fig. 20** Microstructures of SLM fabricated TiC reinforced TMCs processed with laser energy density of: (**a**) 120, (**b**) 180 and (**c**) 360 J/mm$^3$ [75]

### 8.1.3 Morphology of In-Situ Synthesized Reinforcements

The applied energy density can also affect the morphology of in-situ synthesized reinforcements in some systems. As presented in Table 5, one of the systems in which the size and morphology of in-situ synthesized reinforcements are greatly dependent on the applied energy density is the TiC/Ti system having nanoscale TiC powder particles. As being observed in Fig. 20, the morphology of in-situ synthesized TiC precipitates in SLM processing of this system follows successive change from discrete nanoscale lamellar to accumulated whiskers and finally to coarse dendrites with the enhancement of laser energy density [74, 75].

In general, TiC primarily nucleates as NaCl-type hexagonal structure with center-symmetric position of Ti and C atoms in the lattice. The deposition of Ti and C atoms on TiC nuclei is believed to form a lamellar crystal with [111] basal plane (Fig. 21) [74]. The combined effects of evaporative recoil pressure and surface tension of melt induced on [111] plane of TiC crystals especially at high scan speeds (low laser energy density), restricts the growth along [111] direction and results in a desirable nanostructure (Fig. 20a). However, the increased activation of Ti and C atoms caused by elevated temperatures at higher energy densities favors their deposition on [111] planes of the TiC crystal and leads to morphological change from nanoscale lamellar structure to coarsened sub-micrometer scale whisker-like structure

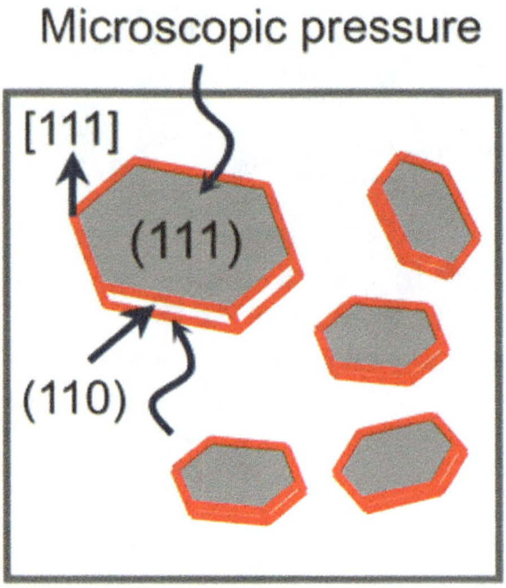


**Fig. 21** Schematic illustration of primarily nucleated TiC structure [75]

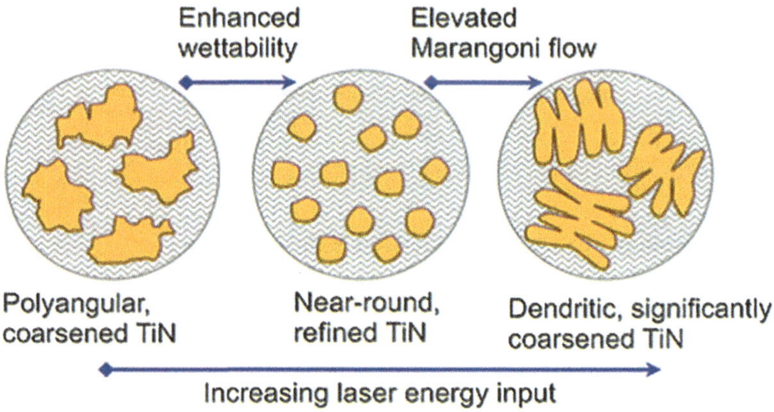


**Fig. 22** Schematic illustration of morphological change experienced by in-situ TiN phase in SLM processed $Si_3N_4$/Ti mixed powder system [79]

(Fig. 20b). The enhanced instability of melt pool caused by the intensified Marangoni flow combined with the extremely high melt pool temperatures at excessive energy densities facilitates the dendritic growth of crystals to form coarsened structure (Fig. 20c).

The improved wettability of solid reinforcements with the surrounding melt caused by the enhanced energy density can also affect the morphology of reinforcements. Figure 22 schematically shows the variation in morphology of in-situ synthesized TiN precipitates as a function of energy density subjected to the $Si_3N_4$/Ti powder mixture during SLM processing [79]. The limited wettability of TiN precipitates with the melt at relatively low energy densities leads to their non-uniform polyangular morphology with some agglomeration. The elevated temperature and improved wettability gained at higher energy densities change the morphology to near-round shape with some refinement. However, at considerably high energy densities, the intensified perturbations and noticeable internal energy caused by significantly reduced dynamic viscosity and enhanced Marangoni flow makes the solid-liquid interface instable and favors the formation of coarse dendritic TiN precipitates.

## *8.2 Characteristics of Powder Mixture*

The effective laser absorption of a powder mixture containing metal, ceramic and pores is defined as the following [119]:

$$A_{eff} = \sum A_i V_i \sqrt{(1 + 0.5P)} \tag{6}$$

where $A_i$ and $V_i$ represent the absorption coefficient and volume fraction of condensed part, respectively, and $P$ corresponds to the fraction of porosity.

Due to the significant effects ceramic reinforcing particles have on laser absorptivity of powder mixture, their features (e.g. laser absorptivity, size and volume fraction) play important roles in microstructural evolution and mechanical properties of laser-based AM fabricated MMCs. The influence of size and volume fraction of reinforcing particles on microstructural characteristics is discussed in the following.

### 8.2.1 Size of Reinforcing Particles

The limited in-situ reaction between the coarse reinforcing particles and the melt results in heterogeneous microstructures containing high volumes of un-melted reinforcing particles along with low amounts of in-situ synthesized precipitates. However, compared with the larger reinforcing particles, the higher effective surface area of finer ones can lead to the improved laser absorption efficiency. This can consequently elevate the melt pool temperature and induce much more in-situ reactions in the system. Therefore, more uniform microstructures containing higher amounts of in-situ reinforcements may be formed. It is worth noting that when employing relatively low energy densities, the fine reinforcing particles may gather to form clusters [47]. This can diminish the efficiency of laser absorption by reducing the effective surface area of ceramic reinforcing particles.

### 8.2.2 Volume Fraction of Reinforcing Particles

Due to the higher laser absorption coefficient of ceramic reinforcing particles than the metallic powder part, the increase in volume fraction of ceramic particles

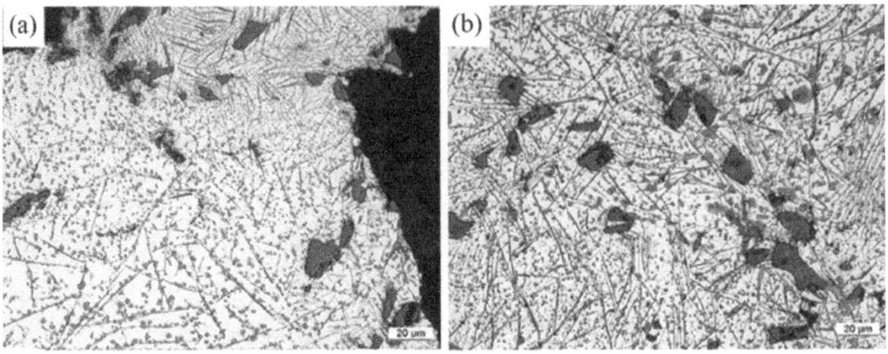


**Fig. 23** Optical micrographs of laser sintered SiC/A356 mixed powder system having: (**a**) 10 and (**b**) 20 vol.% SiC [47]

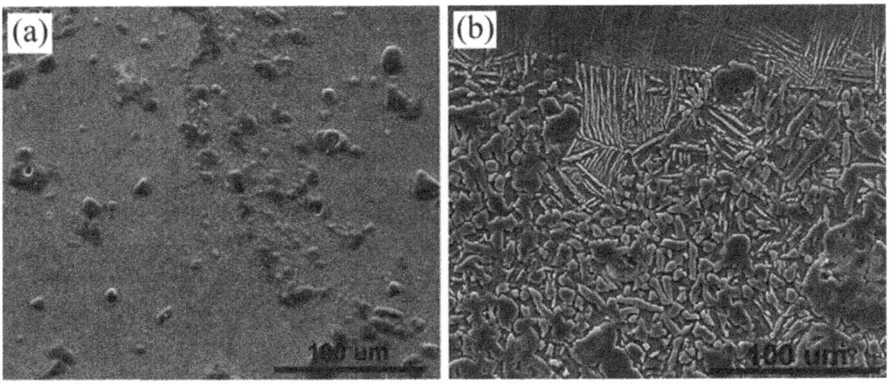


**Fig. 24** SEM micrographs of LENS processed TiN/Ti6Al4V system with: (**a**) 20 and (**b**) 40 wt.% TiN [50]

elevates the effective laser absorptivity of mixed powder system. This in turn elevates the maximum increase in the temperature of laser-induced melt pool as the following [120]:

$$\Delta T_{max} = \frac{2A_{eff}\eta}{k}\left(\sqrt{\frac{\alpha\tau_p}{\pi}}\right) \quad (7)$$

in which $A_{eff}$ signifies the effective laser absorptivity, $\eta$ represents the energy density, $k$ is the thermal conductivity, $\alpha$ is the thermal diffusivity and $\tau_p$ refers to the duration of laser irradiation.

The elevated temperature enhances the chemical reaction in the system and leads to microstructures containing higher amounts of in-situ synthesized reinforcements. This can be confirmed by comparing the amounts of in-situ reaction products (plate-like $Al_4SiC_4$ and extremely fine Si particles) in Fig. 23 as well as the in-situ formed elongated TiN precipitates in Fig. 24.

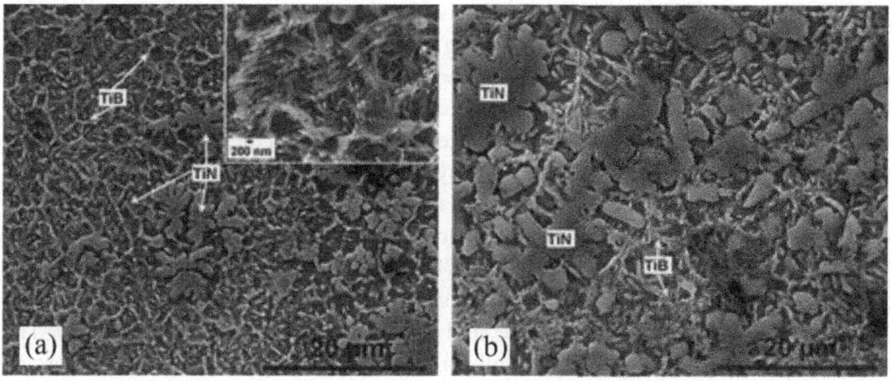


**Fig. 25** Microstructures of LENS processed BN/Ti6Al4V mixed powder system containing: (**a**) 5 and (**b**) 15 wt.% BN. Inset in (**a**) provides high magnification micrograph of nano-scale needle-like TiB phase [89]

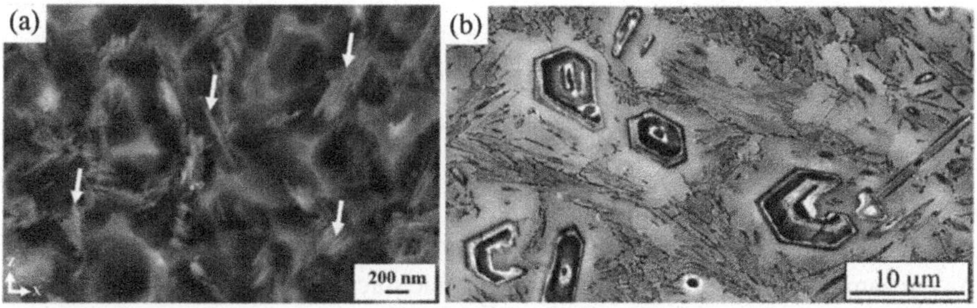


**Fig. 26** SEM micrographs of TiB reinforced TMCs fabricated by: (**a**) SLM processing of 5 wt.% $TiB_2$/Ti [84] and (**b**) LENS processing of 2 wt.% B/TNTZ [123] powder mixtures

In addition to the elevated temperature caused by the increased volume fraction of reinforcing particles, the heat released by the exothermic reaction in some systems can also induce even higher temperatures in the melt pool. Figure 25 shows the effect of BN content on size and distribution pattern of in-situ synthesized precipitates in hybrid (TiB + TiN) in-situ reinforced TMCs. The high laser absorptivity of BN and the exothermic nature of reaction between BN and Ti are believed to be responsible for the elevated melt pool temperature and coarsening of the reaction products [89, 121].

Owing to the changes in melt composition, the amount of reinforcing particles existing in the system may affect the size of in-situ synthesized reinforcements. Figure 26 shows the microstructures of AM processed $TiB_2$/Ti powder mixtures containing different volume fractions of $TiB_2$ constituent. For the hypoeutectic alloys where the B content in the melt pool is less than that of the eutectic composition (1.64 wt.%), the solidification starts with the nucleation of β-Ti grains based on the Ti-B binary phase diagram [122]. Due to the extremely low solid solubility of B in Ti, B atoms are rejected from the growing solidification interface of β-Ti into the melt. The melt enriched in B transforms to a mixture of TiB and β-Ti phases at the eutectic temperature, leading to the microstructure containing colonies of needle-shape TiB whiskers distributed uniformly in the Ti matrix (Fig. 26a) [83, 84]. However, for the hypereutectic compositions (those with B contents higher than the eutectic composition), the solidification starts with the formation of primary coarse TiB precipitates and is followed by the transformation of remaining melt to TiB + β-Ti mixture at the eutectic temperature. Therefore, microstructures containing multi-scale coarse primary and fine eutectic TiB precipitates dispersed in a Ti matrix are developed (Fig. 26b) [78, 123].

As reported by Gu et al. [76], the change in volume fraction of TiC in TiC/Ti system significantly affects the size and morphology of in-situ synthesized TiC reinforcements. While the in-situ synthesized TiC precipitates indicate a nanoscale lamellar-structure in SLM processed part with 7.5 wt.% TiC, they lose their nanoscale and change morphology to a dendritic structure for powder mixtures with increased TiC contents.

# 9 Distribution Pattern of Reinforcements

## *9.1 Non-homogenous Distribution*

### 9.1.1 Bimodal Distribution

As the same as the non-uniform coarse and fine microstructures observed in laser-based additively manufactured Al-Si and Al-Si-Mg alloys [124–129], some of the AM processed MMCs also show non-homogeneities in the size of reinforcements. These non-uniformities are in fact caused by the differences in heat cycles received by the track cores and overlaps. While the track cores experience only a single melting stage, the melting takes place twice in track overlaps. As shown in Figs. 27a, b, the MMC structure obtained by SLM processing of $Al_{85}Nd_8Ni_5Co_2$ alloy consists of finer precipitates along the melt pool boundaries (track overlaps) and coarser ones along the cores of tracks. Depending on the type of alloy and the applied processing parameters, the track cores may contain finer precipitates compared with the track overlaps (Fig. 27c).

### 9.1.2 Agglomeration or Clustering of Reinforcements

The heterogeneous distribution pattern of reinforcements in the matrix and their clustering could occur in two different conditions as follows:

(i) At relatively low laser energy densities [60, 62, 66, 82, 130] or even low volume fractions of reinforcing particles in the mixed powder system [64], the insufficient heat generated in the melt pool results in reduced melt pool temperature.

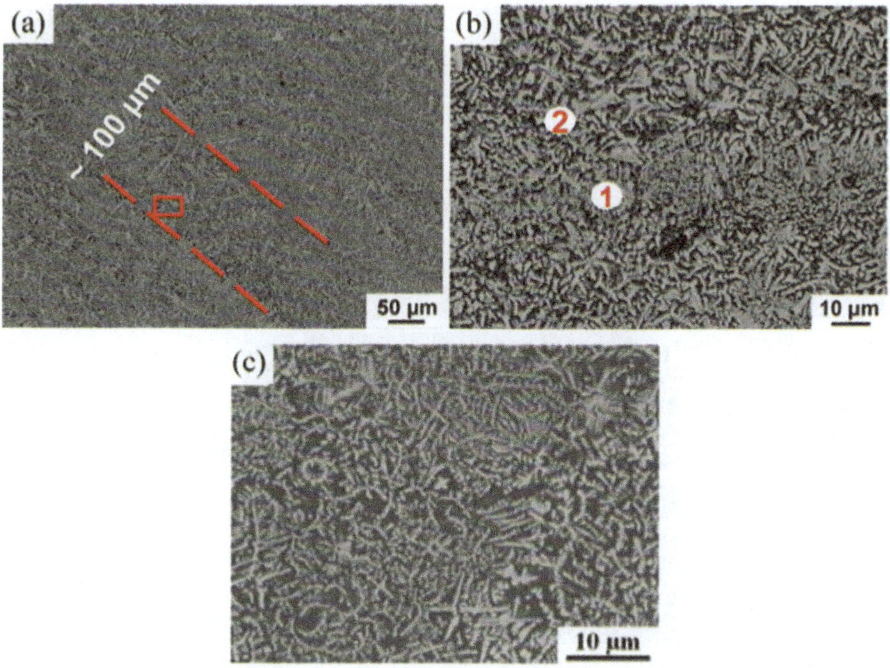


**Fig. 27** (**a**) Low and (**b**) high magnification SEM micrographs of SLM processed $Al_{85}Nd_8Ni_5Co_2$ alloy [70]. The area between dashed lines in (**a**) indicates the track width while (**b**) shows the microstructure for the area marked in (**a**). (**c**) Microstructure of SLM processed Al-12Si-TNM powder system [69]

The combined effects of weak Marangoni flow and high dynamic viscosity obtained at this condition restricts the rearrangement of solid reinforcing particles/reinforcements in the melt pool and leads to their limited distribution and even clustering/agglomeration in the solidified MMC structure, as shown in Fig. 28.

(ii) While the increased heat input could be beneficial to the efficient distribution of reinforcing particles/reinforcements by enhancing the Marangoni flow and reducing the dynamic viscosity, the excessive heat input subjected to the system at extremely high energy densities may be also detrimental to obtaining homogeneous MMCs. As shown in Fig. 29, application of excessive energy densities can cause the aggregation of reinforcing particles to form non-homogenous MMC structures.

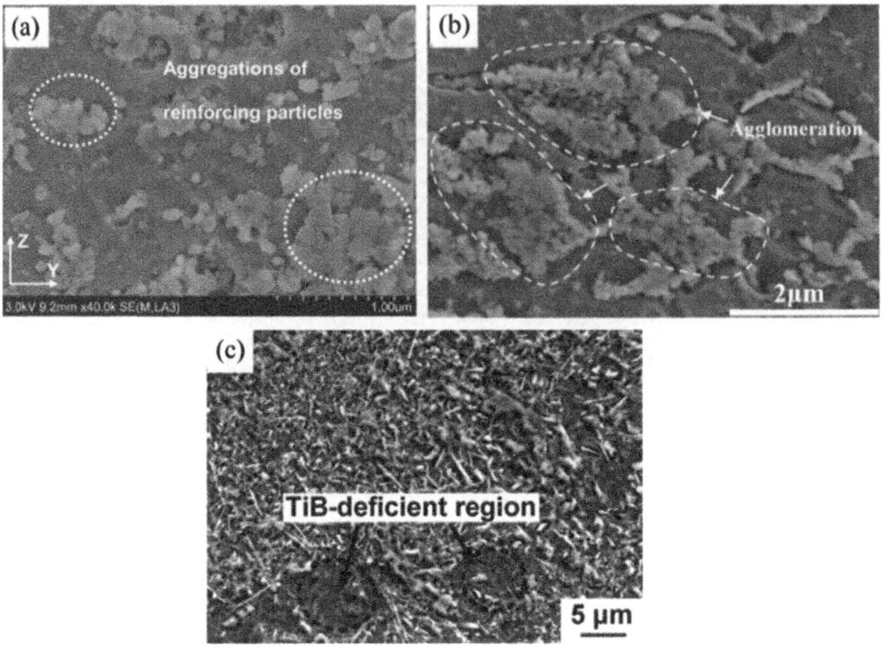


**Fig. 28** SEM micrographs of: (**a**) SLM processed TiC reinforced AMC [61], (**b**) SLM fabricated $Al_2Si_4O_{10}$ reinforced AMC obtained by processing $Al_2O_3$/AlSi10Mg mixed powder system [66], and (**c**) LENS processed TiB reinforced TMC [130]

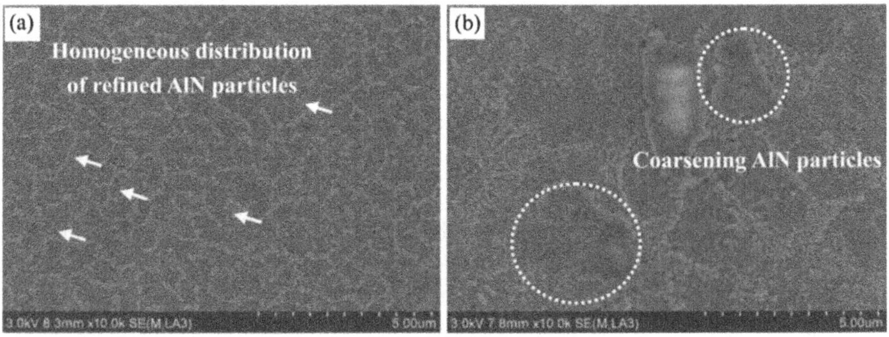


**Fig. 29** Microstructures of SLM produced AlN reinforced AMCs processed with energy densities of: (**a**) 450 and (**b**) 1800 J/m [67]

## *9.2 Homogenous Distribution of Reinforcements*

Typical microstructures of laser-based AM fabricated MMCs with homogenous distribution of reinforcements are shown in Fig. 30. When applying appropriate processing parameters and selecting suitable fraction of reinforcing particles in the

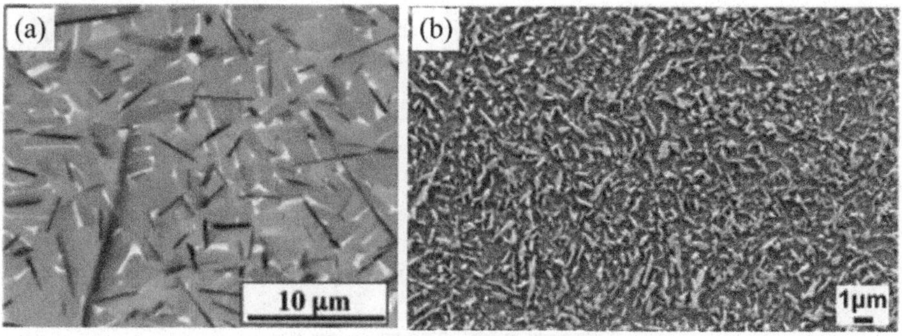


**Fig. 30** Microstructures of SLM processed TiC reinforced: (**a**) AMC [80] and (**b**) TMC [76] with homogenous distribution pattern of reinforcements

mixed powder system, the elevated temperature and reduced surface tension in the melt pool both act to decline the dynamic viscosity and intensify the Marangoni flow. This accelerates the rearrangements of reinforcements and activates microscopic pressure on them [59, 64, 89]. In addition, the decreased dynamic viscosity may decline the friction between the solid particles and the molten material. Moreover, the improved fluidity of melt pool competes with the Van der Walls force between the solid particles (especially those in nanoscale) and favors repulsion forces between them, inhibiting their clustering and agglomeration. Therefore, all these mechanisms contributing to the improved dispersion state of reinforcing particles/reinforcements in the melt pool pave the way for the formation of MMCs with homogenous distribution pattern of reinforcements.

## *9.3 Microstructures with Ring-Like Distribution Pattern of Reinforcements*

Recently, unique microstructures with tailored distribution patterns of reinforcements have been developed in several MMCs including TiC reinforced AMnCs, $Al_2Si_4O_{10}$ reinforced AMCs [66], TiB reinforced TMCs [82, 130], TiC reinforced H13 steel [131], $TiB_2$ reinforced 316 L stainless steel [132, 133], TiC reinforced 316 L stainless steel [134], $TiB_2$ reinforced H13 steel [135] and SLM processed CNT/AlSi10Mg mixture [136]. Figure 31 presents the ring-like structures formed in AM processed AMCs and TMCs. As being observed, the reinforcements in these structures are uniformly distributed along the grain/sub-grain boundaries.

At present, two different mechanisms have been proposed for the formation of such microstructures in the SLM processed MMCs as discussed below:

**First Mechanism**: This mechanism correlates the formation of such microstructures mainly to the effects of forces and flows in the melt pool on the rearrangement of reinforcements [60, 62, 64, 66]. Figure 32 schematically illustrates the mechanism

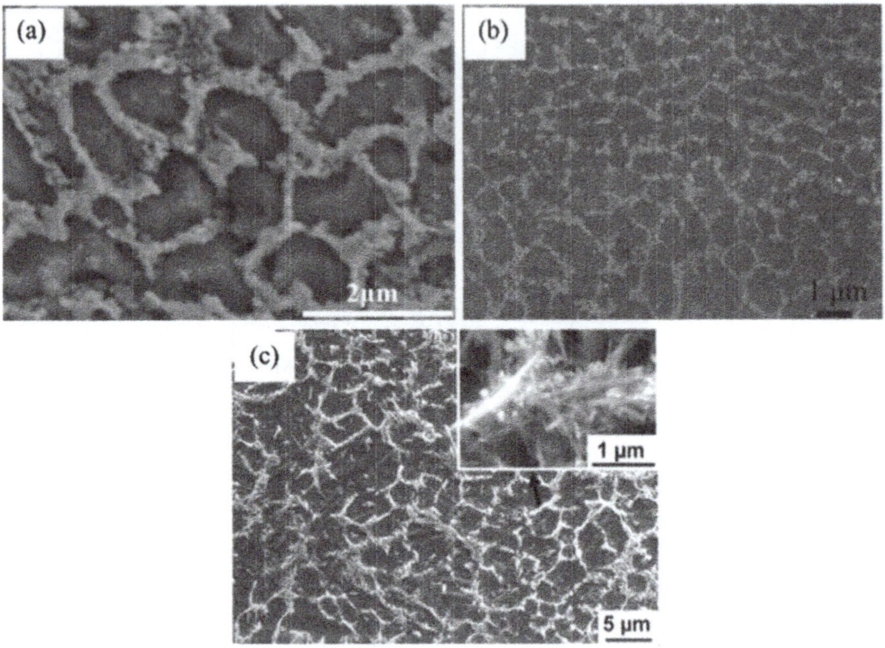


**Fig. 31** SEM images of: (**a**) SLM fabricated $Al_2Si_4O_{10}$ reinforced AMC [66], (**b**) SLM processed CNTs/AlSi10Mg composite [136], and (**c**) LENS processed TiB reinforced TMC [82]

**Fig. 32** Schematic illustration of the mechanism behind the formation of ring-like structure in some of SLM processed systems [66]

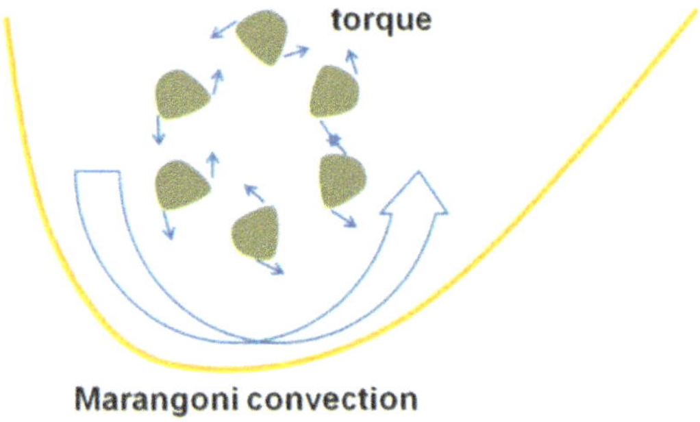


behind the formation of such structure. As being observed, the Marangoni flow drives the material from the center to the edges of the melt pool. These materials descend to the bottom of melt pool under the action of gravity force and then come back to the surface again under the buoyancy force. The repetitive circular flow induced in the melt pool acts on solid reinforcements and improves their dispersion state. Moreover, due to the irregular shape of reinforcing particles and the misalignment of the particle's center, a torque is generated around each particle which tends to rotate it. The final output of these multiple forces existing in the melt

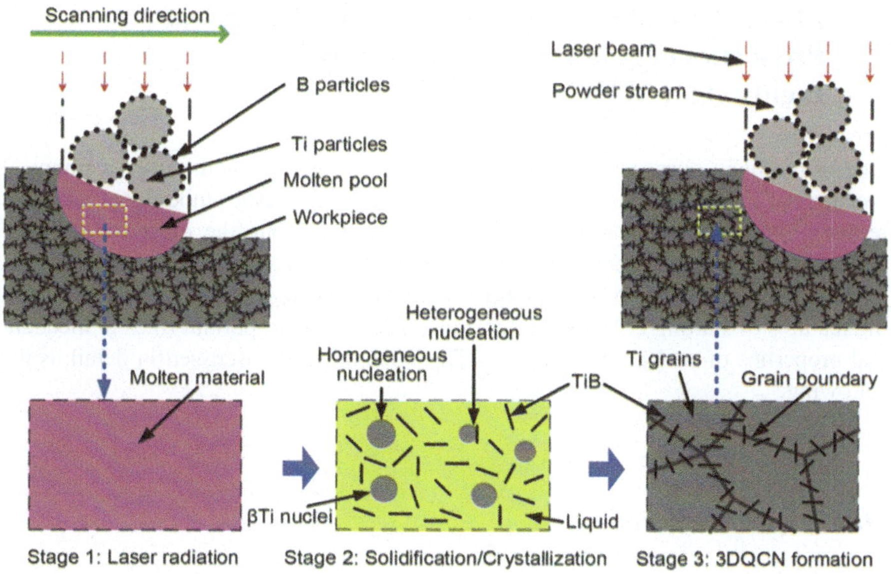


**Fig. 33** The schematic view of mechanism suggested for the formation of ring-like (3DQCN) structure in LENS processed TiB reinforced TMCs [82]

pool is the decoration of grains with reinforcing particles, leading to ring-like structures.

**Second Mechanism**: The mechanism proposed recently by Hu et al. [82] ascribes the formation of ring-like structures to the combined effects of flows in the melt pool and the particle pushing mechanism in the solidification front. The active mechanism behind the formation of these structures is schematically shown in Fig. 33 for TiB reinforced TMCs. Due to the full melting nature of the SLM process, the powder mixture experiences full melting when interacting with laser beam (stage 1). Solidification starts with the in-situ reaction between Ti and B elements, leading to the nucleation of TiB precipitates, followed by their growth as needle-shape morphology. Subsequently, Ti melt starts to solidify by means of heterogeneous nucleation at TiB precipitates and homogenous nucleation in the melt. Due to the extremely low solubility of TiB in β-Ti [89, 137, 138], the TiB precipitates are pushed into the melt in front of growing nuclei-liquid interface (stage 2). Finally, aggregation of TiB precipitates at the grain boundaries results in the formation of three dimensional quasi-continuous network (3DQCN) microstructures (stage 3).

# 10 Effect of Reinforcement Features on Mechanical Properties of Additively Manufactured Lightweight MMCs

One of the main purposes behind the fabrication of MMCs is the improvement of mechanical properties (e.g. hardness, strength and elastic modulus). Ceramic reinforcing particles, because of having high hardness, are the most widely used materials added to the metallic matrix to enhance the mechanical properties.

For all three types of ex-situ, in-situ and hybrid ex-situ/in-situ reinforced MMCs, the features of starting ceramic reinforcing particles play important roles in mechanical properties of AM processed parts. These features are discussed in detail as the following.

## *10.1 Volume Fraction*

The effect of reinforcement volume fraction on the hardness of MMCs can be discussed based on the simple rule of mixture as [139–142, 176, 177]:

$$\mathbf{H_{MMC}}=(\mathbf{H_r}\times\mathbf{V_r})+(\mathbf{H_m}\times\mathbf{V_m}) \tag{8}$$

where $H_{MMC}$ represents the hardness of MMC, while $H_r$ and $H_m$ are the hardnesses of reinforcement and matrix, respectively. The reinforcement and matrix volume fractions are also indicated as $V_r$ and $V_m$, respectively. Due to the higher hardness of ceramic reinforcements than the matrix, MMCs benefitting from higher contents of reinforcements are associated with higher hardness.

Optimization of process parameters using Taguchi method performed by Ghosh et al. [90] indicated that among different variables of AM process, the volume fraction of reinforcing particles (composition of mixed powder system) had the highest contribution to hardness (60.16%). Higher laser powers and lower layer thicknesses also led to better melting of powder mixture and improved the hardness with contributions of 33.35% and 4.74%, respectively. However, the effects of other processing parameters on the hardness were negligible in their study.

As reported by Ghosh et al. [29], enhancement of SiC content from 10 to 30 vol. % in the ex-situ reinforced DMLS processed SiC/Al-4.5Cu-3 Mg AMCs parts can efficiently improve the microhardness from ~1.7 to ~4 GPa. Results presented by Wang et al. [54] also shows that the increased volume fraction of in-situ synthesized TiB precipitates caused by the enhancement of $TiB_2$ content in the mixed powder system (5 to 10 wt.%) can significantly improve the hardness from 560 to 750 HV. This is while the non-reinforced Ti6Al4V alloy had a hardness of 340 HV in their study. Balla et al. [143] have also reported the significant increase in hardness of LENS processed TiN reinforced TMCs from 527 to 1138 HV with the enhancement of TiN content from 10 to 40 wt.%. However, the pure Ti6Al4V fabricated with the same process had a hardness of 394 HV.

The increase in reinforcement volume fraction can also affect the Young's modulus of MMCs. For instance, more contents of in-situ reinforcements formed by using higher volumes of BN in the mixed powder system (15 vs. 5 wt.%), increased the Young's modulus of hybrid in-situ (TiB + TiN) reinforced TMCs from 184 to 204 GPa [87]. These moduli are believed to be higher than those for TMCs having either TiB or TiN as reinforcement [87, 144, 145].

The increase in volume faction of primarily added ceramic reinforcing particles may have some other effects on the mechanical properties of AM processed MMCs as:

(i) The elevated working temperature can induce much more in-situ reactions in the system and consequently enhance the mechanical properties [90]. Moreover, the size of in-situ synthesized reaction products may be also affected. As reported by Wang et al. [54], the length and width of needle-like TiB precipitates in TiB reinforced TMCs increased as the volume fraction of $TiB_2$ powder in the mixed powder system enhanced from 5 to 10 wt.%. This can also enhance the mechanical properties of MMCs. It is also worth mentioning that when using higher contents of reinforcing particles, the greater heat released by probable exothermic in-situ reaction can reduce the energy density required to achieve MMCs with the optimum hardness.
(ii) The enhanced fusion of solids can significantly diminish the amount of interconnected porosities and consequently enhance the mechanical properties [146, 147]. SLM processing of Al/$Fe_2O_3$ powder mixtures containing 5–15 wt. % $Fe_2O_3$ indicates the strong dependence of hardness on microstructural features which were affected by $Fe_2O_3$ content [146, 147]. When using high contents of $Fe_2O_3$, the grain boundaries containing fine, homogenous and well-bounded particulate features of reinforcements were thickened. This is believed to be responsible for the improved hardness through the dispersion strengthening mechanism.

Although the addition in volume faction of reinforcing particles is in most cases associated with the enhanced strength, there may be an optimum content of reinforcing particles for some systems above which the hardness and strength drop. For instance, the increase in TiC content elevates the microhardness of TiC reinforced TMCs and leads to a microhardness of 577 HV for the TMC having 12.5 wt.% TiC, which is noticeably higher than that for the pure Ti fabricated with the same processing parameters (241–287 HV). However, the decreased densification levels as well as the microstructural variations (e.g. change in size and morphology of in-situ synthesized precipitates) occurred at TiC contents higher than 12.5 wt.% decreased the microhardness [76].

It should be noted that the enhanced volume fraction of reinforcing particles may be detrimental to some of the mechanical properties. As reported by Zhang et al. [148], while the addition of TiC content from 10 to 40 wt.% had a significant influence on hardness and wear resistance, it did not have a remarkable influence on tensile strength of hybrid ex-situ/in-situ TiC reinforced TMCs. Furthermore, it noticeably decreased the ductility. Although the TiC precipitates are hard and have a good interfacial bonding with the Ti matrix (especially the in-situ ones), enhancement of TiC content in the mixed powder leads to higher amounts of ex-situ (un-melted) TiC particles in the AM processed part which are prone to cracking [148]. This can make the fabricated MMCs part less ductile when subjected to external loading.

## *10.2 Size*

As the size of ceramic reinforcing particles in the mixed powder system is decreased, several mechanisms contribute to improve the mechanical properties of MMCs as the following:

(i) Compared with the MMCs having larger primary ceramic reinforcing particles, those containing smaller ones lead to higher densification levels and consequently better mechanical properties.
(ii) The enhanced efficiency of laser absorption induces more in-situ reactions between the ceramic reinforcing particles and the matrix. The in-situ reaction products may have much more coherent interface and better bonding with the matrix. Compared with the non-coherent interface, the coherent interface transfers the load from the matrix to the reinforcement more efficiently.
(iii) Due to the enhanced nucleation sites for the molten metal matrix during the solidification process, MMCs with much refined matrix are formed. Based on the grain boundary strengthening mechanism, higher strength is attained since larger stresses are required for the deformation of these structures [149, 178, 179].

When using coarse ceramic reinforcing particles, the mechanical properties of fabricated MMCs may be even worse than the non-reinforced parts. For instance, the increase in the mean size of SiC particles from 5 to 50 μm can significantly decline the microhardness of SLM processed hybrid ex-situ/in-situ ($SiC+Al_4SiC_4$) reinforced AMCs from ~218 to 127 HV [46]. This is while the non-reinforced part processed with SLM is believed to have a maximum microhardness of ~145 HV [150]. The significant decrease in hardness and strength by application of relatively coarse reinforcing particles could be due to the combined effects of decreased density and weakened interfacial bonding between the un-melted/partially melted large-size reinforcing particles and the matrix [46].

## *10.3 Distribution Pattern*

The distribution pattern of reinforcing particles/reinforcements in the matrix significantly affects the mechanical properties of MMCs. When the reinforcements are distributed in the matrix as a non-homogenous pattern, some regions of the matrix are left non-reinforced while some others contain clusters or agglomerates of reinforcements. Therefore, the optimum improvements in the mechanical properties by the addition of reinforcements may not be attained. The improved dispersion of reinforcements obtained by optimizing the process parameters can significantly enhance the mechanical properties of MMCs. For instance, the change in distribution pattern of reinforcements from heterogeneous to homogenous in AM processed TiC reinforced AlSi10Mg matrix composites containing 5 wt.% TiC is believed to increase the microhardness from 157.3 to 177.6 HV [59]. Uniform distribution of nanoscale in-situ synthesized TiC reinforcements in the Ti matrix can also significantly enhance the maximum nanohardness and elastic modulus, e.g. ~22.7-fold and ~2.4-fold, respectively compared to the non-reinforced Ti parts [75]. The improved distribution pattern and refinement of in-situ synthesized TiN precipitates in SLM processed $TiN/Ti_5Si_3$ composites achieved at optimal energy density is also believed to be responsible for improved mean microhardness with less fluctuations [75]. Under this condition, the fabricated TiN reinforced $Ti_5Si_3$ matrix composites show at least a four-fold increase in microhardness compared with the non-reinforced Ti parts [79].

Thanks to the outstanding capabilities of laser-based AM technology, MMCs microstructures with unique distribution pattern of reinforcements and improved mechanical properties have been developed in recent years. For instance, the microhardness of in-situ synthesized TiC reinforced TMCs can be significantly increased from 156.6 to 184.7 HV by changing the distribution pattern of reinforcements from individually dispersed particulates to uniform ring-like structure. In fact, the ring-like structures with uniform distribution pattern of reinforcements more efficiently constrain the local deformation of the matrix during microhardness indentation and consequently improve the hardness [62].

## 11 Applications of Additively Manufactured Lightweight MMCs

AM has massive implications not only in part design, but also in the reduction of manual labor, and the elimination of supply chains and inventories. This is particularly applicable in the automotive industry. The ability to create parts in-house and on demand allows for reduction of inventory risk, reduced shipping costs and material procurement costs. In 2014, according to the National Automotive Dealers Association (NADA), the average dealership had over $300,000 of inventory which presents a potential risk that could be removed by fully utilizing AM production techniques [151]. A unique feature of using AM in the automotive sector is assuring the part dealer that they constantly have the most up-to-date part designs [40]. When producing automotive parts, the ability to create complex structures has a huge weight reduction benefit. The weight of parts can be reduced significantly by leveraging the ability of AM processes to produce parts with complex geometries while maintaining relative strengths. Operating expenses and the fuel consumption rate is reduced when weight is reduced, making it extremely attractive to automotive companies. Complex automotive parts are often built using expensive casting and machining processes. Nowadays, AM has the ability to create complex parts for direct end use in production using MMCs [152, 153].

Biomedical industry is another application of AM of MMCs. AM is a very promising technology in the medical industry due to its ability to quickly produce highly custom-designed components. It is now possible to manufacture bones, ears, exoskeletons, windpipes, jaw bones, eyeglasses, cell cultures, stem cells, blood vessels, vascular networks, tissue and organs with AM techniques. Currently, medical applications only represent a small fraction of the 3D printing market. The medical sector of AM represents only 1.6% of the current $700 million AM market, but is predicted to grow by 21% in the next 10 years [154]. The main types of AM systems used in medical applications are SLS, Inkjet 3DP and FDM [155, 156]. Recently, AM is being used more frequently in the production of customized implants and prostheses. There is a great need for bespoke models in the medical industry since standard implants are not always suitable for some patients. Before AM was available, surgeons had to modify standard implants themselves in order to obtain the desired shape. There is no doubt that printing custom prototypes is more accurate than the previous method [157–159]. AM is also commonly used to produce custom-made hearing aids. In fact, 99% of this type of product is now 3D printed. Since all ear canals are uniquely shaped, it is very difficult to produce one standard prototype. Therefore, AM is an efficient way to manufacture hearing aids by combining a great fit with cost effectiveness [157]. The most commonly used materials in SLS for tissue engineering and biomedical applications are semi-crystalline and amorphous thermoplastics, glass and ceramics, metals, and composites.

MMCs could be used in AM of aerospace components. The main challenge in the aerospace industry is to produce light complex geometries with reasonable mechanical properties. This makes AM a very efficient manufacturing method for aircraft and spacecraft manufacturers. Additionally, the aerospace industry manufactures parts in small quantities. These reasons make it beneficial to use AM technology for aerospace applications [160]. The most common AM technologies used in the aerospace industry are DED [161], SLM [162], SLS [163], FDM [164], and EBM [165]. AM is used for low volume production of complex aerospace parts, aircraft wings, and replacement parts in the aerospace industry along with fabrication of specialized parts, lightweight structures, parts with minimal waste, on-demand parts, and replacement parts to support space exploration [166].

Compared to traditional metals and alloys, lightweight MMCs are used for applications where weight reduction is scoped such as aerospace, automotive, and biomedical industries. In general, MMCs have been developed for the following objectives [2]:

1. High yield and ultimate tensile/compressive strengths at high temperatures.
2. High creep and fatigue resistances at high temperatures.

**Table 6** Applications of additively manufactured lightweight MMCs in various industrial sectors

| Matrix | Reinforcement | Effect of reinforcement | References |
|---|---|---|---|
| AlSi10Mg<br>Al-7Si-0.3Mg<br>Al-4.5Cu-3Mg<br>Al4C3 | SiC | Adding SiC improves hardness and enhances wear resistance of the material. It also increases the densification rate of Al matrix during laser melting. This allows the potential application of the material in automotive industry. | [167] |
| AlSi10Mg | TiC | Adding TiC reinforcement with a uniform distribution of the nanoscale TiC particles lowers the wear rate and the coefficient of friction of the parts produced. This will open potential applications for the produced composite in the automotive and aerospace industries. | [59, 168–170] |
| AlSi10Mg | $TiB_2$ | Adding $TiB_2$ to the Al matrix increases the tensile hardness, strength and wear resistance of the material. The obtained composite could be potentially used in automotive industry. | [73] |
| AlSi10Mg | $Fe_2O_3$ | Adding $Fe_2O_3$ results in significant superior microhardness for composites than that of Al alloys. This promising enhancement allows the use of the composite in the automotive industry. | [171] |
| Ti6Al4V | TiB | The modulus of elasticity, yield, ultimate strength and wear resistance were increased. However, the ductility was decreased. The obtained composites could be used in aerospace and biomedical industries. | [54, 172–175] |

3. Good thermal shock and corrosion resistances.
4. Low thermal elongation and high life time.

The AM technology offers high degree of freedom in design and fabrication of complex geometries and customized products. Hence, integrating the AM technology in lightweight MMCs production would allow maintaining the above objectives in complex lightweight structures. Incorporation of reinforcements into the matrix material fulfills a specific function and property in the MMCs produced. Table 6 provides the improved properties and applications of additively manufactured lightweight MMCs.

## References

1. A. I. Mertens and J. Lecomte-Beckers, "On the Role of Interfacial Reactions, Dissolution and Secondary Precipitation During the Laser Additive Manufacturing of Metal Matrix Composites: A Review," in *New Trends in 3D Printing*, I. V. Shishkovsky, Ed. Rijeka: InTech, 2016, p. Ch. 09.

2. K. U. Kainer, "Basics of Metal Matrix Composites," in *Metal Matrix Composites*: Wiley-VCH Verlag GmbH & Co. KGaA, 2006, pp. 1–54.
3. S. Suresh, *Fundamentals of metal-matrix composites*. Elsevier, 2013.
4. M. Rosso, "Ceramic and metal matrix composites: Routes and properties," *Journal of Materials Processing Technology,* vol. 175, no. 1–3, pp. 364–375, 2006.
5. C. W. Hull, "Apparatus for Production of Three-Dimensional Objects by Stereolithography," Unites States Patent Patent US4575330 A, 1986.
6. S. S. Crump, "Apparatus and Method for Creating Three-Dimensional Objects," United States Patent Patent US5121329 A, 1992.
7. T. Wohlers and T. Gornet, "Wohlers Report (2014), History of Additive Manufacturing," Wohlers Associates Inc.2014.
8. Z. Quan *et al.*, "Additive manufacturing of multidirectional preforms for composites: opportunities and challenges," *Materials Today,* vol. 18, no. 9, pp. 503–512, 2015.
9. M. K. Thompson *et al.*, "Design for Additive Manufacturing: Trends, opportunities, considerations, and constraints," *CIRP Annals – Manufacturing Technology,* vol. 65, pp. 737–760, 2016.
10. Y. Zhang, Y. Liu, X. Zhao, and Y. Tang, "The interface microstructure and tensile properties of direct energy deposited TC11/Ti2AlNb dual alloy," *Materials & Design,* vol. 110, pp. 571–580, 2016.
11. W. Liu and J. DuPont, "Fabrication of functionally graded TiC/Ti composites by laser engineered net shaping," *Scripta Materialia,* vol. 48, no. 9, pp. 1337–1342, 2003.
12. D. Herzog, V. Seyda, E. Wycisk, and C. Emmelmann, "Additive manufacturing of metals," *Acta Materialia,* vol. 117, no. Supplement C, pp. 371–392, 2016/09/15/ 2016.
13. D. D. Gu, W. Meiners, K. Wissenbach, and R. Poprawe, "Laser additive manufacturing of metallic components: materials, processes and mechanisms," *International Materials Reviews,* vol. 57, no. 3, pp. 133–164, 2012/05/01 2012.
14. S. Kumar and J. P. Kruth, "Composites by rapid prototyping technology," *Materials & Design,* vol. 31, no. 2, pp. 850–856, 2010/02/01/ 2010.
15. W. E. Frazier, "Metal additive manufacturing: a review," *Journal of Materials Engineering and Performance,* vol. 23, no. 6, pp. 1917–1928, 2014.
16. *Standard Terminology for Additive Manufacturing – General Principles – Terminology*, 2015.
17. A. Hehr and M. J. Dapino, "Interfacial shear strength estimates of NiTi–Al matrix composites fabricated via ultrasonic additive manufacturing," *Composites Part B: Engineering,* vol. 77, no. Supplement C, pp. 199–208, 2015/08/01/ 2015.
18. T. DebRoy *et al.*, "Additive manufacturing of metallic components – Process, structure and properties," *Progress in Materials Science,* vol. 92, no. Supplement C, pp. 112–224, 2018/03/01/ 2018.
19. M. Yakout, A. Cadamuro, M. A. Elbestawi, and S. C. Veldhuis, "The selection of process parameters in additive manufacturing for aerospace alloys," *The International Journal of Advanced Manufacturing Technology,* journal article vol. 92, no. 5, pp. 2081–2098, September 01 2017.
20. K.-H. Shin, H. Natu, D. Dutta, and J. Mazumder, "A Method for the Design and Fabrication of Heterogeneous Objects," *Materials & Design,* vol. 24, no. 5, pp. 339–353, 2003.
21. M. Zhong and W. Liu, "Laser Surface Cladding: the State of the Art and Challenges," in *Proceedings of the Institution of Mechanical Engineers*, London, England, 2010.
22. R. P. Mudge and N. R. Wald, "Laser Engineered Net Shaping Advances Additive Manufacturing and Repair," *Welding Journal,* vol. 86, pp. 44–48, 2007.
23. L. Xue and M. Ul-Islam, "Laser Consolidation – A Novel One-Step Manufacturing Process for Making Net-Shape Functional Components," in *Cost Effective Manufacture via Net-Shape Processing (RTO-MP-AVT-139)*, France, 2006, pp. (15) 1–14: NATO Research and Technology Organization (RTO).

24. R. Anandkumar, A. Almeida, and R. Vilar, "Microstructure and sliding wear resistance of an Al–12wt.% Si/TiC laser clad coating," *Wear,* vol. 282–283, no. Supplement C, pp. 31–39, 2012/04/05/ 2012.
25. L. Bian, S. M. Thompson, and N. Shamsaei, "Mechanical properties and microstructural features of direct laser-deposited Ti-6Al-4V," *Jom,* vol. 67, no. 3, pp. 629–638, 2015.
26. H. Zhang, J. Xu, and G. Wang, "Fundamental Study on Plasma Deposition Manufacturing," in *International Conference on Open Magnetic Systems for Plasma Confinement*, Jeju Island, Korea, 2003.
27. B. Baufeld, O. V. d. Biest, and R. Gault, "Additive Manufacturing of Ti-6Al-4V Components by Shaped Metal Deposition: Microstructure and Mechanical Properties," *Materials & Design,* vol. 31, pp. S106-S111, 2010.
28. S. K. Ghosh, P. Saha, and S. Kishore, "Influence of size and volume fraction of SiC particulates on properties of ex situ reinforced Al–4.5Cu–3Mg metal matrix composite prepared by direct metal laser sintering process," *Materials Science and Engineering: A,* vol. 527, no. 18, pp. 4694–4701, 2010/07/15/ 2010.
29. S. K. Ghosh and P. Saha, "Crack and wear behavior of SiC particulate reinforced aluminium based metal matrix composite fabricated by direct metal laser sintering process," *Materials & Design,* vol. 32, no. 1, pp. 139–145, 2011.
30. A.Gåård, P.Krakhmalev, and J.Bergström, "Microstructural characterization and wear behavior of (Fe,Ni)–TiC MMC prepared by DMLS," *Journal of Alloys and Compounds,* vol. 421, no. 1–2, pp. 166–171, 2006.
31. H. A. Hegab, "Design for additive manufacturing of composite materials and potential alloys: a review," *Manufacturing Review,* vol. 3, no. 11, 2016.
32. Z. Quan *et al.*, "Additive manufacturing of multi-directional preforms for composites: opportunities and challenges," *Materials Today,* vol. 18, no. 9, pp. 503–512, 2015.
33. S. Guessasma, W. Zhang, J. Zhu, S. Belhabib, and H. Nouri, "Challenges of additive manufacturing technologies from an optimisation perspective," *International Journal for Simulation and Multidisciplinary Design Optimization,* vol. 6, 2015.
34. P. Hanzl, M. Zetek, T. Bakša, and T. Kroupa, "The Influence of Processing Parameters on the Mechanical Properties of SLM Parts," *Procedia Engineering,* vol. 100, pp. 1405–1413, 2015.
35. F. Górski, W. Kuczko, and R. Wichniarek, "Influence of Process Parameters on Dimensional Accuracy of Parts Manufactured Using Fused Deposition Modelling Technology," *Advances in Science and Technology Research Journal,* vol. 7, no. 19, pp. 27–35, 2013.
36. D. L. Bourell, M. C. Leu, and D. W. Rosen, "Roadmap for Additive Manufacturing: Identify the Future of Freeform Processing," in "International Solid Freeform Fabrication Symposium," USA2009
37. N. Hrabe and T. Quinn, "Effects of Processing on Microstructure and Mechanical Properties of a Titanium Alloy (Ti–6Al–4V) Fabricated using Electron Beam Melting (EBM), Part 1: Distance from Build Plate and Part Size," *Materials Science & Engineering A,* vol. 375, pp. 264–270, 2013.
38. N. Guo and M. C. Leu, "Additive Manufacturing: Technology, Applications and Research Needs," *Frontiers of Mechanical Engineering,* vol. 8, no. 3, pp. 215–243, 2013.
39. J. Scott, N. Gupta, C. Weber, S. Newsome, T. Wohlers, and T. Caffrey, "Additive Manufacturing: Status and Opportunities," IDA Science and Technology Policy Institute, Washington, DC, USA 2012.
40. C. Beyer, "Strategic Implications of Current Trends in Additive Manufacturing," *Journal of Manufacturing Science and Engineering,* vol. 136, no. 6, pp. 064701 (1–6), 2014.
41. W. Gao *et al.*, "The Status, Challenges, and Future of Additive Manufacturing in Engineering," *Computer-Aided Design,* 2015.
42. T. Wohlers. (2011) Making Products By Using Additive Manufacturing. *Manufacturing Engineering Magazine.* 70–77.
43. S. Kumar and J.-P. Kruth, "Composites by rapid prototyping technology," *Materials & Design,* vol. 31, no. 2, pp. 850–856, 2010.

44. S. C. Tjong and Z. Ma, "Microstructural and mechanical characteristics of in situ metal matrix composites," *Materials Science and Engineering: R: Reports,* vol. 29, no. 3, pp. 49–113, 2000.
45. S. K. Ghosh, P. Saha, and S. Kishore, "Influence of size and volume fraction of SiC particulates on properties of ex situ reinforced Al–4.5 Cu–3Mg metal matrix composite prepared by direct metal laser sintering process," *Materials Science and Engineering: A,* vol. 527, no. 18, pp. 4694–4701, 2010.
46. F. Chang, D. Gu, D. Dai, and P. Yuan, "Selective laser melting of in-situ Al4SiC4+ SiC hybrid reinforced Al matrix composites: Influence of starting SiC particle size," *Surface and Coatings Technology,* vol. 272, pp. 15–24, 2015.
47. A. Simchi and D. Godlinski, "Densification and microstructural evolution during laser sintering of A356/SiC composite powders," *Journal of materials science,* vol. 46, no. 5, pp. 1446–1454, 2011.
48. A. Simchi and D. Godlinski, "Effect of SiC particles on the laser sintering of Al–7Si–0.3 Mg alloy," *Scripta Materialia,* vol. 59, no. 2, pp. 199–202, 2008.
49. J. Jue, D. Gu, K. Chang, and D. Dai, "Microstructure evolution and mechanical properties of Al-Al 2 O 3 composites fabricated by selective laser melting," *Powder Technology,* vol. 310, pp. 80–91, 2017.
50. V. K. Balla, A. Bhat, S. Bose, and A. Bandyopadhyay, "Laser processed TiN reinforced Ti6Al4V composite coatings," *Journal of the mechanical behavior of biomedical materials,* vol. 6, pp. 9–20, 2012.
51. M. Das, V. K. Balla, D. Basu, S. Bose, and A. Bandyopadhyay, "Laser processing of SiC-particle-reinforced coating on titanium," *Scripta Materialia,* vol. 63, no. 4, pp. 438–441, 2010.
52. M. Das *et al.*, "Microstructure, mechanical and wear properties of laser processed SiC particle reinforced coatings on titanium," *Surface and Coatings Technology,* vol. 205, no. 19, pp. 4366–4373, 2011.
53. F. Wang, J. Mei, and X. Wu, "Compositionally graded Ti6Al4V+ TiC made by direct laser fabrication using powder and wire," *Materials & design,* vol. 28, no. 7, pp. 2040–2046, 2007.
54. F. Wang, J. Mei, and X. Wu, "Direct laser fabrication of Ti6Al4V/TiB," *Journal of materials processing technology,* vol. 195, no. 1, pp. 321–326, 2008.
55. S. C. Tjong, "Novel nanoparticle-reinforced metal matrix composites with enhanced mechanical properties," *Advanced engineering materials,* vol. 9, no. 8, pp. 639–652, 2007.
56. H. Wang, Q. Jiang, X. Li, and J. Wang, "In situ synthesis of TiC/Mg composites in molten magnesium," *Scripta Materialia,* vol. 48, no. 9, pp. 1349–1354, 2003.
57. N. Saheb *et al.*, "Spark plasma sintering of metals and metal matrix nanocomposites: a review," *Journal of Nanomaterials,* vol. 2012, p. 18, 2012.
58. M. Das and V. K. Balla, "Additive Manufacturing and Innovation in Materials World," *Additive Manufacturing,* p. 297, 2015.
59. D. Gu *et al.*, "Densification behavior, microstructure evolution, and wear property of TiC nanoparticle reinforced AlSi10Mg bulk-form nanocomposites prepared by selective laser melting," *Journal of Laser Applications,* vol. 27, no. S1, p. S17003, 2015.
60. D. Gu, H. Wang, and D. Dai, "Laser additive manufacturing of novel aluminum based nanocomposite parts: tailored forming of multiple materials," *Journal of Manufacturing Science and Engineering,* vol. 138, no. 2, p. 021004, 2016.
61. P. Yuan, D. Gu, and D. Dai, "Particulate migration behavior and its mechanism during selective laser melting of TiC reinforced Al matrix nanocomposites," *Materials & Design,* vol. 82, pp. 46–55, 2015.
62. D. Gu, H. Wang, D. Dai, P. Yuan, W. Meiners, and R. Poprawe, "Rapid fabrication of Al-based bulk-form nanocomposites with novel reinforcement and enhanced performance by selective laser melting," *Scripta Materialia,* vol. 96, pp. 25–28, 2015.
63. D. Dai and D. Gu, "Tailoring surface quality through mass and momentum transfer modeling using a volume of fluid method in selective laser melting of TiC/AlSi10Mg powder," *International Journal of Machine Tools and Manufacture,* vol. 88, pp. 95–107, 2015.

64. D. Gu and P. Yuan, "Thermal evolution behavior and fluid dynamics during laser additive manufacturing of Al-based nanocomposites: Underlying role of reinforcement weight fraction," *Journal of Applied Physics,* vol. 118, no. 23, p. 233109, 2015.
65. D. Gu, F. Chang, and D. Dai, "Selective laser melting additive manufacturing of novel aluminum based composites with multiple reinforcing phases," *Journal of Manufacturing Science and Engineering,* vol. 137, no. 2, p. 021010, 2015.
66. L. Wang, J. Jue, M. Xia, L. Guo, B. Yan, and D. Gu, "Effect of the Thermodynamic Behavior of Selective Laser Melting on the Formation of In situ Oxide Dispersion-Strengthened Aluminum-Based Composites," *Metals,* vol. 6, no. 11, p. 286, 2016.
67. D. Dai and D. Gu, "Influence of thermodynamics within molten pool on migration and distribution state of reinforcement during selective laser melting of AlN/AlSi10Mg composites," *International Journal of Machine Tools and Manufacture,* vol. 100, pp. 14–24, 2016.
68. P. Ma, Y. Jia, K. G. Prashanth, S. Scudino, Z. Yu, and J. Eckert, "Microstructure and phase formation in Al–20Si–5Fe–3Cu–1Mg synthesized by selective laser melting," *Journal of Alloys and Compounds,* vol. 657, pp. 430–435, 2016.
69. K. G. Prashanth *et al.*, "Processing of Al–12Si–TNM composites by selective laser melting and evaluation of compressive and wear properties," *Journal of Materials Research,* vol. 31, no. 1, pp. 55–65, 2016.
70. K. Prashanth *et al.*, "Production of high strength Al85Nd8Ni5Co2 alloy by selective laser melting," *Additive Manufacturing,* vol. 6, pp. 1–5, 2015.
71. S. Dadbakhsh and L. Hao, "Effect of Al alloys on selective laser melting behaviour and microstructure of in situ formed particle reinforced composites," *Journal of alloys and compounds,* vol. 541, pp. 328–334, 2012.
72. D. Manfredi *et al.*, "Additive manufacturing of Al alloys and aluminium matrix composites (AMCs)," in *Light metal alloys applications*: InTech, 2014.
73. X. P. Li *et al.*, "Selective laser melting of nano-TiB2 decorated AlSi10Mg alloy with high fracture strength and ductility," *Acta Materialia,* vol. 129, no. Supplement C, pp. 183–193, 2017/05/01/ 2017.
74. D. Gu, W. Meiners, Y.-C. Hagedorn, K. Wissenbach, and R. Poprawe, "Bulk-form TiCx/Ti nanocomposites with controlled nanostructure prepared by a new method: selective laser melting," *Journal of Physics D: Applied Physics,* vol. 43, no. 29, p. 295402, 2010.
75. D. Gu, Y.-C. Hagedorn, W. Meiners, K. Wissenbach, and R. Poprawe, "Nanocrystalline TiC reinforced Ti matrix bulk-form nanocomposites by Selective Laser Melting (SLM): Densification, growth mechanism and wear behavior," *Composites Science and Technology,* vol. 71, no. 13, pp. 1612–1620, 2011.
76. D. Gu, G. Meng, C. Li, W. Meiners, and R. Poprawe, "Selective laser melting of TiC/Ti bulk nanocomposites: Influence of nanoscale reinforcement," *Scripta Materialia,* vol. 67, no. 2, pp. 185–188, 2012.
77. Y. Zhang, J. Sun, and R. Vilar, "Characterization of (TiB+ TiC)/TC4 in situ titanium matrix composites prepared by laser direct deposition," *Journal of Materials Processing Technology,* vol. 211, no. 4, pp. 597–601, 2011.
78. S. Nag, S. Samuel, A. Puthucode, and R. Banerjee, "Characterization of novel borides in Ti–Nb–Zr–Ta+ 2B metal-matrix composites," *Materials characterization,* vol. 60, no. 2, pp. 106–113, 2009.
79. D. Gu, C. Hong, and G. Meng, "Densification, microstructure, and wear property of in situ titanium nitride-reinforced titanium silicide matrix composites prepared by a novel selective laser melting process," *Metallurgical and Materials Transactions A,* vol. 43, no. 2, pp. 697–708, 2012.
80. R. Banerjee, P. Collins, A. Genc, and H. Fraser, "Direct laser deposition of in situ Ti–6Al–4V–TiB composites," *Materials Science and Engineering: A,* vol. 358, no. 1, pp. 343–349, 2003.
81. A. Genç, R. Banerjee, D. Hill, and H. Fraser, "Structure of TiB precipitates in laser deposited in situ, Ti-6Al-4V–TiB composites," *Materials Letters,* vol. 60, no. 7, pp. 859–863, 2006.

82. Y. Hu, W. Cong, X. Wang, Y. Li, F. Ning, and H. Wang, "Laser deposition-additive manufacturing of TiB-Ti composites with novel three-dimensional quasi-continuous network microstructure: Effects on strengthening and toughening," *Composites Part B: Engineering,* vol. 133, pp. 91–100, 2018.
83. H. Attar *et al.*, "Effect of powder particle shape on the properties of in situ Ti–TiB composite materials produced by selective laser melting," *Journal of Materials Science & Technology,* vol. 31, no. 10, pp. 1001–1005, 2015.
84. H. Attar, M. Bönisch, M. Calin, L.-C. Zhang, S. Scudino, and J. Eckert, "Selective laser melting of in situ titanium–titanium boride composites: processing, microstructure and mechanical properties," *Acta Materialia,* vol. 76, pp. 13–22, 2014.
85. H. Attar *et al.*, "Mechanical behavior of porous commercially pure Ti and Ti–TiB composite materials manufactured by selective laser melting," *Materials Science and Engineering: A,* vol. 625, pp. 350–356, 2015.
86. D. Gu, D. Dai, G. Zhang, and H. Wang, "Growth mechanisms of in situ TiC in laser melted Ti–Si–C ternary system," *Applied Physics Letters,* vol. 101, no. 17, p. 171603, 2012.
87. M. Das *et al.*, "In situ synthesized TiB–TiN reinforced Ti6Al4V alloy composite coatings: microstructure, tribological and in-vitro biocompatibility," *Journal of the mechanical behavior of biomedical materials,* vol. 29, pp. 259–271, 2014.
88. D. Gu, Z. Wang, Y. Shen, Q. Li, and Y. Li, "In-situ TiC particle reinforced Ti–Al matrix composites: powder preparation by mechanical alloying and selective laser melting behavior," *Applied Surface Science,* vol. 255, no. 22, pp. 9230–9240, 2009.
89. M. Das, V. K. Balla, D. Basu, I. Manna, T. S. Kumar, and A. Bandyopadhyay, "Laser processing of in situ synthesized TiB–TiN-reinforced Ti6Al4V alloy coatings," *Scripta Materialia,* vol. 66, no. 8, pp. 578–581, 2012.
90. S. K. Ghosh, K. Bandyopadhyay, and P. Saha, "Development of an in-situ multi-component reinforced Al-based metal matrix composite by direct metal laser sintering technique—Optimization of process parameters," *Materials Characterization,* vol. 93, pp. 68–78, 2014.
91. B. Vrancken, L. Thijs, J.-P. Kruth, and J. Van Humbeeck, "Microstructure and mechanical properties of a novel β titanium metallic composite by selective laser melting," *Acta Materialia,* vol. 68, pp. 150–158, 2014.
92. S. L. Sing, W. Y. Yeong, and F. E. Wiria, "Selective laser melting of titanium alloy with 50 wt % tantalum: Microstructure and mechanical properties," *Journal of Alloys and Compounds,* vol. 660, pp. 461–470, 2016.
93. T. A. Campbell and O. S. Ivanova, "3D printing of multifunctional nanocomposites," *Nano Today,* vol. 8, no. 2, pp. 119–120, 2013.
94. M. Chu and M. Premkumar, "Mechanism of TiC formation in Al/TiCin situ metal-matrix composites," *Metallurgical and Materials Transactions A,* vol. 24, no. 12, pp. 2803–2805, 1993.
95. D. Gu *et al.*, "Selective laser melting additive manufacturing of TiC/AlSi10Mg bulk-form nanocomposites with tailored microstructures and properties," *Physics Procedia,* vol. 56, pp. 108–116, 2014.
96. E. Zhang, S. Zeng, Q. Li, Y. Bo, and M. Ma, "A study on the kinetic process of reaction synthesis of TiC: Part I. Experimental research and theoretical model," *Metallurgical and Materials Transactions A,* vol. 30, no. 4, pp. 1147–1151, 1999.
97. D. C. Van Aken, P. Krajewski, G. Vyletel, J. Allison, and J. Jones, "Recrystallization and grain growth phenomena in a particle-reinforced aluminum composite," *Metallurgical and Materials Transactions A,* vol. 26, no. 6, pp. 1395–1405, 1995.
98. G. Vyletel, J. Allison, and D. C. Van Aken, "The effect of matrix microstructure on cyclic response and fatigue behavior of particle-reinforced 2219 aluminum: Part I. Room temperature behavior," *Metallurgical and Materials Transactions A,* vol. 26, no. 12, pp. 3143–3154, 1995.
99. G. S. P. Kumar, P. G. Koppad, R. Keshavamurthy, and M. Alipour, "Microstructure and mechanical behaviour of in situ fabricated AA6061–TiC metal matrix composites," *Archives of Civil and Mechanical Engineering,* vol. 17, no. 3, pp. 535–544, 2017.

100. D. Zhou, F. Qiu, and Q. Jiang, "The nano-sized TiC particle reinforced Al–Cu matrix composite with superior tensile ductility," *Materials Science and Engineering: A,* vol. 622, pp. 189–193, 2015.
101. H. Yin, S. Felicelli, and L. Wang, "Fluid flow, heat and mass transfer in the molten pool of the LENS process," in *TMS Annual Meeting*, 2008, pp. 261–270.
102. H. Niu and I. Chang, "Selective laser sintering of gas and water atomized high speed steel powders," *Scripta Materialia,* vol. 41, no. 1, pp. 25–30, 1999.
103. M. Thompson and J. Szekely, "The transient behavior of weldpools with a deformed free surface," *International journal of heat and mass transfer,* vol. 32, no. 6, pp. 1007–1019, 1989.
104. K. Arafune and A. Hirata, "Thermal and solutal Marangoni convection in In–Ga–Sb system," *Journal of Crystal Growth,* vol. 197, no. 4, pp. 811–817, 1999.
105. Y. Lei, H. Murakawa, Y. Shi, and X. Li, "Numerical analysis of the competitive influence of Marangoni flow and evaporation on heat surface temperature and molten pool shape in laser surface remelting," *Computational Materials Science,* vol. 21, no. 3, pp. 276–290, 2001.
106. A. Kumar and S. Roy, "Effect of three-dimensional melt pool convection on process characteristics during laser cladding," *Computational Materials Science,* vol. 46, no. 2, pp. 495–506, 2009.
107. S. A. Khairallah, A. T. Anderson, A. Rubenchik, and W. E. King, "Laser powder-bed fusion additive manufacturing: Physics of complex melt flow and formation mechanisms of pores, spatter, and denudation zones," *Acta Materialia,* vol. 108, pp. 36–45, 2016.
108. L. Han, F. W. Liou, and S. Musti, "Thermal behavior and geometry model of melt pool in laser material process," *Journal of Heat Transfer,* vol. 127, no. 9, pp. 1005–1014, 2005.
109. X. He, J. Norris, P. Fuerschbach, and T. DebRoy, "Alloying element vaporization and liquid metal expulsion during laser microjoining of stainless steel with short pulse," in *Trends Weld. Res. Proc. 7th Int. Conf*, 2005, pp. 109–114.
110. M. Simonelli *et al.*, "A study on the laser spatter and the oxidation reactions during selective laser melting of 316L stainless steel, Al-Si10-Mg, and Ti-6Al-4V," *Metallurgical and Materials Transactions A,* vol. 46, no. 9, pp. 3842–3851, 2015.
111. B. Rabin, G. Smolik, and G. Korth, "Characterization of entrapped gases in rapidly solidified powders," *Materials Science and Engineering: A,* vol. 124, no. 1, pp. 1–7, 1990.
112. S. Khairallah, A. Anderson, A. Rubenchik, J. Florando, S. Wu, and H. Lowdermilk, "Simulation of the main physical processes in remote laser penetration with large laser spot size," *AIP Advances,* vol. 5, no. 4, p. 047120, 2015.
113. A. C. Perkins, *Mechanisms of instability in Rayleigh-Benard convection.* Georgia Institute of Technology, 2011.
114. L. Lü, J. Fuh, and Y.-S. Wong, *Laser-induced materials and processes for rapid prototyping.* Springer Science & Business Media, 2013.
115. R. M. German, "Supersolidus liquid phase sintering. I: Process review," *International journal of powder metallurgy (1986),* vol. 26, no. 1, pp. 23–34, 1990.
116. M. Agarwala, D. Bourell, J. Beaman, H. Marcus, and J. Barlow, "Direct selective laser sintering of metals," *Rapid Prototyping Journal,* vol. 1, no. 1, pp. 26–36, 1995.
117. T. Iida and R. I. Guthrie, "The physical properties of liquid metals," *Clarendon Press, Walton Street, Oxford OX 2 6 DP, UK, 1988.,* 1988.
118. D. Gu, "Novel Aluminum Based Composites by Selective Laser Melting (SLM) Additive Manufacturing (AM): Tailored Formation of Multiple Reinforcing Phases and its Mechanisms," in *Laser Additive Manufacturing of High-Performance Materials*: Springer, 2015, pp. 201–222.
119. A. Grabowski, M. Nowak, and J. Śleziona, "Optical and conductive properties of AlSi-alloy/SiCp composites: application in modelling CO2 laser processing of composites," *Optics and Lasers in Engineering,* vol. 43, no. 2, pp. 233–246, 2005.
120. P. Fischer, V. Romano, H.-P. Weber, N. Karapatis, E. Boillat, and R. Glardon, "Sintering of commercially pure titanium powder with a Nd: YAG laser source," *Acta Materialia,* vol. 51, no. 6, pp. 1651–1662, 2003.

121. S. P. Gordienko, "Reaction of titanium with boron nitride under self-propagating high-temperature synthesis conditions," *Powder Metallurgy and Metal Ceramics,* vol. 40, no. 1–2, pp. 58–60, 2001.
122. K. Morsi and V. Patel, "Processing and properties of titanium–titanium boride (TiBw) matrix composites—a review," *Journal of materials science,* vol. 42, no. 6, pp. 2037–2047, 2007.
123. S. Samuel, S. Nag, T. W. Scharf, and R. Banerjee, "Wear resistance of laser-deposited boride reinforced Ti-Nb–Zr–Ta alloy composites for orthopedic implants," *Materials Science and Engineering: C,* vol. 28, no. 3, pp. 414–420, 2008.
124. L. Thijs, K. Kempen, J.-P. Kruth, and J. Van Humbeeck, "Fine-structured aluminium products with controllable texture by selective laser melting of pre-alloyed AlSi10Mg powder," *Acta Materialia,* vol. 61, no. 5, pp. 1809–1819, 2013.
125. D. Manfredi, F. Calignano, M. Krishnan, R. Canali, E. P. Ambrosio, and E. Atzeni, "From powders to dense metal parts: Characterization of a commercial AlSiMg alloy processed through direct metal laser sintering," *Materials,* vol. 6, no. 3, pp. 856–869, 2013.
126. F. Trevisan *et al.*, "On the Selective Laser Melting (SLM) of the AlSi10Mg Alloy: Process, Microstructure, and Mechanical Properties," *Materials,* vol. 10, no. 1, p. 76, 2017.
127. G. Dinda, A. Dasgupta, and J. Mazumder, "Evolution of microstructure in laser deposited Al–11.28% Si alloy," *Surface and Coatings Technology,* vol. 206, no. 8, pp. 2152–2160, 2012.
128. H. Asgari, C. Baxter, K. Hosseinkhani, and M. Mohammadi, "On microstructure and mechanical properties of additively manufactured AlSi10Mg_200C using recycled powder," *Materials Science and Engineering: A,* vol. 707, pp. 148–158, 2017.
129. M. Mohammadi and H. Asgari, "Achieving low surface roughness AlSi10Mg_200C parts using direct metal laser sintering," *Additive Manufacturing,* 2017.
130. Y. Hu, B. Zhao, F. Ning, H. Wang, and W. Cong, "In-situ ultrafine three-dimensional quasi-continuous network microstructural TiB reinforced titanium matrix composites fabrication using laser engineered net shaping," *Materials Letters,* vol. 195, pp. 116–119, 2017.
131. B. AlMangour, D. Grzesiak, and J.-M. Yang, "Nanocrystalline TiC-reinforced H13 steel matrix nanocomposites fabricated by selective laser melting," *Materials & Design,* vol. 96, pp. 150–161, 2016.
132. B. AlMangour, D. Grzesiak, and J.-M. Yang, "Selective laser melting of TiB 2/316L stainless steel composites: The roles of powder preparation and hot isostatic pressing post-treatment," *Powder Technology,* vol. 309, pp. 37–48, 2017.
133. B. AlMangour, D. Grzesiak, and J.-M. Yang, "Rapid fabrication of bulk-form TiB 2/316L stainless steel nanocomposites with novel reinforcement architecture and improved performance by selective laser melting," *Journal of Alloys and Compounds,* vol. 680, pp. 480–493, 2016.
134. B. AlMangour and D. Grzesiak, "Selective laser melting of TiC reinforced 316L stainless steel matrix nanocomposites: Influence of starting TiC particle size and volume content," *Materials & Design,* vol. 104, pp. 141–151, 2016.
135. B. AlMangour, D. Grzesiak, and J.-M. Yang, "Selective laser melting of TiB 2/H13 steel nanocomposites: Influence of hot isostatic pressing post-treatment," *Journal of Materials Processing Technology,* vol. 244, pp. 344–353, 2017.
136. X. Zhao, B. Song, W. Fan, Y. Zhang, and Y. Shi, "Selective laser melting of carbon/AlSi10Mg composites: microstructure, mechanical and electronical properties," *Journal of Alloys and Compounds,* vol. 665, pp. 271–281, 2016.
137. S. Tamirisakandala, R. Bhat, J. Tiley, and D. Miracle, "Grain refinement of cast titanium alloys via trace boron addition," *Scripta Materialia,* vol. 53, no. 12, pp. 1421–1426, 2005.
138. M. Li, J. Huang, Y.-y. Zhu, Z.-g. Li, and Y.-x. Wu, "Growth morphology and solidification behavior of in-situ synthesized TiN and TiB," *MATERIALS TRANSACTIONS,* vol. 57, no. 1, pp. 15–19, 2016.
139. E. Fereiduni and S. G. Banadkouki, "Reliability/unreliability of mixture rule in a low alloy ferrite–martensite dual phase steel," *Journal of alloys and compounds,* vol. 577, pp. 351–359, 2013.

140. C. T. Lynch, *Metal matrix composites*. CRC press, 2018.
141. A. Agarwal, S. R. Bakshi, and D. Lahiri, *Carbon nanotubes: reinforced metal matrix composites*. CRC press, 2016.
142. M. Taya and R. J. Arsenault, *Metal matrix composites: thermomechanical behavior*. Elsevier, 2016.
143. R. Mahamood and E. T. Akinlabi, "Laser metal deposition of functionally graded Ti6Al4V/TiC," *Materials & Design*, vol. 84, pp. 402–410, 2015.
144. A. Biswas, L. Li, U. Chatterjee, I. Manna, and J. D. Majumdar, "Diode laser assisted surface nitriding of Ti-6Al-4V: Properties of the nitrided surface," *Metallurgical and Materials Transactions A*, vol. 40, no. 12, p. 3031, 2009.
145. R. Banerjee, A. Genc, P. Collins, and H. Fraser, "Comparison of microstructural evolution in laser-deposited and arc-melted in-situ Ti-TiB composites," *Metallurgical and Materials Transactions A*, vol. 35, no. 7, pp. 2143–2152, 2004.
146. S. Dadbakhsh and L. Hao, "Effect of hot isostatic pressing (HIP) on Al composite parts made from laser consolidated Al/Fe 2 O 3 powder mixtures," *Journal of Materials Processing Technology*, vol. 212, no. 11, pp. 2474–2483, 2012.
147. S. Dadbakhsh, L. Hao, P. Jerrard, and D. Zhang, "Experimental investigation on selective laser melting behaviour and processing windows of in situ reacted Al/Fe 2 O 3 powder mixture," *Powder technology*, vol. 231, pp. 112–121, 2012.
148. Y. Zhang, Z. Wei, L. Shi, and M. Xi, "Characterization of laser powder deposited Ti–TiC composites and functional gradient materials," *journal of materials processing technology*, vol. 206, no. 1, pp. 438–444, 2008.
149. E. Hall, "The deformation and ageing of mild steel: III discussion of results," *Proceedings of the Physical Society. Section B*, vol. 64, no. 9, p. 747, 1951.
150. D. Buchbinder, H. Schleifenbaum, S. Heidrich, W. Meiners, and J. Bültmann, "High power selective laser melting (HP SLM) of aluminum parts," *Physics Procedia*, vol. 12, pp. 271–278, 2011.
151. NADA, "NADA Data 2014: ANNUAL FINANCIAL PROFILE OF AMERICA'S FRANCHISED NEW-CAR DEALERSHIPS," Available: nada.org
152. D. Cooper, J. Thornby, N. Blundell, R. Henrys, M. A. Williams, and G. Gibbons, "Design and Manufacture of High Performance Hollow Engine Valves by Additive Layer Manufacturing," *Materials & Design*, vol. 69, pp. 44–55, 2015.
153. S. Biamino, B. Klöden, T. Weißgärber, B. Kieback, and U. Ackelid, "Properties of a TiAl Turbocharger Wheel Produced by Electron Beam Melting," in *Fraunhofer Direct Digital Manufacturing Conference (DDMC)*, Berlin, Germany, 2014.
154. C. Schubert, M. C. v. Langeveld, and L. A. Donoso, "Innovations in 3D Printing: A 3D Overview from Optics to Organs," *Br J Ophtalmol*, vol. 98, no. 2, pp. 159–161, 2014.
155. X. Cui, T. Boland, D. D. D'Lima, and M. K. Lotz, "Thermal Inkjet Printing in Tissue Engineering and Regenerative Medicine," *Recent Pat Drug Deliv Formul*, vol. 6, no. 2, pp. 149–155, 2012.
156. M. B. Hoy, "3D Printing: Making Things at the Library," *Medical Reference Services Quarterly*, vol. 32, no. 1, pp. 94–99, 2013.
157. J. Banks, "Adding Value in Additive Manufacturing: Researchers in the United Kingdom and Europe Look to 3D Printing for Customization," *IEEE Pulse*, vol. 4, no. 6, pp. 22–26, 2013.
158. S. Bartlett, "Printing Organs on Demand," *Lancet Respiratory Medicine*, vol. 1, no. 9, p. 684, 2013.
159. H. Lipson, "New World of 3-D Printing Offers "Completely New Ways of Thinking:" Q & A with Author, Engineer, and 3-D Printing Expert Hod Lipson," *IEEE Pulse*, vol. 4, no. 6, pp. 12–14, 2013.
160. K. V. Wong and A. Hernandez, "A Review of Additive Manufacturing," *International Scholarly Research Network (ISRN) Mechanical Engineering*, vol. 2012, p. 10 Pages, 2012.
161. N. K. Dey, "Masters Theses, Additive Manufacturing Laser Deposition of Ti-6Al-4V for Aerospace Repair Application," Missouri University of Science and Technology 2014.

162. T. Wohlers, "Making Products By Using Additive Manufacturing," *Manufacturing Engineering Magazine,* pp. 70–77, April 2011.
163. D. Eyers and K. Dotchev, "Technology Review for Mass Customisation Using Rapid Manufacturing," *Assembly Automation,* vol. 30, no. 1, pp. 39–46, 2010.
164. D. Yagnik, "Fused Deposition Modeling – A Rapid Prototyping Technique for Product Cycle Time Reduction Cost Effectively in Aerospace Applications," *IOSR Journal of Mechanical and Civil Engineering (IOSR-JMCE),* vol. International Conference on Advances in Engineering & Technology – 2014 (ICAET-2014), Volume 5, pp. 62–68, 2014.
165. P. Michaleris, "Modeling Metal Deposition in Heat Transfer Analyses of Additive Manufacturing Processes," *Finite Elements in Analysis and Design,* vol. 86, pp. 51–60, 2014.
166. J. Coykendall, M. Cotteleer, J. Holdowsky, and M. Mahto, "3D Opportunity in Aerospace and Defense: Additive Manufacturing Takes Flight," Deloitte University Press, UK2014.
167. F. Chang, D. Gu, D. Dai, and P. Yuan, "Selective laser melting of in-situ Al4SiC4+SiC hybrid reinforced Al matrix composites: Influence of starting SiC particle size," *Surface and Coatings Technology,* vol. 272, no. Supplement C, pp. 15–24, 2015/06/25/ 2015.
168. D. Gu *et al.*, "Selective Laser Melting Additive Manufacturing of TiC/AlSi10Mg Bulk-form Nanocomposites with Tailored Microstructures and Properties," *Physics Procedia,* vol. 56, no. Supplement C, pp. 108–116, 2014/01/01/ 2014.
169. D. Gu, H. Wang, D. Dai, P. Yuan, W. Meiners, and R. Poprawe, "Rapid fabrication of Al-based bulk-form nanocomposites with novel reinforcement and enhanced performance by selective laser melting," *Scripta Materialia,* vol. 96, no. Supplement C, pp. 25–28, 2015/02/01/ 2015.
170. P. Yuan, D. Gu, and D. Dai, "Particulate migration behavior and its mechanism during selective laser melting of TiC reinforced Al matrix nanocomposites," *Materials & Design,* vol. 82, no. Supplement C, pp. 46–55, 2015/10/05/ 2015.
171. S. Dadbakhsh and L. Hao, "In situ formation of particle reinforced Al matrix composite by selective laser melting of Al/Fe2O3 powder mixture," *Advanced Engineering Materials,* vol. 14, no. 1–2, pp. 45–48, 2012.
172. R. Banerjee, A. Genç, D. Hill, P. C. Collins, and H. L. Fraser, "Nanoscale TiB precipitates in laser deposited Ti-matrix composites," *Scripta Materialia,* vol. 53, no. 12, pp. 1433–1437, 2005/12/01/ 2005.
173. R. Banerjee, A. Genç, P. C. Collins, and H. L. Fraser, "Comparison of microstructural evolution in laser-deposited and arc-melted In-Situ Ti-TiB composites," *Metallurgical and Materials Transactions A,* journal article vol. 35, no. 7, pp. 2143–2152, July 01 2004.
174. S. Pouzet *et al.*, "Additive layer manufacturing of titanium matrix composites using the direct metal deposition laser process," *Materials Science and Engineering: A,* vol. 677, no. Supplement C, pp. 171–181, 2016/11/20/ 2016.
175. R. Banerjee, P. C. Collins, A. Genç, and H. L. Fraser, "Direct laser deposition of in situ Ti–6Al–4V–TiB composites," *Materials Science and Engineering: A,* vol. 358, no. 1, pp. 343–349, 2003/10/15/ 2003.
176. Fereiduki, E., & Banadkouki, S. G. (2014). Ferrite hardening response in a low alloy ferrite–martensite dual phase steel. Journal of Alloys and Compounds, 589, 288–294.
177. Fereiduni, E., & Banadkouki, S. G. (2014). Improvement of mechanical properties in a dual-phase ferrite–martensite AISI4140 steel under tough-strong ferrite formation. Materials & Design (1980–2015), 56, 232–240.
178. Eskandar Fereiduni, Mojtaba Movahedi, Amir Hossein Kokabi, Hossein Najafi, (2017) Effect of Dwell Time on Joint Interface Microstructure and Strength of Dissimilar Friction Stir Spot-Welded Al-5083 and St-12 Alloy Sheets. Metallurgical and Materials Transactions A 48 (4):1744–1758.
179. E. Fereiduni, M. Movahedi, A. Baghdadchi, (2017) Ultrahigh-strength friction stir spot welds of aluminium alloy obtained by Fe O nanoparticles. Science and Technology of Welding and Joining 23 (1):63–70.

180. Mostafa Yakout, M. A. Elbestawi, Stephen C. Veldhuis, (2018) On the characterization of stainless steel 316L parts produced by selective laser melting. The International Journal of Advanced Manufacturing Technology 95 (5-8):1953–1974.
181. Yakout, M., & Elbestawi, M. A. (2017). Additive Manufacturing of Composite Materials: An Overview. In 6th International Conference on Virtual Machining Process Technology.

# Process-Structure-Property Relationships in Additively Manufactured Metal Matrix Composites

**Eskandar Fereiduni and Mohamed Elbestawi**

## 1 Introduction

During the past few decades, there has been a significant interest in development and fabrication of near net-shape and complex parts made from novel materials such as bi-materials, composites and complex materials. The parts produced from these materials could have outstanding combinations of properties, introducing them as potential candidates meeting the requirements of automotive, aerospace, rapid tooling and biomedical industries.

Thanks to the significant advances in additive manufacturing (AM) of single materials as well as the optimized process parameters prescribed for fabrication of parts made of commercially available materials, near net-shape components with high qualities could be fabricated from these materials using AM processes. However, processing of composites and complex materials is rather challenging in some of the AM techniques. Moreover, AM processing of these materials is relatively new and still needs extensive research to be fully explored.

Besides the flexibility in fabricating the components with complex geometries, recent advances in AM technology have paved the way for the design and manufacturing of parts with tailored properties including composites, functionally graded materials (FGMs) and multifunctional materials. When combined with the basics of composite fabrication, AM technology can open new opportunities to develop composites with outstanding properties.

The ASTM-recognized AM methods involve materials extrusion, material jetting, sheet lamination, vat-photo-polymerization, binder jetting, directed energy deposition (DED) and powder bed fusion (PBF) [1]. Among various AM technologies available, only a few of them have shown their capability in composite

E. Fereiduni · M. Elbestawi (✉)
Department of Mechanical Engineering, McMaster University, Hamilton, ON, Canada
e-mail: elbestaw@mcmaster.ca



B. AlMangour (ed.), *Additive Manufacturing of Emerging Materials*,
https://doi.org/10.1007/978-3-319-91713-9_4

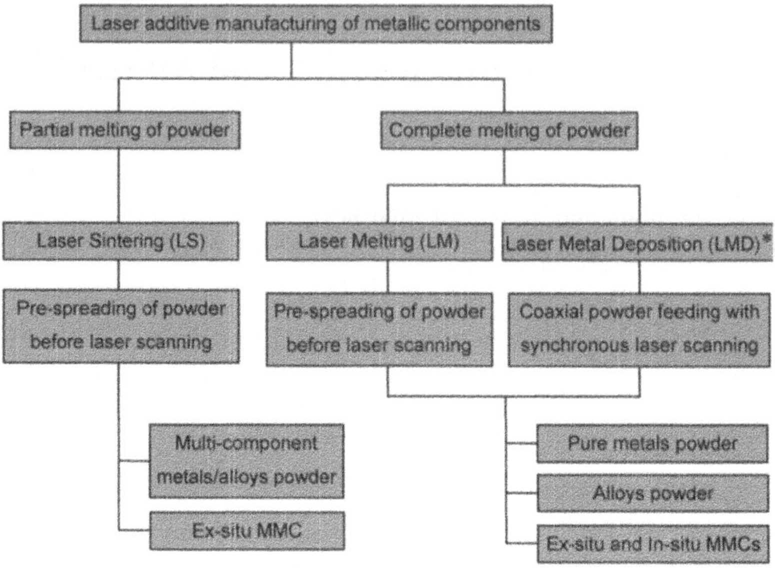


**Fig. 1** Classification of AM processes according to the laser/material interaction mechanism [2]

fabrication. While major efforts have been made focusing on fabrication of polymer matrix composites using AM technology, limited numbers of researches have been performed employing this technology to produce metal matrix composites (MMCs).

Although the electron beam has been used in a few works to fabricate MMCs, most of the AM processed coatings or three-dimensional (3D) MMC structures have been manufactured by employing laser beam as the heat source. Based on Fig. 1, the AM processes used to fabricate MMCs from starting powder particles can be classified into two major groups with partial and complete melting nature of particles. The process with partial melting of powder particles is laser sintering (LS), also known as SLS[1] and DMLS.[2] This is while the processes with complete melting nature of powder particles are categorized as laser melting (LM) and laser metal deposition (LMD). LM is a general term for SLM[3] and DMLR[4] processes. However, LMD processes are also named as DMD,[5] LENS,[6] DLF[7] and DLD.[8]

[1]Selective laser sintering.

[2]Direct metal laser sintering.

[3]Selective laser melting.

[4]Direct metal laser remelting.

[5]Direct metal deposition.

[6]Laser engineered net shaping.

[7]Direct laser fabrication.

[8]Direct laser deposition.

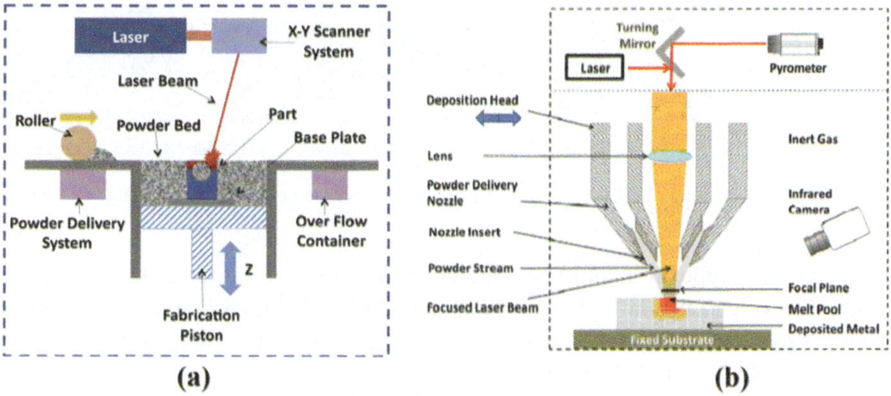


**Fig. 2** Schematic of: (**a**) SLM and (**b**) DLD processes [3]

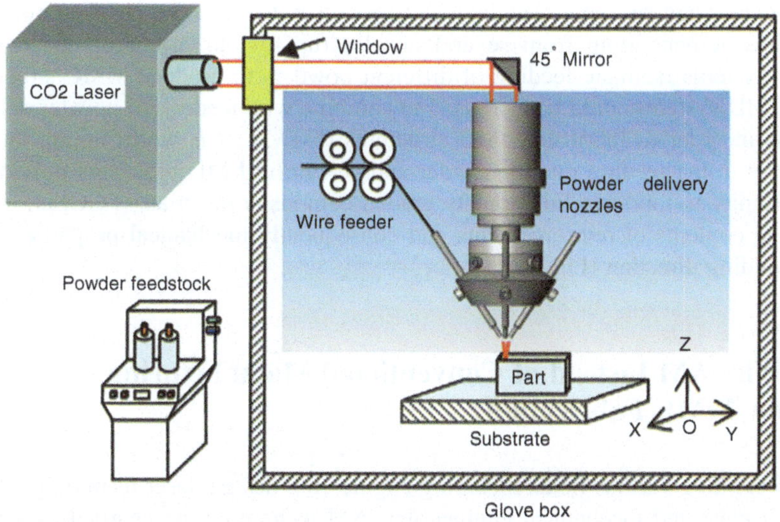


**Fig. 3** Schematic illustration of a DLF system with simultaneous powder and wire feeding [4]

The LS and LM processes are powder bed fusion (PBF), meaning that the powder is spread on a bed prior to the laser scanning process. However, in the LMD processes with a powder-fed nature, the powder is coaxially blown with synchronous laser scanning. The schematic illustrations of SLM and DLD as the most famous processes with LM and LMD nature, respectively, are provided in Fig. 2.

In case of SLM process, a uniform layer of powder is first deposited on the base plate. Then, the predefined regions of this layer are melted by the laser beam, resulting in the formation of a single layer of the desired part. Next, the powder bed is lowered, and a new layer of deposited powder is subjected to laser beam.

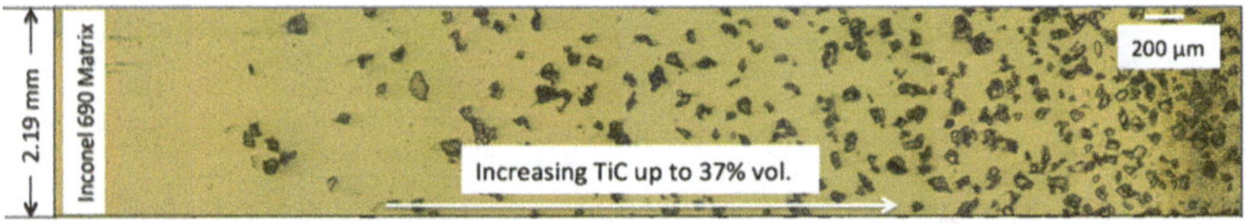


**Fig. 4** Typical structure of a LENS processed functionally graded composite material (FGCM) [5]

These powder deposition and laser melting procedures are repeated, leading to the fabrication of part in a layer-by-layer manner. Unlike SLM, DLD process combines both the material delivery and laser systems to simultaneously deposit the material and fabricate the part in the same region.

When it comes to the fabrication of MMCs using AM processes, the desired powder blends are mixed using powder pre-processing techniques. While pre-mixing of powder blends prior to the process is essential for fabricating MMCs using powder bed fusion AM processes (LS and LM), quite different techniques may be employed in LMD processes to fabricate such components. Because of benefitting from several nozzles (up to four or more), the existing LMD systems facilitate feeding of different powders to produce MMC structures. The LMD systems could be also used to combine wire feeder and powder delivery nozzle in order to fabricate MMC structures (Fig. 3). It is worth noting that the freedom in the feeding rate of powder and/or wire in LMD processes provide the opportunity to fabricate functionally graded composite materials (FGCMs) having varying contents of reinforcements and consequently mechanical properties along the building direction (Fig. 4).

## 2 Why AM Instead of Conventional Manufacturing for MMC Fabrication?

By integrating various fields including engineering design, laser technology, materials science and mechanical engineering, AM technology is regarded as a major revolution in manufacturing, introducing it as a "next generation" technology. AM technology competes with conventional manufacturing processes in many aspects such as fabrication cost and time, design freedom, component accuracy and part quality. The most significant advantages of AM technology over conventional manufacturing techniques employed to fabricate MMCs components are as follows:

- Because of the need for expensive and dedicated tools such as molds and dies, most of the conventional manufacturing processes are not suitable candidates for small volume production and fabricating components with complex geometries. However, besides the capability for small volume production, the unique features associated with AM processes enable the technological opportunities to fabricate small quantities of near net-shape custom-designed components in a rapid and efficient way.

- The need for processing temperatures significantly higher than the melting point of the matrix combined with the relatively slow cooling rates involved in most of conventional manufacturing processes may lead to the noticeable grain growth and inversely affect the mechanical properties of fabricated MMCs. However, the extremely high cooling rates of AM processes results in substantially refined microstructures and improved mechanical properties compared with the conventionally processed counterparts. It is worth noting that the outstanding features of AM technology can also provide the fabrication of metal matrix nano-composites (MMnCs) with significantly improved mechanical properties.
- Fabrication of MMCs with conventional processing techniques has been always encountered with some major processing problems such as gas entrapment, reinforcement/matrix interfacial cracking, limited densification levels and non-uniform size distribution of reinforcements as well as their segregation at the grain boundaries. Combinations of these phenomena result in premature failure, reduced strength and decreased ductility of parts. The outstanding properties associated with AM processes can significantly improve the performance of MMCs by reducing the porosity and developing larger numbers of refined precipitates with more homogenous distribution throughout the matrix.
- While some of conventional manufacturing processes need additional steps to fabricate MMCs with in-situ synthesized reinforcements, AM technology facilitates the fabrication of these composites in a single step.

## 3 Additively Manufactured MMCs (Challenges, Opportunities and Existing Literature)

### *3.1 Aluminum-Matrix Composites (AMCs)*

The poor wear resistance as well as low hardness of aluminum (Al) is a serious impediment which limits its further development in industrial applications [6, 7]. Because of their outstanding combination of properties including low density, low coefficient of thermal expansion (CTE), high hardness, improved wear resistance and comprehensive mechanical properties, Al matrix composites (AMCs) reinforced with ceramic particles are capable of meeting most of the industrial requirements. Accordingly, these materials are considered as promising candidates to be used widely in automotive, aerospace, microelectronics and other related industries [8–10].

Manufacturing of Al alloy parts by AM processes is challenging due to three major reasons: (i) high laser reflectivity of the Al powder, (ii) high heat conductivity away from the melt pool through the already solidified material, and (iii) elevated affinity to oxygen, resulting in the formation of oxide layers on top of the melt pool [11, 12].

Among different Al alloys used as the matrix in AM processed AMCs, the most frequently investigated ones are Al-Si alloys (especially Al-Si-Mg) because of their unique properties such as high fluidity, high weldability, low CTE and good corrosion resistance. The binary Al-Si system is a eutectic system with eutectic

**Table 1** Mixed powder systems and AM techniques employed to fabricate various types of AMCs

| References | Mixed powder system | Matrix | Reinforcement | Process |
|---|---|---|---|---|
| [13] | TiC/AlSi10Mg | AlSi10Mg | TiC | SLM |
| [14] | | | | |
| [15] | | | | |
| [16] | | | | |
| [17] | | | | |
| [18] | SiC/AlSi10Mg | AlSi10Mg | Hybrid ($Al_4SiC_4$+SiC) | SLM |
| [19] | | | $Al_4SiC_4$ | |
| [20] | | | $Al_4C_4$ | DMLS |
| [21, 22] | SiC/A356 | A356 | $Al_4SiC_4$ | LS |
| [23] | SiC/Al-4.5Cu-3Mg | Al-4.5Cu-3Mg | SiC | DMLS |
| [24] | | | | |
| [25] | $Al_2O_3$/Al | Al | $Al_2O_3$ | SLM |
| [26] | $Al_2O_3$/AlSi10Mg | AlSi10Mg | $Al_2Si_4O_{10}$ | |
| [27] | AlN/AlSi10Mg | AlSi10Mg | AlN | |
| [28] | Al-20Si-5Fe-3Cu-1Mg | Al-20Si-5Fe-3Cu-1Mg | $Al_4FeSi_2$ | |
| [29] | Al-12Si-TNM | $Al_{SS\text{-}Si\text{-}Ti}$ solid solution | $Al_6MoTi$ | |
| [30] | Al85Nd8Ni5Co2 | Al | $AlNd_3$, $Al_4CoNi$, $AlNdNi_4$ | |
| [31] | $Fe_2O_3$/Al | Al | Al-Fe intermetallics, $\alpha$-$Al_2O_3$ | |
| | $Fe_2O_3$/AlMg1SiCu | AlMg1SiCu | Al-Fe intermetallics, $\alpha$-$Al_2O_3$ | |
| | $Fe_2O_3$/AlSi10Mg | AlSi10Mg | Al-Fe intermetallics, $Al_{0.5}Fe_3Si_{0.5}$ | |

composition at ~12 wt.% Si. Through the addition of other alloying elements such as Mg (e.g. AlSi10Mg) and Cu, the Al-Si alloys can be strengthened either by means of heat treating or by employing rapid solidification techniques (e.g. melt spinning or AM) which leads to microstructural refinement. In case of AlSi10Mg alloy, the formation of fine $Mg_2Si$ precipitates can further improve the hardness and strength.

Among various AMCs systems (Table 1), those containing TiC and SiC as the starting ceramic reinforcing particle have attracted a great attention in recent years. Table 1 summarizes the works performed to fabricate AMCs using AM processes.

## 3.2 *Titanium-Matrix Composites (TMCs)*

Because of having high specific strength, sufficient stiffness, elevated high temperature application, good fatigue and wear behavior as well as outstanding corrosion resistance, titanium (Ti) and its alloys are widely used in various industrial sections such as chemical, aeronautical and defense among others [32, 33]. A major concern with Ti alloys is their limited wear performance especially in applications demanding high resistance to abrasion and erosion. Accordingly, noticeable numbers of researches have been performed in recent decades to incorporate ceramic reinforcing particles into the matrix for fabricating TMCs parts with improved mechanical properties and wear resistance [34–37]. These attempts include highly efficient surface modification techniques such as laser melt injection [35], laser cladding [38] and laser surface alloying [39], as well as manufacturing techniques like casting, powder metallurgy and combustion synthesis [40] to fabricate bulk-form TMCs parts.

TiC and TiB, due to their high modulus, high thermal stability as well as close density and chemical compatibility with Ti, are generally known as the most appropriate reinforcements for TMCs. TiC reinforced TMCs and Ti matrix nanocomposites (TMnCs) have been fabricated in recent years from different powder mixtures such as TiC/Ti, SiC/Ti and Ti/Al/graphite using various AM processes (Table 2). This is while TiB reinforced ones have been manufactured mostly using LENS and SLM processes (Table 2). The TiB reinforcement is formed in these composites through the in-situ reaction between Ti and B elements in various mixed powder systems.

**Table 2** Mixed powder systems and AM techniques employed to fabricate various types of TMCs

| References | Mixed powder system | Matrix | Reinforcement | Process |
|---|---|---|---|---|
| [41–44] | TiC/Ti | Ti | TiC | SLM |
| [45] | | | | LPD |
| [46] | Ti, Al, graphite elemental powder | Ti(Al) solid solution | | SLM |
| [47] | SiC/Ti | Ti | | SLM |
| [48] | $TiB_2$/Ti | | TiB | SLM |
| [49] | | | | |
| [50] | Elemental B/Ti | | | LENS |
| [51] | Elemental B/Ti6Al4V | Ti6Al4V | | DLD |
| [52] | | | | LENS |
| [37] | $TiB_2$/Ti6Al4V | | | DLF |
| [53] | $TiB_2$/TNZT(elemental Ti +Nb+Zr+Ta) | TNZT | | LENS |
| [54] | $B_4C$/Ti | Ti | TiB+TiC | DLD |
| [55] | SiC/Ti | | (unmelted SiC+TiC +$TiSi_2$+$Ti_5Si_3$) | LENS |
| [56] | | | | |
| [57] | $Si_3N_4$/Ti | | TiN | SLM |
| [58] | $TiB_2$/Ti | | TiB | |
| [59] | TiN/Ti6Al4V | Ti6Al4V | TiN | LENS |
| [60] | BN/Ti6Al4V | | (TiB+TiN) | |
| [61] | | | | |
| [62] | Ta+Ti | Ti | Ta | SLM |
| [63] | Mo+Ti6Al4V | Ti6Al4V | Mo | |

## 3.3 Nickel-Based Matrix Composites

Because of having outstanding combinations of high temperature workability and mechanical properties, high fatigue strength and good oxidation and corrosion resistance, high performance nickel-based superalloys are noticeably used in power generation, aerospace, marine, chemical and petrochemical industries [64–68]. Inconel series of Ni-based superalloys, both the precipitation and/or solution-hardened, are the most widely known ones with fantastic combination of properties, introducing them as excellent choices for miscellaneous industrial applications. The high hardness, low thermal conductivity and high work hardening rates associated with the nickel-based superalloys are the most important features introducing them as "difficult to machine" alloys [69]. Due to their tool-less nature, their ability to fabricate net or near net-shape components and their noticeably high freedom in geometry design, AM processes are considered as fascinating processing routes for the fabrication of superalloys.

Most of superalloys desired for high temperature applications are in precipitation hardened state in which the gamma prime ($\gamma'$) phase consisting of $Ni_3Al$ or $Ni_3Ti$ precipitates is formed by the presence of Ti and/or Al in the alloy [70]. The increasing of Ti and Al concentration in the superalloy is one of the approaches for the enhanced content of $\gamma'$ phase and consequently the improved high temperature properties of nickel-based superalloys. However, this technique is believed to be detrimental to the weldability/printability of the superalloy, since it increases the susceptibility to cracking [71], as shown in Fig. 5. There exists a limit of weldability beyond which the superalloy is considered as "unweldable". The sum of Al, Ti and Nb elements of this limit is reported to be ~4 at.%. The alloys exceeding this limit suffer from strain age cracking either during the welding/AM process or when subjected to post-processing heat treatment cycles [72]. Accordingly, fabrication of high-quality parts from such alloy systems with AM processes would be challenging or even impossible.

In order to avoid the cracking and further enhance the high temperature performance of nickel-based superalloys, hard and temperature-resistant ceramic reinforcing particles can be incorporated into the nickel-based matrix to produce MMCs.

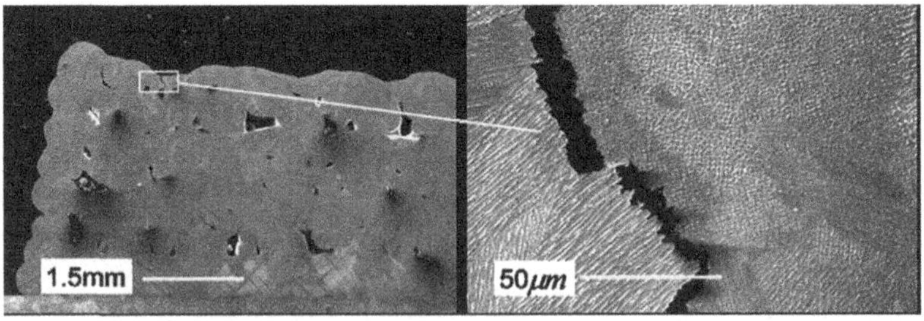


**Fig. 5** Cross-sectional view of SLM processed Waspalloy showing the formation of cracks [73]

**Table 3** AM techniques employed to fabricate various types of nickel-based matrix composites

| References | Matrix | Reinforcement | Applied AM process |
|---|---|---|---|
| [76] | Inconel 625 | $Al_2O_3$ | LM |
| | | SiC | |
| | | TiC | |
| [5] | Inconel 690 | TiC | DLD |
| [77] | Ni | | DLF |
| [78] | Inconel 718 | | SLM |
| [79] | | | |
| [80, 81] | | | LMD |
| [82] | Inconel 625 | | |
| [83, 84] | | | |
| [85] | | | |
| [86] | Inconel 625 | | LENS |
| [87] | NiTi | TiC | |
| [88] | Ni | CNT | |
| [89] | Inconel 718 | WC | SLM |
| [90] | | BN | |

Because of having promising properties such as high hardness, outstanding thermal stability and good wear and corrosion resistance [74, 75], TiC is the most frequently used ceramic reinforcing particle incorporated into the nickel-based superalloys to fabricate MMCs employing AM processes. Table 3 summarizes the works performed on AM processing of nickel-based superalloy matrix composites.

## *3.4 Copper-Matrix Composites*

Because of benefiting from outstanding combinations of high electrical and thermal conductivity as well as high resistance to fatigue and corrosion, copper (Cu) is extensively used in many applications, especially in electrical contacts [91, 92]. However, the limited wear resistance of pure Cu restricts its application in the environments demanding high wear resistance [93, 94]. To overcome this imperfection, secondary hard reinforcing particles could be incorporated into the Cu matrix to develop Cu-matrix composites with significantly enhanced mechanical properties (e.g. improved wear resistance).

Cu-matrix composites have found their applications in many industrial parts such as high voltage electrical applications, arcing tips, current-carrying contacts and electrodes of electric discharge machining (EDM) machine among others [93, 95]. These components are conventionally fabricated by various processes including powder metallurgy [92], squeeze casting [96], infiltration [93] and self-propagating high temperature synthesis (SHS) [97].

Table 4 summarizes the works performed to fabricate Cu-matrix composites with various types of ceramic reinforcing particles using AM processes. Tungsten (W) is

**Table 4** Mixed powder systems and AM techniques employed to fabricate various types of Cu-matrix composites

| References | Mixed powder system | Matrix | Reinforcement | Reinforcement content | Process |
|---|---|---|---|---|---|
| [106] | Cu/Ti/C, Cu/Ni/Ti/C | Cu, CuNi | In-situ TiC | 10 vol.% | SLM |
| [98, 107] | W/Cu | Cu | W | 28, 40, 52 wt.% | DMLS |
| [105] | WC/Cu | | WC | 60 wt% | |
| [101, 103] | WC-Co/Cu | | WC-Co | 30 wt.% | |
| [102, 108] | | | | 20, 30, 40 wt.% | |
| [104] | | | | 50 wt.% | |
| [109] | CNT/Cu-10Sn | Cu-10Sn | CNT | 0–20 vol.% | LENS |

a refractory metal that has been widely incorporated into the Cu matrix to produce composites. However, the mutual insolubility of Cu and W as well as the poor wettability of W particles with the Cu melt are the major concerns with these MMCs, limiting their densification response [98]. Tungsten carbide (WC), because of having lower density, higher hardness and higher stability than W, could be considered as substituting reinforcement to W. In order to further improve the mechanical properties; i.e. hardness, strength, wear resistance and fracture toughness, the well-known WC-Co hard metal is also used as reinforcement to the Cu matrix in AM processed composite [99–104]. Rare earth (RE) elements have been also added in small amounts to AM processed Cu-matrix composites to improve the processability, to control the microstructural evolution and to enhance the mechanical properties of fabricated Cu matrix composites [104, 105].

## *3.5 Iron-Based Matrix Composites*

As far as the AM technology has come to existence, iron-based alloys have been among the mostly investigated materials due to their wide range of application in various industries. Tool steels and stainless steels are two categories of iron-based alloys which have attracted a great attention in the field of AM, especially when it comes to the fabrication of MMCs having these materials as the matrix.

H13 is a hot work tool steel with a medium carbon content which is extensively used to make tools for cutting, forming and shaping purposes because of having outstanding properties such as high strength, high wear resistance as well as sound stability at high temperatures [110]. However, when needing materials capable of withstanding harsh environments with extremely high loads and temperature gradients, non-reinforced H13 may not be the best candidates. Because of having improved strength, wear resistance, reduced cost and lower density, MMCs are believed to be the materials being able to serve in these environments [111].

Using AM processes, especially SLM, TiC and $TiB_2$ have been used recently as reinforcing particles to fabricate H13 matrix nano-composites. These reinforcing particles, due to their high elastic modulus (EM), high thermal stability, low density and high thermodynamic stability with the matrix, are among the best ceramic materials used as reinforcements to the steel matrices [112, 113]. While SiC has been also used as reinforcement to iron-based composites, its strong reactivity with the transition metals of the matrix may lead to the formation of detrimental phases at high temperature and deteriorate the mechanical properties [114].

Stainless steels are also an important grade of iron-based alloys with a wide variety of applications in many industrial fields, especially those needing materials with good mechanical properties and excellent corrosion resistivity. One of the major problems associated with stainless steels is their limited hardness and wear resistance.

Incorporation of reinforcing particles into the stainless steel matrix to form MMCs has been considered as an effective approach to produce parts with enhanced physical and mechanical properties which meet the requirements of high-tech industries (e.g. aerospace and biomedical) [115, 116]. Among various types of reinforcements (e.g. carbides, nitrides, borides and metal oxides) incorporated into steel by conventional processes, $TiB_2$, TiC and hydroxyapatite (HA) are those added to the stainless steel matrices using AM processes. The main purpose behind the addition of HA to stainless steel is to combine the relatively high strength of stainless steel with outstanding biocompatibility of HA, leading to MMCs with excellent load-bearing and bioactive properties. These MMCs could be employed as internal, hip and maxillofacial implants in the human body. Based on the fact that finer HA particles have higher surface to volume ratio and are associated with higher bioactivity [117], attempts have been also performed to replace the micron-sized HA particles by nanoparticles [118, 119]. Compared to the significantly sharp change in

**Table 5** Mixed powder systems and AM techniques employed to fabricate various types of iron-based matrix composites

| References | Mixed powder system | Matrix | Reinforcement | Process |
|---|---|---|---|---|
| [120] | TiC/H13 steel MMnC | H13 steel | TiC | SLM |
| [121] | TiC/316L SS[a] MMnC | 316L SS | | |
| [122] | | | | |
| [123] | $TiB_2$/316L SS MMnC | 316L SS | $TiB_2$ | |
| [124] | | | | |
| [125] | TiB2/H13 steel MMnC | H13 steel | | |
| [126] | TiC/Invar 36 | Invar 36 | TiC | DMLS |
| [127] | HA/316L SS MMC and MMnC | 316L SS | Hydroxyapatite (HA) | SLM |
| [119] | | | | |
| [128] | SiC/iron MMnC | Iron | SiC | |
| [129] | SiC/iron (with hybrid micro- and nano-size SiC) | | | |

[a]SS refers to stainless steel

mechanical properties at the interface of coated implant parts, the uniform metallurgical bonding between the matrix and the reinforcements in MMCs can provide much more functional parts to be used in biomedical applications. Table 5 presents the iron-based matrix composites made using AM technology.

# 4 Pre-processing of Mixed Powder System

Since powder is used as the starting material to fabricate parts in most of AM processes, its perquisite conditions are of a crucial importance and significantly affect the quality of AM processed parts. The morphology of powder particles (e.g. size and shape) defines the extent of particles packing and consequently influences the thickness of deposited layers as well as the surface roughness. Due to the improved flowability and enhanced quality of parts fabricated by spherical powder particles, the gas atomized powders having such morphology are widely used in AM processes, especially the PBF ones [130]. However, the limited flowability as well as mechanical interlocking and entangling of irregular-shaped

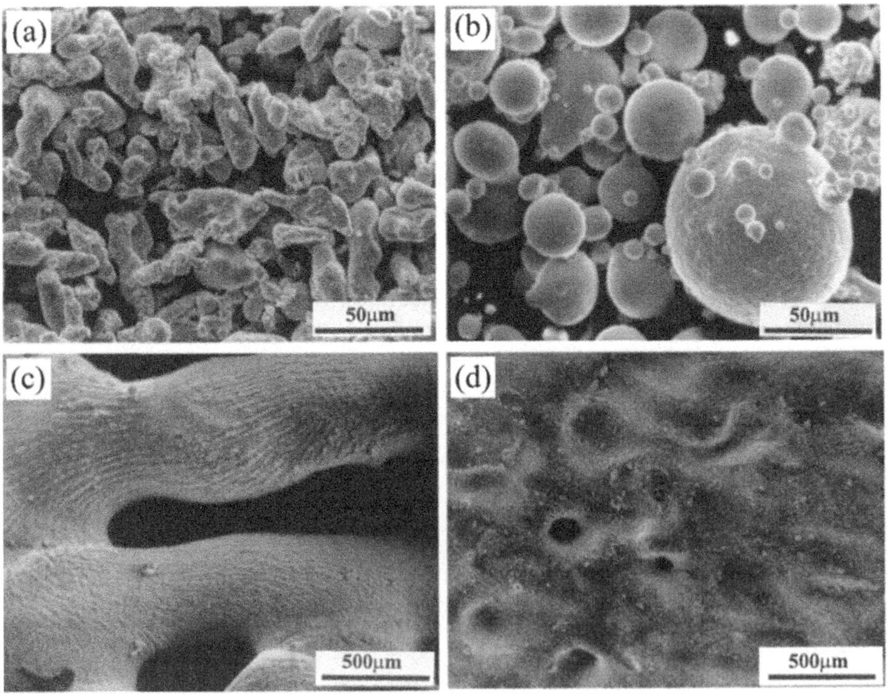


**Fig. 6** Scanning electron microscopy (SEM) images of (**a**) water atomized and (**b**) gas atomized high speed steel powder particles. (**c**) and (**d**) SEM micrographs of laser sintered powders shown in (**a**) and (**b**), respectively [130]

powder particles limits the flow of powder, makes the melt pool tracks non-homogeneous and consequently favors the formation of porosities in the solidified part. Figure 6 clearly compares the surface of selective laser sintered parts fabricated using irregular-shaped water atomized high speed steel powders with that of spherical gas atomized powders.

Despite the noticeable technological advances in AM technology, there are still some limitations on the compositions of available powders. Accordingly, when MMCs are considered, desired powders should be blended and even pre-processed and then fed to fabricate parts.

Besides the direct mechanical mixing, ball milling has been regarded as one of the most applicable processes for mixing of powders. During this process, the powder particles are imparted to tremendous energy induced by the balls, leading to fragmentation of brittle powder particles and induction of severe deformation or even fragmentation in the ductile ones [131].

The milling time has a significant effect on the morphology and distribution pattern of particles during powder pre-processing stage. Figure 7 shows the spherical commercially pure (CP)-Ti and irregular-shaped $TiB_2$ starting powders used for fabricating TiB reinforced TMCs.

The morphology of these powder mixtures ball milled for 1–4 h are provided in Fig. 8. As being observed, relatively short milling time of 1 h led to inhomogeneous distribution of $TiB_2$ powder particles in the mixture with their limited connectivity to the Ti particles (Fig. 8a). Enhancement of milling time to 2 h provided the improved dispersion of $TiB_2$ ceramic particles around the Ti particles in a uniform manner (Fig. 8b). While the prolonged milling time could further improve the distribution of fragmented $TiB_2$ particles, it may transform the desired spherical shape of mixed powder system to a flattened and irregular shape, which is not considered as an optimal condition for the PBF AM processes (Fig. 8c, d).

The applied milling time can also affect the chemical reactions that may be activated in the powder system during mixing. Due to the short-term action of mechanical force and the resultant decreased energy subjected to the powder particles at relatively short milling times, the system may not undergo solid state chemical reactions between the elements to form new phases. For instance, ball

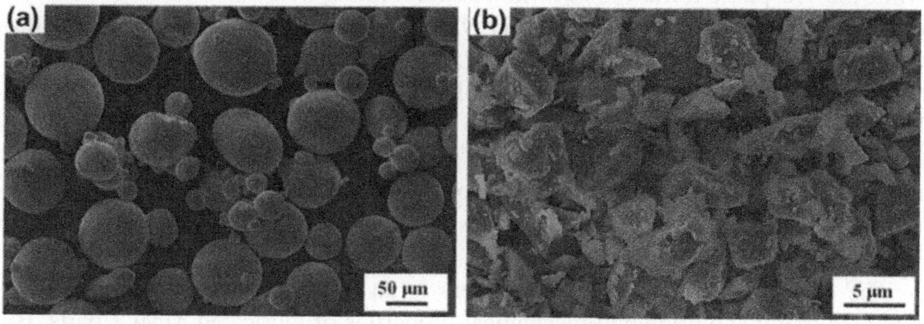


**Fig. 7** SEM micrographs of: (**a**) CP-Ti and (**b**) $TiB_2$ powder particles [49]

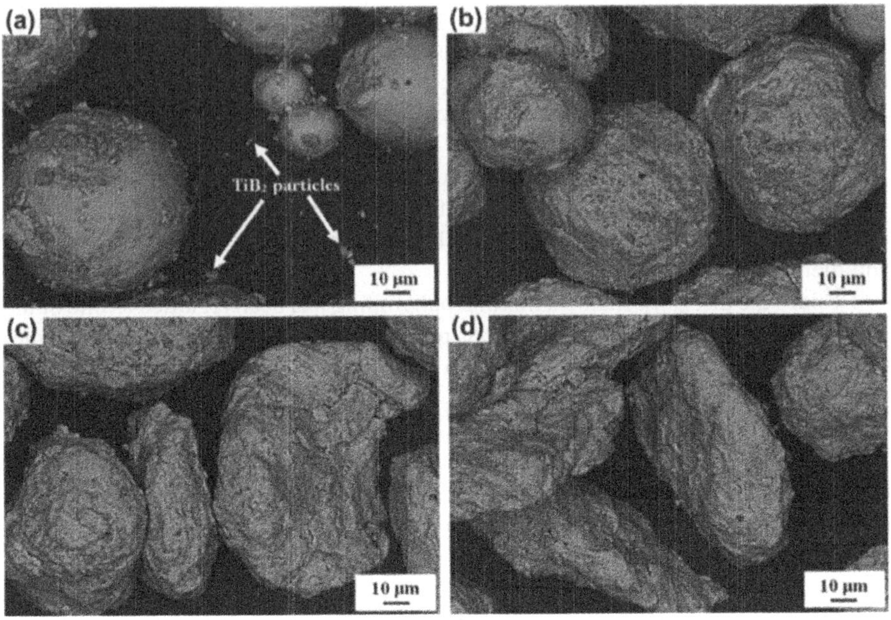


**Fig. 8** SEM images of CP-Ti and $TiB_2$ powders in Fig. 7 mixed using ball milling for different times of: (**a**) 1 h, (**b**) 2 h, (**c**) 3 h and (**d**) 4 h [49]

milling of the Ti and $TiB_2$ powder mixtures for the times as long as 4 h does not activate the metallurgical bonding between the powders to form in-situ TiB phase from the reaction between Ti and B elements [49, 58]. However, the prolonged and continuous microscopic action of the mechanical force by ball-powder-ball collisions may favor such chemical reactions. For example, ball milling of Ti+SiC powder mixture for relatively long milling time of 15 h favors the formation of TiC and $Ti_5Si_3$ phases [47].

In addition to the synthesis of new phases, the ball milling process may also provide some systems with the formation of solid solution. Mechanical alloying of Ti, Al and graphite elemental powder mixtures prior to the SLM process reveals that while the TiC phase forms in-situ within a short milling time due to its large negative enthalpy and Gibbs free formation energy, the Ti(Al) solid solution appears gradually with the increase in milling time [46].

The size of powder particles during ball milling is determined by the competition between two mechanisms of fracturing and cold-welding. While the fracturing mechanism tends to decrease the particle size, the cold-working mechanism favors the attachment of powder particles to form larger-sized particles. Therefore, the refining or coarsening of powder particles depends on whether the fracture mechanism or the cold-welding is predominant.

The category of mixed powder system is an important issue influencing the microstructural evolution and mechanical properties of fabricated MMCs parts. By changing the milling time from 4 to 10 h, Gu et al. [42] prepared two different TiC/Ti powder blends named directly mixed and ball milled, respectively. The mechanism of laser interaction with two different categories of powder blends is schematically presented in Fig. 9. While the direct action of laser beam on TiC particles for the directly mixed system results in simultaneous melting of TiC and Ti powder particles, the TiC particles in the ball milled category are not directly exposed to the laser irradiation. Accordingly, the dissolution of TiC powder particles occurs after the melting of Ti in the ball milled system. These differences in the laser interaction affected the dissolution and re-precipitation of TiC particles as well as the microstructural evolution of laser processed MMCs.

In addition to the direct mechanical mixing and ball milling processes, blending of powders in some systems may be performed using dilute solutions followed by drying. To prepare suitable premixed composite powder, Das et al. [60] blended Ti6Al4V and BN powders with dilute polyvinyl alcohol solution in a tubular mixer for 24 h and then dried the mixed powders. The obtained mixture of powders in

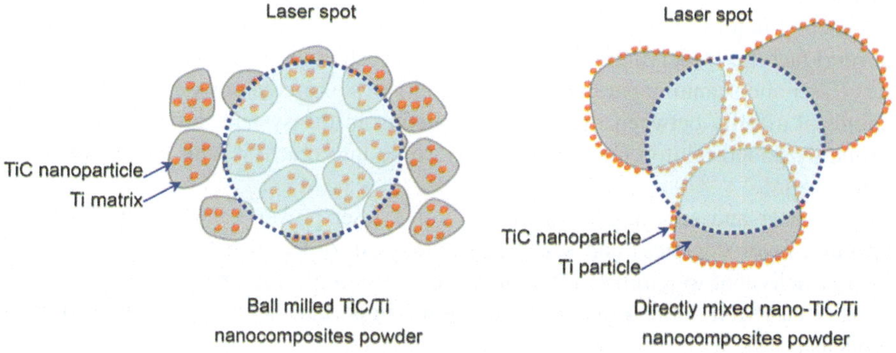


**Fig. 9** Schematic illustrations of laser beam interaction with two different categories of TiC/Ti mixed powder system [42]

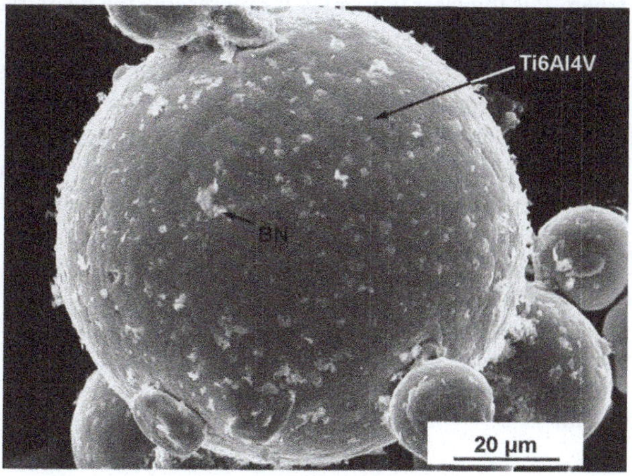


**Fig. 10** SEM image of Ti6Al4V and BN powder particles premixed with diluting followed by drying [60]

Fig. 10 reveals that the fine BN particles are uniformly coated on the surface of Ti6Al4V alloy powder particles. This type of rearrangement can facilitate the homogenous distribution of reaction products in the solidified microstructure.

## 5 Microstructural Evolution in Additively Manufactured MMCs

As the laser beam interacts with a mixed powder system containing ceramic and metallic constituents, part of the mixed powder with lower melting point melts first while the part with noticeably higher melting point (usually ceramic reinforcing particles) may remain mostly unaffected and partially and/or completely melted. Therefore, the system may contain solid reinforcements distributed in a molten matrix.

Microstructural evolution in MMCs fabricated by AM processes is affected by several factors. Depending on the applied AM technique, selected process parameters (laser and scanning parameters), characteristics of mixed powder system and the chemical affinity between the elements existing in the system, a part or all of the primarily added reinforcing particles may come into reaction with the surrounding liquid metal.

The laser and scanning parameters involve laser power, scanning speed, hatch spacing, layer thickness and scanning strategy. In order to specify the amount of energy delivered to a unit volume of powder system during processing, the energy density has been defined by combining some important laser and scanning parameters as [132]:

$$E_\rho = \frac{P}{\pi r^2} \frac{2r}{\upsilon} \frac{2r}{h} \tag{1}$$

where $P$ is the laser power (W), $r$ represents the beam radius (mm), $\upsilon$ is the scanning speed, $h$ signifies the hatch spacing, and $E_\rho$ is the energy density (J/mm$^2$).

Referring to Eq. 1, the energy density has three terms. The first term is related to the laser intensity. The second term signifies the time that a circle of radius $r$ is exposed to the beam while the third term indicates the number of laser exposures received by any point on the surface [132]. The volumetric energy density is the most commonly used energy density in the literature, defined as the energy delivered to a unit volume [133]:

$$\eta = \frac{P}{\upsilon h t} \tag{2}$$

in which $t$ is the layer thickness, and $\eta$ signifies the energy density per volume (J/mm$^3$).

The characteristics of mixed powder system include the features of powder particles prior to and after the pre-processing stage. When the mixed powder system contains ceramic reinforcing particles added to the metallic powder, the size and volume fraction of these ceramic particles have major effects on laser absorptivity of the system and consequently the quality and mechanical properties of AM processed MMCs parts. The applied procedure employed to mix the powder particles may also influence the size, morphology and distribution pattern of powder particles and consequently play role in the properties of fabricated parts.

Generally, the microstructural features of AM processed MMCs can be categorized as:

(i) Characteristics of reinforcements distributed in the matrix,
(ii) Reinforcement/matrix reactions, and
(iii) Microstructural evolutions in the matrix induced by the presence of reinforcements.

These microstructural features have been discussed in the following for various AM processed MMCs systems.

## *5.1 Characteristics of Reinforcements Distributed in the Matrix*

### 5.1.1 Size and Morphology of Reinforcements

The applied energy density and the characteristics of mixed powder system are the two parameters playing the most important roles in the size and morphology reinforcements have in AM processed MMCs. Effects of these parameters on the features of reinforcements can be discussed in terms of ex-situ or in-situ reinforced nature of MMCs.

#### Ex-Situ Reinforced MMCs

For the ex-situ reinforced MMCs where the primarily added reinforcing particles play the reinforcing role to the matrix in the solidified MMC structure, the elevated melt pool temperature and the reduced cooling rate associated with the enhanced energy density can lead to partial melting or dissolution of reinforcing particles and consequently decrease their size in the solidified MMC microstructure [19]. Moreover, the shape of primarily added reinforcing particles in ex-situ reinforced MMCs may be affected by the applied energy density. The research performed by Gu et al. [77] indicated that enhancement of energy density in LMD processed TiC reinforced Inconel 718 matrix composites changed the shape of TiC particles from primary irregular polyangular to near round shape with some degrees of smoothening and refinement. In fact, the improved wettability of reinforcing particles by the surrounding melt induced at elevated temperatures is believed to dissolve the edges of reinforcing particles and result in their smoothening [81]. It is worth noting that while the elevated energy density could make the ex-situ reinforcements smooth and homogenize their distribution in the microstructure [81, 82], the excessive energy density applied to the system may coarsen the microstructure of matrix [82].

### In-Situ Reinforced MMCs

Based on the type of MMC system under investigation, the elements of reinforcing particles dissolved into the matrix may precipitate as newly developed in-situ synthesized phase during solidification. When it comes to such systems, the used mixed powder system and the applied process parameters need to be thoroughly examined to control the characteristics of reinforcements (i.e. size and morphology).

By having the following consequences, the enhanced energy density may change the size and morphology of in-situ synthesized reinforcements:

(i) The elevated temperature and increased thermallization in the melt pool which favours the coarsening of in-situ reaction products [41, 43, 83].
(ii) The intensified surface tension and resultant Marangoni flow in the melt pool.
(iii) The improved wettability of solid reinforcements by the surrounding melt [57].
(iv) The reduced cooling rate and prolonged lifetime of the liquid molten material, facilitating the growth of reinforcements.

The research performed by Li et al. [77] on DLF of Ni/Ti/C mixed powder system reveals a strong relationship between the size and volume fraction of in-situ synthesized TiC precipitates in the microstructure. As it is well-known, ceramic reinforcements have usually higher laser absorptivity than the metallic powder acting as the matrix in MMCs. The enhanced volume fraction of in-situ synthesized TiC reinforcements is believed to increase the melt pool temperature, favouring the coarsening of in-situ synthesized reinforcements (Fig. 11). Moreover, the increased

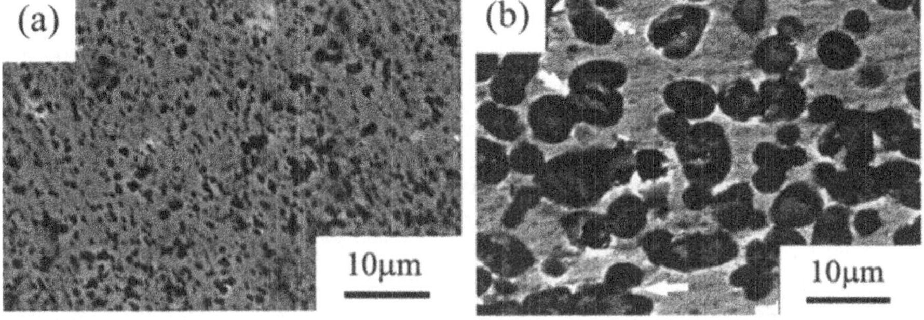


**Fig. 11** Microstructures of DLF processed TiC reinforced Ni matrix composites containing: (**a**) 20 and (**b**) 60 vol.% in-situ synthesized TiC [77]

content of solid reinforcements reduces the amount of molten material in the system and encourages the coalescence of reinforcements to form larger-sized precipitates.

Among various features of the mixed powder system, the size of starting powder particles is of a great importance in the obtained MMCs structures. Because of benefiting from higher surface to volume ratio, the finer primarily added reinforcing particles can enhance the laser absorptivity of the system and induce much more heat in the melt pool. This may affect the melting efficiency of starting reinforcing particles and the characteristics of subsequent in-situ synthesized precipitates.

### 5.1.2 Distribution Pattern of Reinforcements

The distribution pattern of reinforcements in AM processed MMCs is generally controlled by the mutual interaction between reinforcements and the solidification interface. Accordingly, the features of reinforcements and the properties of solidification interface are the main factors affecting the distribution pattern of reinforcements. The solidification mode and growth velocity are two features of the interface having major effects on distribution pattern of reinforcements in MMCs. The growth velocity ($V_s$) of the solidification interface is mainly determined by the applied scanning speed. However, the solidification mode is affected by several factors including the processing parameters and alloy type. Microstructural characterization of the AM processed MMCs reveals a non-planar solidification mode in most of these materials due to the following reasons:

(i) The solid reinforcements incorporated into the system may have quite different thermal conductivity and specific heat from the matrix. This may locally destabilize the planar solidification interface,
(ii) The laser-induced turbulence in the melt pool may demolish the planar solidification interface,
(iii) The dissolution of reinforcements into the surrounding melt may provide constitutional undercooling for the matrix. This can change the solidification mode of the matrix from planar to non-planar.

Therefore, the distribution pattern of reinforcements in the matrix of MMCs needs to be discussed in terms of the interaction between the reinforcements and the non-planar solidification interface (usually dendritic).

The following describes the most important factors affecting the distribution pattern of reinforcements in AM processed MMCs.

#### Effect of Scanning Speed

At relatively low solidification velocities (low scanning speeds), the reinforcements are pushed by the dendrite tips (solidification interface) into the adjacent melt, resulting in local pile-up of the majority of reinforcements and their aggregation along the boundaries of the solidified MMC structure (Fig. 12a, b) [101]. While most

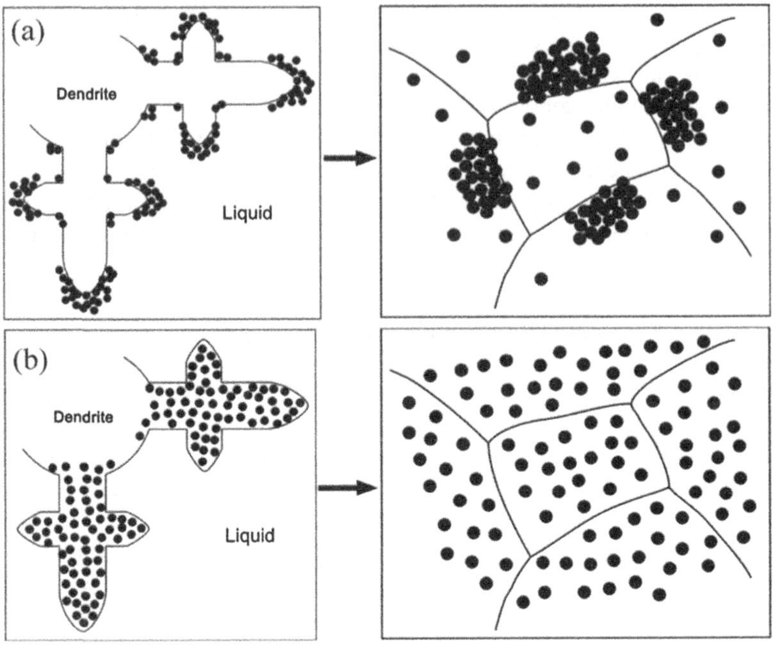


**Fig. 12** The reinforcement/dendrite interactions in: (**a**) particle pushing and (**b**) particle engulfment mechanisms [101]

of the reinforcements may locate at the boundaries, a few of them may be also found within the grains [120, 124]. When particle pushing mechanism is dominant, the reinforcements are not capable of restricting the crystal growth to a great extent [134].

In case of particle engulfment mechanism where the solidification velocity ($V_s$) exceeds a critical velocity ($V_{cr}$), the solid reinforcements are kept stationery and are entrapped in the growing dendrites, bringing about a uniform distribution pattern of reinforcements in the matrix (Fig. 12c, d) [101]. $V_{cr}$ is defined as [135]:

$$V_{cr} = \frac{\Delta\sigma_o d_0}{6(n-1)\eta r}\left(2 - \frac{k_p}{k_l}\right) \tag{3}$$

in which $d_0$ is the interatomic distance of the matrix alloy, $n = 2$ to 7, $\eta$ is the melt viscosity, $r$ is the radius of solid reinforcement which is assumed to be spherical, and $k_p$ and $k_l$ refer to thermal conductivities of reinforcement and liquid, respectively. $\Delta\sigma_o = \sigma_{ps} - (\sigma_{pl} + \sigma_{sl})$, in which $\sigma_{ps}$, $\sigma_{pl}$ and $\sigma_{sl}$ are reinforcement-interface, reinforcement-liquid and interface-liquid surface energies, respectively.

## Effect of Energy Density

The amount of energy density applied to the system is of a great importance in determining the distribution pattern of reinforcements. Generally, relatively low energy densities are associated with clustering of reinforcements and their inhomogeneous dispersion [80, 83]. However, the appropriate selection of energy density results in much more homogenous microstructures in terms of distribution pattern of reinforcements due to: (i) decreased melt viscosity, (ii) improved wettability of reinforcements and (iii) the intensified convective flow in the melt pool.

## Effects of Size and Volume Fraction of Reinforcements

The size and volume fraction of reinforcements also play important roles in their distribution pattern as well as the crystal growth of the obtained microstructures. Using two different starting micro- and nano-sized TiC powder particles, AlMangour et al. [121] found the noticeable effect of reinforcement size on microstructural features of SLM processed TiC reinforced 316L stainless steel matrix composites. While the fabricated MMCs both showed cellular dendritic structures containing TiC reinforcements aggregated along the cell boundaries, those with nano-sized TiC particles had thinner walls of cellular structure.

In case of MMCs with large-size reinforcements, the solidification interface may not be able to push the particles, leading to their engulfment in dendrites or their entrapment into inter-dendritic regions. When the diameter of reinforcements is larger than a critical value ($\lambda_{cr}$), they are capable of avoiding crystal growth. The critical diameter of reinforcements is defined as [136]:

$$\lambda_{cr} = 2\lambda_1 R_d \tag{4}$$

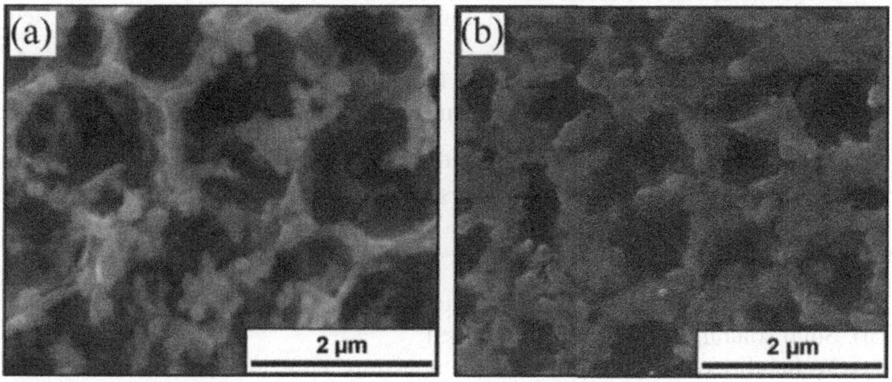


**Fig. 13** SEM micrographs of SLM processed $TiB_2$ reinforced 316L stainless steel matrix composites containing: (**a**) 2.5 and (**b**) 15 vol.% reinforcement [123]

where $\lambda_1$ refers to the primary dendrite arm spacing while $R_d$ represents the radius of dendrite tip.

Although high volume fractions of reinforcements enhance the chance for restriction of crystal growth and the subsequent grain refinement of the matrix, the particle pushing mechanism activated at low solidification velocities may promote the clustering of reinforcements to form agglomerates with larger effective sizes or greater effective thicknesses. Microstructural characterization of the SLM processed $TiB_2$ reinforced 316L stainless steel matrix nanocomposites presented in Fig. 13 reveals that by the increased volume fraction of $TiB_2$ reinforcements from 2.5 to 15 vol.%, clustering of much more reinforcements along the cell boundaries leads to coarsening of walls of cellular structure [123].

## 5.2 *Reinforcement/Matrix Reactions*

### 5.2.1 Reaction Mechanisms

Generally, in-situ reaction products in MMCs are formed through five different mechanisms as the following:

(i) Solution precipitation: The in-situ formation of precipitates from the saturated solution through nucleation and the subsequent growth.
(ii) Solid-liquid interfacial reaction: The reaction between the melt and outer surface of solid particles dispersed in it. This mechanism results in the formation of reaction layer around the solid particles. The thickness of this reaction layer is dictated by the reaction progress.
(iii) Combined solution precipitation and solid-liquid reaction: The existence of both solution precipitation and solid-liquid interfacial reaction for some systems.
(iv) Solid-solid interfacial reaction in the melt: The interfacial in-situ reaction between solid phases existing in a melt which leads to the formation of new phase.
(v) Solid diffusion reaction: The in-situ formation of new phase from the induced solid state diffusion reaction between solid particles.

Among different mechanisms discussed above, due to the presence of melt in the system, the in-situ reaction between the solid reinforcing particles and the molten material in AM processing mainly follows the first three mechanisms and is affected by the following factors:

(i) The amount and temperature of molten material formed during processing,
(ii) Solid solubility of particles in the melt,
(iii) Wettability of the particles with the molten material, and
(iv) Solid-phase diffusion in the melt.

### 5.2.2 Reinforcements/Matrix Interfacial Reaction

The physical and mechanical properties of the interface between the reinforcements and the matrix are the parameters playing significant roles in densification level and mechanical properties of AM processed MMCs. Generally, the interfaces between the ceramic reinforcements and the metallic matrix in particulate reinforced MMCs are weak, introducing them as sensitive and potential locations for crack nucleation and propagation during sliding. However, the researches have recently shown the formation of good interfacial bondings between the reinforcements and the matrix in AM processed TiC reinforced Ni matrix [77], $TiB_2$ reinforced 316L stainless steel matrix [123], TiC reinforced Al matrix [28] and TiC reinforced titanium matrix [45] composites.

While the formation of interfacial reaction layers between reinforcements and matrix in some systems can modulate the difference between the deformation behaviors and improve the bonding coherence and consequently improve the mechanical properties of the fabricated MMCs, they may be detrimental to the mechanical properties in some other systems. Therefore, thorough efforts have been allocated in recent years for determining the characteristics of reinforcement/matrix interfaces in AM processed MMCs.

In addition to the powder particles considered to be melted and solidified as the metallic matrix, the high amount of energy associated with the AM processes can lead to the partial/complete melting and dissolution of the primarily added ceramic reinforcing particles with high melting points. The extent of these melting and dissolution is dictated by the energy density [137, 138]. In cases where a chemical potential exists for reaction between the solid reinforcing particles and the molten matrix, reaction layers may be formed at the particle/matrix interfaces.

Depending on the type of elements existing in the mixed powder system, the in-situ reaction between elements may lead to the formation of new or the same phase as the starting ceramic reinforcing particle. The in-situ reaction between the solid reinforcing particles and the melt for two AM processed MMCs showing combined solution precipitation and solid-liquid reaction mechanisms are provided in Figs. 14 and 15. Figure 14 shows the microstructures for LENS processed TiN/Ti6Al4V mixed powder system with 10 and 20 wt.% TiN, respectively. Microstructural observations around a relatively large-sized SiC particle in LENS processed SiC/Ti powder mixture is also presented in Fig. 15. The reaction layers around the TiN particles are believed to be TiN. However, that surrounding the SiC particles is detected as TiC, confirmed by energy dispersive X-ray spectroscopy (EDS) line scan as well as compositional EDS map analysis results. In fact, the higher affinity of Ti to C compared with Si is the main driving force for the formation of TiC through SiC + Ti $\rightarrow$ TiC + Si exothermic reaction [139].

The features (e.g. thickness and chemical composition) of reaction layers around the reinforcing particles are affected by the applied processing parameters as well as the size of starting reinforcing particles. Referring to Fig. 14, it can be found that reaction layers are thinner around the coarser TiN particles. The solubility of a solid particle in the surrounding liquid is negatively dependent on its size. Therefore,

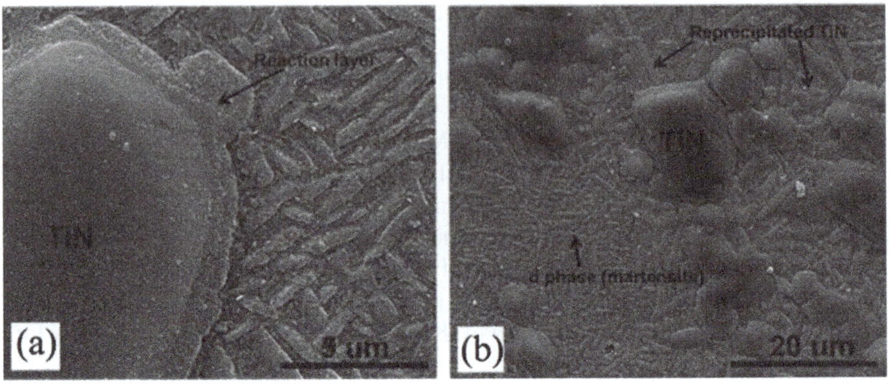


**Fig. 14** SEM micrographs showing reaction layers between: (**a**) coarse TiN particle and the matrix in 10 wt.%+Ti6Al4V and (**b**) finer TiN particles and the matrix in 20 wt.%+Ti6Al4V TMCs fabricated by LENS process [59]

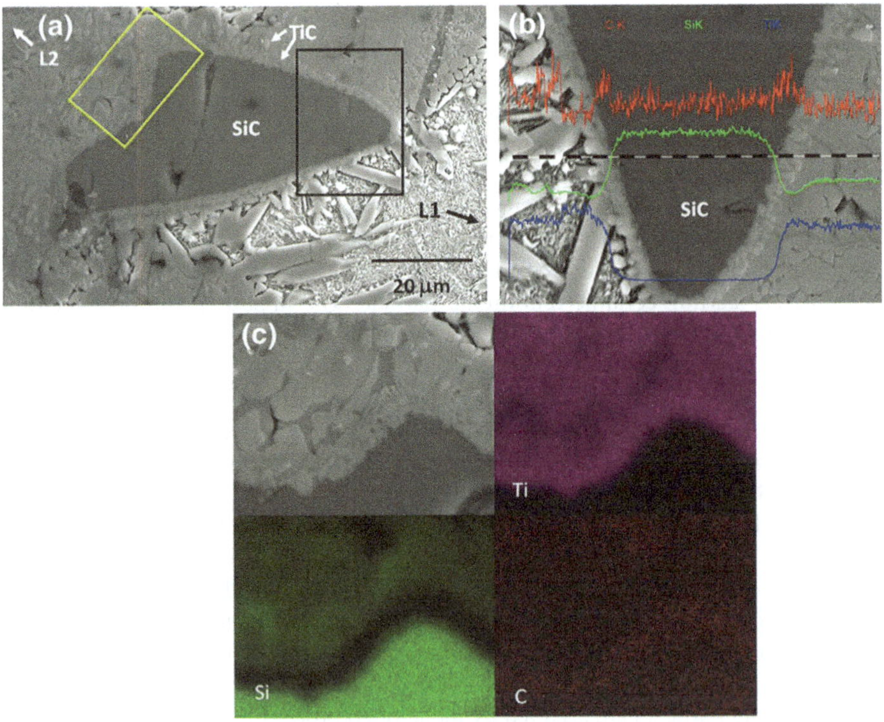


**Fig. 15** (**a**) SEM micrograph showing an undissolved SiC particle in the matrix of LENS processed SiC/Ti system. (**b**) EDS line scan and (**c**) EDS elemental map analysis results from the reaction layer surrounding SiC particle in (**a**) [56]

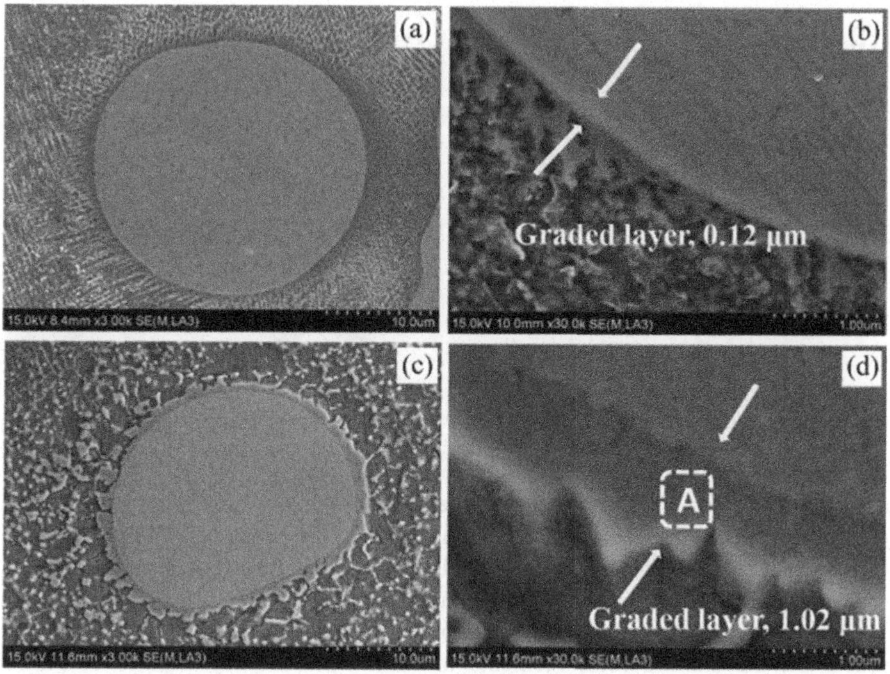


**Fig. 16** SEM micrographs showing the characteristics of reinforcement/matrix interfacial reaction layer formed in WC reinforced Inconel 718 matrix composites at linear laser energy densities of: (**a**), (**b**) 173 and (**c**), (**d**) 303 J/m [89]

compared with the larger particles, the smaller ones have higher energy and solubility in the liquid. This size-dependency in solubility can generate a concentration gradient in the liquid during which the materials migrate from the small particles to the larger ones through diffusion in the liquid.

Figure 16 reveals the effect of laser energy density on the characteristics of reaction layer formed at the interface of WC reinforcements and the Inconel 718 matrix. In this case, the strong carbide forming elements (e.g. Ni, Cr, Fe) existing in the molten matrix come into reaction with W and C elements on the surface of WC particles, leading to the formation of graded (W, M)$C_x$ (M = Ni, Cr, Fe) interfacial layers. The higher working temperature of the melt pool and the subsequent accelerated diffusivity of elements induced at higher energy densities are the main driving forces for the increase in the thickness of reaction layers [80, 81, 89].

Coating of primarily added reinforcing particles could be considered as one of the strategies to promote the wettability of reinforcing particles and simultaneously restrict, modify or avoid unwanted interfacial reactions between the reinforcing particles and the matrix [86, 138, 140]. Using two different Ni-coated and uncoated TiC particles as staring reinforcing particle, Zheng et al. [86] employed LENS process for fabricating TiC reinforced Inconel 625 matrix composites. Microstructural observations revealed that while a part of Ni coating on TiC particles melted

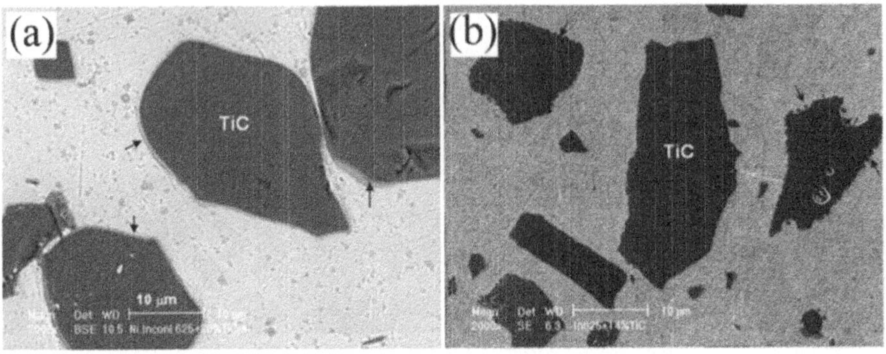


**Fig. 17** SEM images of LENS processed TiC reinforced Inconel 625 matrix composites having: (**a**) Ni-coated and (**b**) un-coated TiC as starting reinforcing particle [86]

and dissolved into the matrix due to the convective flow in the melt pool, TiC particles remained unaffected with a thin layer of well-bonded Ni still remaining on their surface. On the other hand, uncoated TiC particles experienced partial melting when subjected to the same processing parameters (Fig. 17).

Despite the high energy delivered to the mixed powder system during AM processes, the starting reinforcements in some systems may experience partial melting or even remain as unmelted particles in the solidified microstructure. The interfacial region of reinforcements and the matrix in these kinds of MMCs may be almost free of reaction products. The transmission electron microscopy (TEM) characterization of the reinforcement/matrix interface in SLM processed ex-situ BN reinforced Inconel 718 matrix composites in Fig. 18 indicated that while no reaction layers were formed at the interface, some small-sized particles of Inconel 718 matrix with spherical morphology existed at the interfacial regions. The same kind of interface has been also observed in SLM processed TiC reinforced H13 matrix [120] and $TiB_2$ reinforced 316L stainless steel matrix nano-composites [123].

The formation of reinforcement/matrix interface free of reaction layers may be ascribed to the following phenomena:

(i) The high activation energy barrier for breaking the bonds in the primarily added reinforcements [90, 141]
(ii) The combined effects of high diffusion and sudden cooling rates induced at the reinforcement/matrix interface of AM processed MMCs
(iii) The insufficient energy imparted to the powder system which may not be high enough for breaking the bonds in primarily added reinforcements
(iv) The limited solubility of reinforcements into the matrix [142]
(v) The minimized diffusion of matrix alloying elements into the solid reinforcements [90, 142].

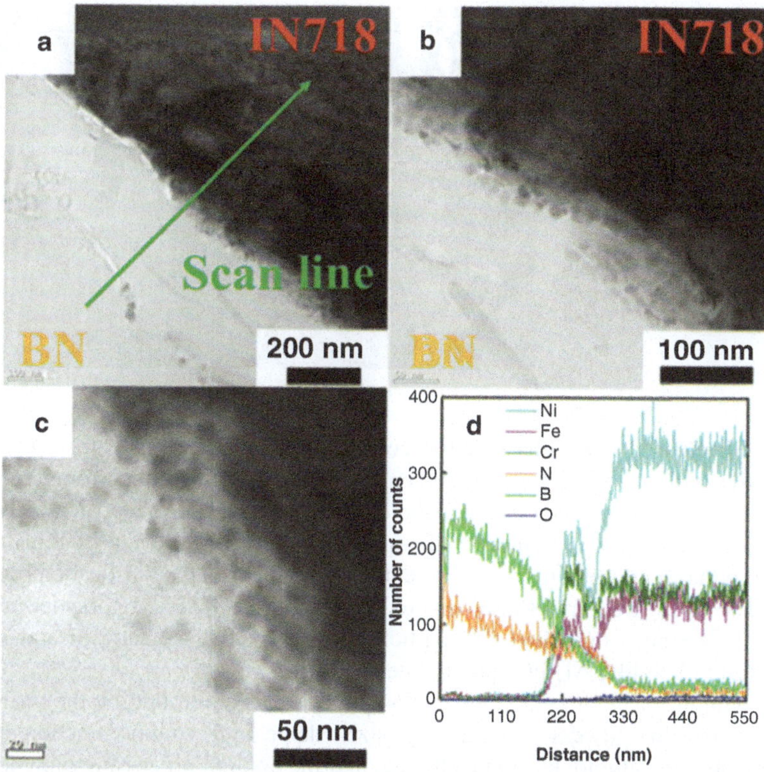


**Fig. 18** (**a–c**) TEM images of the reinforcement/matrix interface in SLM fabricated BN reinforced Inconel 718 matrix composites and (**d**) line scan EDS analysis results showing the distribution of elements along the line in (**a**) [90]

### 5.2.3 Formation of In-Situ Reaction Products

The part of primarily added powder particles melted and dissolved into the matrix can enrich and saturate the surrounding melt from the elements existing in the reinforcing particles. These elements may come into reaction with the elements existing in the molten matrix to form in-situ reaction products through solution-precipitation mechanism. For instance, N atoms dissolved or diffused into the surrounding melt during laser processing of TiN/Ti system reprecipitated as in-situ TiN particles during solidification, especially in parts of the matrix where higher supersaturation of N was formed due to the closer spacing among TiN particles (Fig. 14). Another common mixed powder system showing solution precipitation mechanism is SiC/Al system. The in-situ reaction between the Al melt and SiC ceramic reinforcing particles ($4Al(l) + 4SiC(s) \rightarrow Al_4SiC_4(s) + 3Si$) results in the formation of plate-like $Al_4SiC_4$ phase adjacent to the SiC particles.

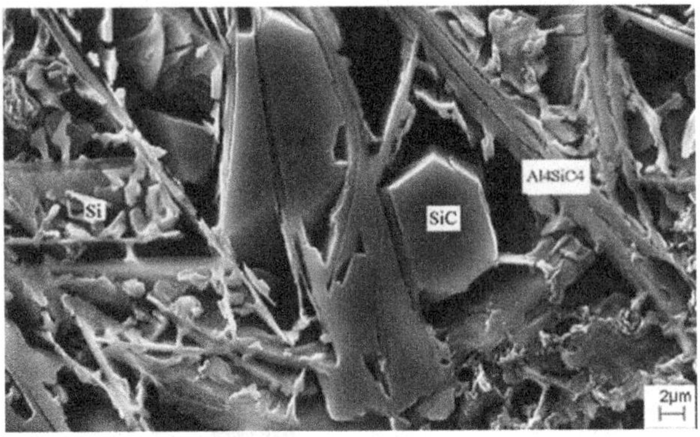


**Fig. 19** Microstructure of AMC fabricated by laser sintering of 20 wt.%SiC/$Al_7Si_{0.3}Mg$ powder mixture [22]

The Si released from this reaction combines with the Si came from the fully dissolved small-sized SiC particles and enhances the Si content of surrounding melt. If the Si content of the melt exceeds the equilibrium eutectic composition of Al-Si binary phase diagram, pro-eutectic silicon will be formed, followed by transformation of remaining melt to Al + Si eutectic as the final solidification product [136]. The microstructure of laser sintered 20 wt.%SiC/$Al_7Si_{0.3}Mg$ powder mixture in Fig. 19 contains unmelted SiC particles as well as pro-eutectic Si and in-situ synthesized plate-like $Al_4SiC_4$ precipitates.

The amount of in-situ reactions in AM processes is controlled by the characteristics of mixture powder system (e.g. size, shape and volume fraction of the constituents), the employed AM technique and the applied processing parameters.

Microstructures of TiC reinforced Inconel 718 matrix composites presented in Fig. 20 clearly shows the effect of energy density on microstructural evolution of the reinforcements. At relatively low energy density, the primarily added TiC particles are remained unmelted, bringing about an ex-situ reinforced structure (Fig. 20a). Enhancement of the applied energy density forces the TiC particles to melt ad dissolve into the matrix. The Ti and C elements dissolved into the matrix reprecipitate again as in-situ synthesized TiC phase in the microstructure (Fig. 20b). At relatively high energy densities, full melting of primarily added TiC particles may lead to the formation of MMCs containing only in-situ synthesized TiC precipitates which are noticeably finer than the ex-situ particles (Fig. 20c). Compared to the ex-situ added TiC particles, the superior interfacial bonding of in-situ ones with the matrix can contribute more to the strength of MMCs [59, 77].

When employing excessive energy densities, higher amounts and coarser reaction products may be formed in the MMCs. This could be attributed to the combinations of: (i) enhanced melting and dissolution of primarily added reinforcing particles, (ii) increased diffusion rate of alloying elements in the system, and (iii) elevated heat accumulation and significant internal energy around the in-situ synthesized precipitates.

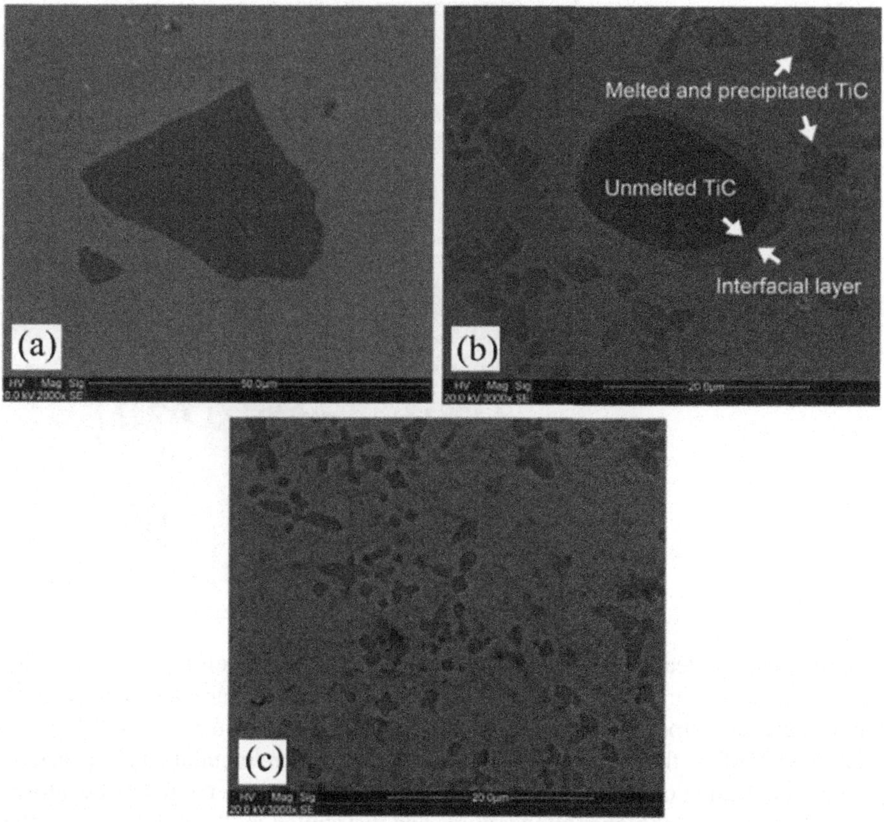


**Fig. 20** Microstructures of LMD processed TiC reinforced Inconel 718 matrix composites processed with energy densities of: (**a**) 80, (**b**) 120 and (**c**) 160 kJ/m [81]

## *5.3 Microstructural Evolutions in the Matrix Induced by the Presence of Reinforcements*

### 5.3.1 Microstructural Refinement of the Matrix

Compared to the conventional manufacturing methods such as casting, the higher solidification and cooling rates associated with the AM processes lead to more significant microstructural refinement. The rapid solidification nature of AM processes is in fact due to the very local heat input as well as noticeably small volumes of the molten material.

When using AM processes to fabricate MMCs, even finer microstructures can be achieved due to the presence of reinforcements in the system. Similar to other processing techniques such as casting and welding, solid reinforcing particles incorporated into the melt can promote the heterogeneous nucleation during

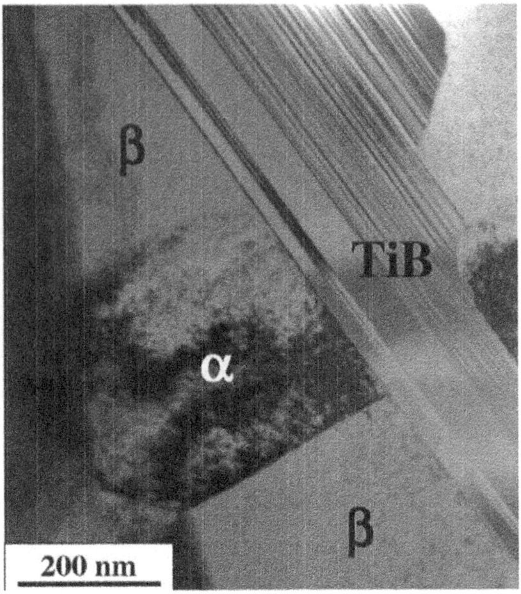


**Fig. 21** TEM image indicating the growth of α phase from in-situ synthesized TiB precipitate in LENS processed TiB reinforced Ti6Al4V matrix composite [51]

solidification and lead to grain refinement of the matrix. Therefore, the size, weigh fraction, distribution pattern and melting efficiency of these particles directly affect the solidification process and microstructural evolution of the matrix. For the ex-situ reinforced MMCs, the un-melted solid particles act as preferential sites for heterogeneous nucleation of grains. This is while for the in-situ ones, the newly developed precipitates formed at higher temperatures and distributed in the molten matrix play this role. The TEM image in Fig. 21 shows an in-situ synthesized TiB precipitate acting as nucleation site for α phase in the Ti6Al4V matrix of LENS processed TiB reinforced TMC structure. Beside the role reinforcements play as nucleation sites, their grain growth inhibiting effects caused by grain boundary pinning can also lead to matrix grain refinement [189, 193].

As an example of the matrix refinement, microstructural observations revealed that while the SLM processed CP-Ti had grains with average width and length of 2.1 and 15.3 μm, respectively, those for the matrix grains of TiB reinforced TMCs processed with the same parameters were noticeably decreased to 0.6 and 1.1 μm, respectively [50, 143]. In addition to the microstructural characterizations, the grain refinement of the matrix can be also identified by broadening and decreased intensity of the matrix peaks in X-ray diffraction (XRD) patterns of the AM processed MMCs [81, 82, 121, 128].

The applied energy density is one of the major factors controlling the degree of microstructural refinement in AM processed MMCs. While maintaining the columnar dendritic structure, enhancement of energy density from 30 to 100 kJ/m decreased the width of matrix dendrites from 3.8 to 1.3 μm in LMD processed TiC reinforced Inconel 625 matrix nano-composites. Such microstructural refinement was ascribed mainly to the increased dissolution of TiC particles and the subsequent larger degrees of constitutional undercooling provided for the melt at elevated energy densities [83].

In cases where the dissolved elements have a limited solubility in the matrix (e.g. B in Ti or Si in Al), the increased constitutional undercooling induced by partitioning of rejected solutes in front of the solid/liquid interface can make the solidification interface unstable and consequently give rise to the driving force for nucleation of much more grains at the solid/liquid interface. The rejected atoms segregated in front of solidification interface can restrict the growth of nuclei and act as grain refiners as well [49]. In addition to the matrix grain refining, alloying elements existing in the system may also help to the refinement of in-situ synthesized precipitates. For instance, much finer Si precipitates have been achieved for SLM processed Al-20Si-5Fe-3Cu-1Mg alloy compared with that of Al-20Si [28, 144]. This has been attributed to the hindering effects of Fe, Cu and Mg alloying elements on growth of Si phase by two possible mechanisms: (i) a portion of Fe atoms are dissolved into Si precipitates, leading to the reduced activation of Si, and (ii) the $Al_4FeSi_2$ phase formed in the microstructure due to the presence of Fe retards the coarsening of Si precipitates by acting as barriers to the diffusion of Si atoms [28].

By increasing the number of nucleation sites for the matrix, higher amounts of reinforcements existing in the system can intensify the level of grain refinement experienced by the matrix [85, 123]. Fig. 22 shows the microstructures for SLM processed pure 316L stainless steel and $TiB_2$ reinforced 316L stainless steel matrix composites containing 2.5 and 5 vol.% reinforcement, respectively. While the addition of 2.5 vol.% $TiB_2$ had a negligible effect on grain refinement of the matrix (Fig. 22a), a remarkable decrease in the grain size was observed for the MMCs containing higher amount of $TiB_2$ (5 vol.%) (Fig. 22c).

Since powder mixtures with different arrangement of particles have quite different interactions with the laser/electron beam, the category of mixed powder system can significantly affect the microstructural characteristic of matrix in AM processed composites. Figure 23a, b show the microstructures for the SLM processed 15 vol.% $TiB_2$/316 L stainless steel mixed powder systems prepared by direct mixing and ball milling, respectively. As being observed, the ball milled powder category shows relatively finer cells of cellular structure.

### 5.3.2 Texture of the Matrix

One of the most famous features of AM processed parts is the anisotropy of the microstructure. This phenomenon is in fact due to the large directional thermal gradients or high values of thermal gradient (G) to solidification rate (R) ratio (G/R) generated by the characteristic layer-by-layer nature of AM processes. Since the heat conduction in the building direction is typically higher than other directions, the crystals grow preferentially along this direction. The anisotropy in microstructure can be associated with the anisotropy in mechanical properties [145–147]. The

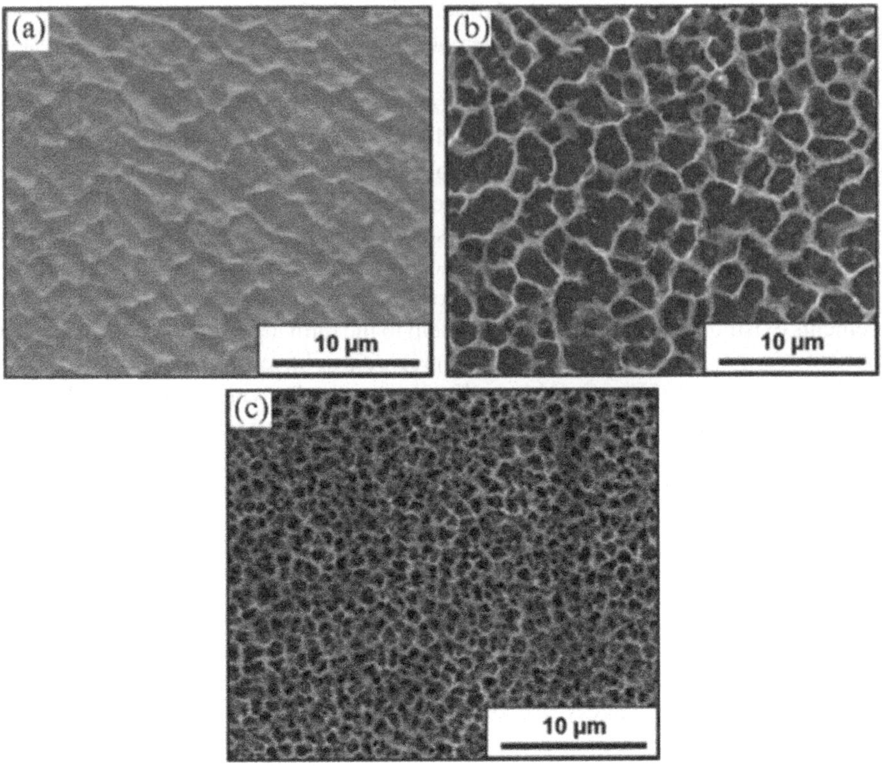


**Fig. 22** The microstructures of SLM processed: (**a**) non-reinforced 316L stainless steel, and $TiB_2$ reinforced 316L stainless steel matrix composites containing: (**b**) 2.5 and (**c**) 5 vol.% $TiB_2$ [123]

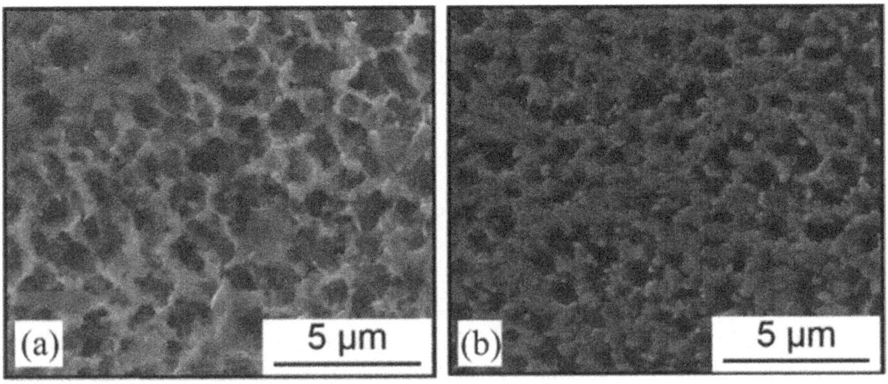


**Fig. 23** Microstructures of SLM processed $TiB_2$ reinforced 316L stainless steel composites with starting: (**a**) directly mixed and (**b**) ball milled powder systems [124]

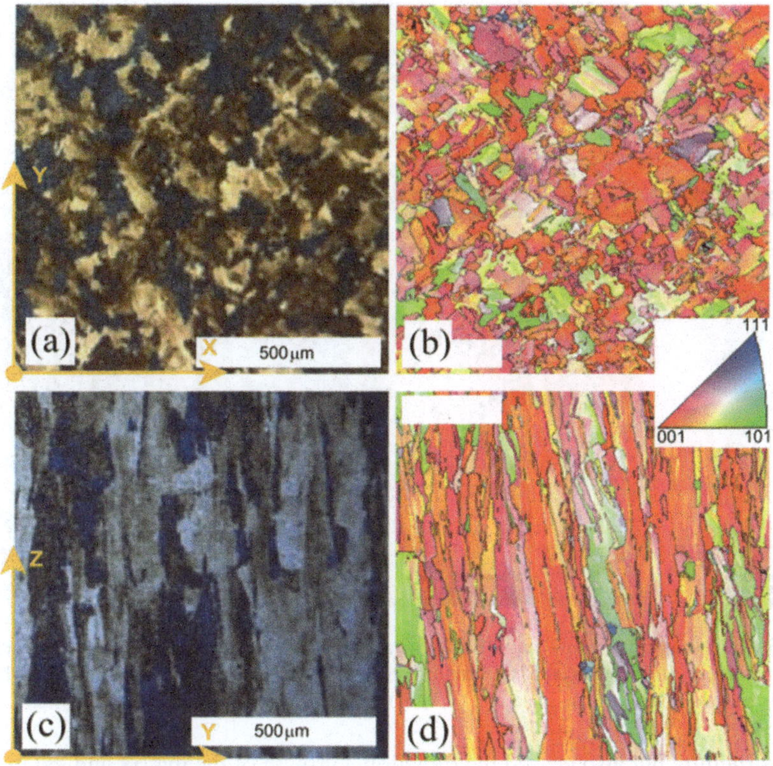


**Fig. 24** (**a**, **c**) Optical micrographs and (**b**, **d**) EBSD crystallographic orientation maps of SLM processed Inconel 738LC alloy. (**a**) and (**b**) are sections perpendicular to the building direction while (**c**) and (**d**) are parallel to the building direction [149]

anisotropy in microstructure has been reported in the literature for many materials fabricated by AM processes [145, 147–149]. Figure 24 shows the optical micrographs and electron back scattered diffraction (EBSD) orientation maps of SLM processed Inconel 738LC alloy obtained from sections perpendicular and parallel to the building direction. The grain shape anisotropy is obvious in the micrographs, with the long axis of grains oriented along the building direction.

Application of scan strategies with rotating scan directions is an applicable technique employed to break up such anisotropic microstructures in AM processed parts. While the rapid solidifications rates associated with the AM processes enhance R, the basic concept behind most of these scan strategies is shortening of scan tracks to decrease the thermal gradient (G) and consequently decline G/R ratio [150]. Incorporation of reinforcements into the melt pool may also affect the texture of the matrix in AM processed MMCs parts. Besides acting as preferential sites for nucleation of grains, these reinforcements can retard the directional growth of grains during the solidification process [82, 129, 147, 151, 152]. As shown in SEM and EBSD micrographs in Fig. 25, the TiC reinforcements not only refined the

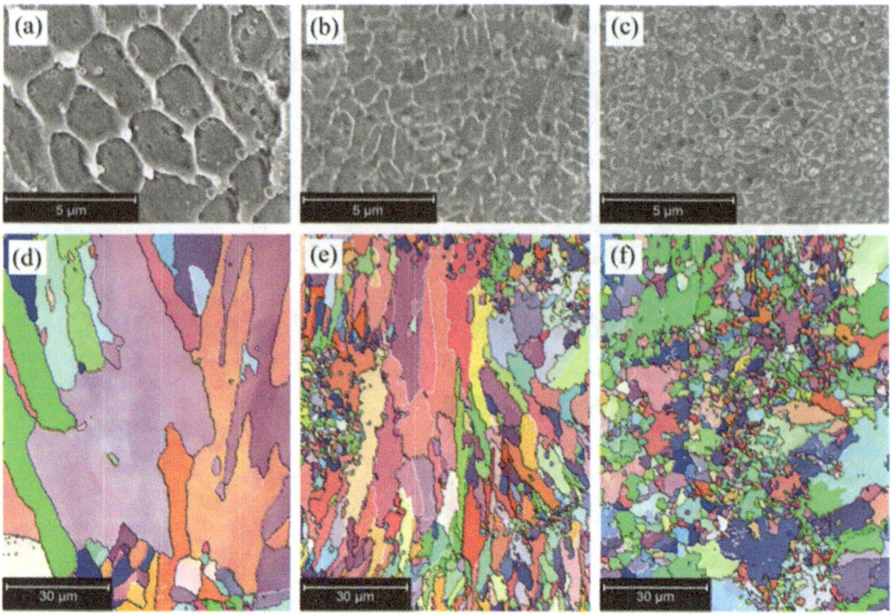


**Fig. 25** SEM images (first row) and EBSD crystallographic orientation maps (second row) of SLM processed: (**a**, **d**) non-reinforced 316L stainless steel; and TiC reinforced 316L stainless steel matrix composites with: (**b**, **e**) micro-scale and (**c**, **f**) nano-scale TiC reinforcements [152]

microstructure, but also ceased the directional growth of the matrix and led to much more isotropic structures. Both the grain refining and directionality prohibiting were much pronounced for the MMCs containing nano-scale reinforcements.

### 5.3.3 Microstructural Evolution and Phase Transformation

Due to the variations in multiple modes of heat, mass and momentum transfer as well as the probable chemical reactions, addition of reinforcements to the system may activate different consolidation mechanisms and microstructural evolutions in the matrix of AM processed MMCs [123]. The microstructures obtained from laser remelting of pure A356 powder and A356/SiC powder mixture in Fig. 26 clearly reveals the major roles reinforcements play in the matrix structure. While the laser sintered pure A356 part shows a directional solidification of columnar grains, incorporation of SiC reinforcements into the system leads to an equiaxed grain structure.

The transition in grain growth mode by addition of reinforcements has been reported in the literature for various systems. Even if the volume fraction of reinforcements is not as high enough to completely transit the solidification mode, such transition can be induced at least in reinforcement-containing regions [154]. Using LENS process to fabricate functionally graded metal matrix composites

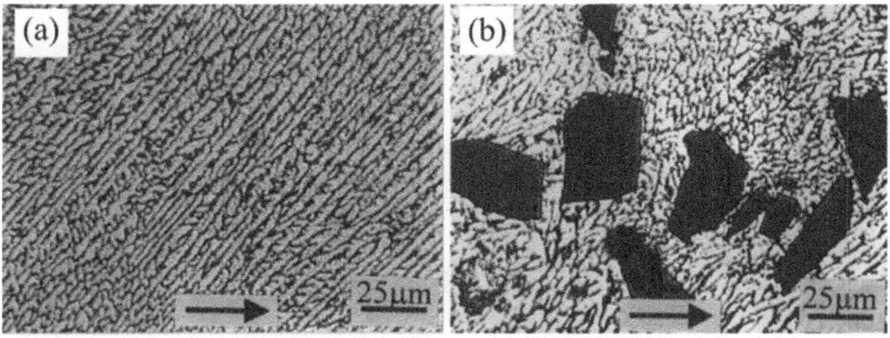


**Fig. 26** Optical micrographs of laser processed: (**a**) non-reinforced A356 Al alloy and (**b**) SiC reinforced A356 matrix composites [153]

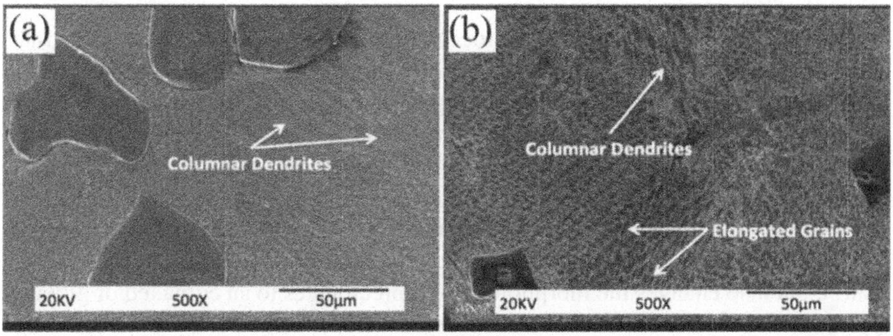


**Fig. 27** Microstructures of LENS processed TiC reinforced Inconel 690 matrix composites containing: (**a**) less than 15 and (**b**) 15–30 vol.% TiC as reinforcement [5]

(FGMMCs), Wilson and Shin [5] studied the effect of TiC content on microstructural evolution of the Inconel 690 matrix. When the TiC content was less than 15 vol. %, the matrix was mainly composed of columnar dendritic structure. However, a transition from columnar dendritic structure to finer elongated grains occurred for the MMCs with 15–30 vol.% TiC, leading to the microstructure having groups of elongated grains as its major fraction (Fig. 27).

For the composite systems in which the reinforcements do not act as heterogeneous nucleation sites for the matrix, the grain boundary pinning and grain growth restriction caused by reinforcements may create sufficient undercooling and provide enough time for new grains to develop. Besides grain refinement, this mechanism also contributes to structural transition [155]. By acting as barriers to solute diffusion and heat transfer, the solid reinforcements can cause disturbance in solute field and temperature field in the solidification front [156]. As a consequence, irregular growth modes may be activated for the matrix of MMCs during solidification.

The reinforcements can also influence the phase transformation in the matrix of AM processed MMCs. Microstructural characterizations of the non-reinforced

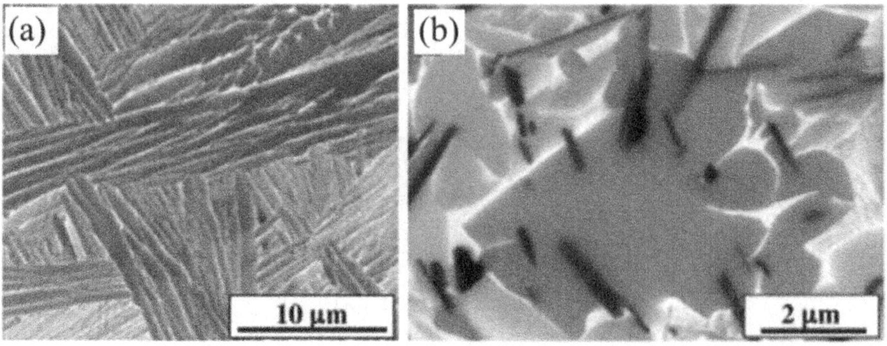


**Fig. 28** Microstructures of LENS processed: (**a**) non-reinforced Ti6Al4V alloy and (**b**) TiB reinforced Ti6Al4V composite [51]

Ti6Al4V alloy fabricated by AM processes indicates that the fast cooling rates associated with these processes favors the transformation of β phase to α′ martensite, leading to microstructures containing Widmanstatten laths of martensitic α′ phase in a β phase matrix [51, 63]. When considering TMCs having Ti6Al4V alloy as the matrix, microstructural features of the matrix may differ from the non-reinforced Ti6Al4V fabricated with the same process and similar processing parameters. For instance, as shown in Fig. 28, the formation of primary TiB precipitates in TiB reinforced Ti6Al4V matrix composites not only refines the α/β microstructure of the matrix, but also changes the morphology of α precipitates to an equiaxed or globular type [37, 51]. This could be due to the roles TiB precipitates play as additional heterogeneous nucleation sites for α phase in the β matrix. In addition to the changes in morphology, the volume fraction of phases formed in the matrix can be affected by the reinforcements. As reported by Balla et al. [59], although having the same acicular morphology, the matrix of TiN reinforced Ti6Al4V matrix composites contained larger amounts of α phase than the non-reinforced LENS processed Ti6Al4V alloy.

The elements dissolved into the matrix in AM processed MMCs can make changes to the microstructural evolution and phase transformation of the matrix. For instance, the stabilizing effect of dissolved N atoms in the matrix and the rapid cooling rates associated with the AM processes are believed to be the main reasons responsible for the formation of needle-like martensitic α phase in TiN reinforced Ti6Al4V matrix composites [59, 157]. The dissolution of SiC reinforcements into the surrounding melt during SLM processing of SiC/iron mixed powder system also resulted in major microstructural transition in the matrix [128, 129]. Due to the extremely low content of carbon in the iron powder (~0.03%), a ferritic (α-Fe) microstructure was obtained for the non-reinforced parts even at extremely high cooling rates of SLM process (Fig. 29a). However, the partial melting and dissolution of SiC particles to C and Si in the SiC reinforced iron-based MMCs led to the increased carbon content of the melt and paved the way for the formation of pearlite and martensite constituents in the matrix (Fig. 29b).

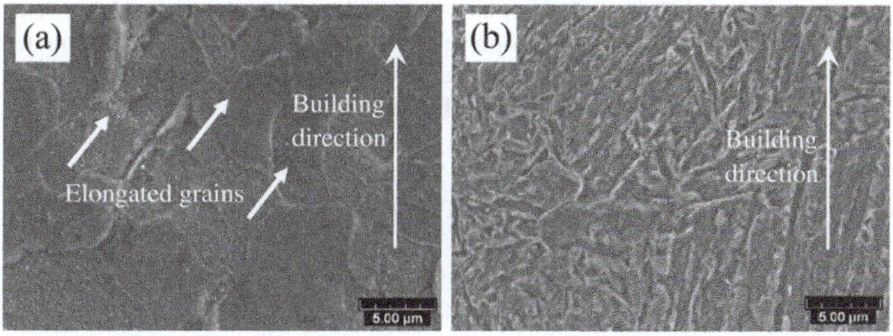


**Fig. 29** SEM micrographs of SLM fabricated: (**a**) non-reinforced pure iron and (**b**) SiC reinforced iron matrix composite [129]

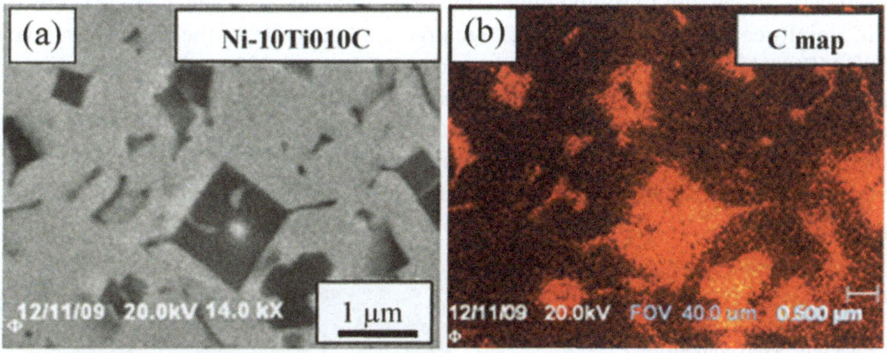


**Fig. 30** (**a**) SEM image of LENS processed TiC reinforced Ni matrix composite and (**b**) AES analysis map of C in the microstructure shown in (**a**). The thermal pseudocolor reveals the relative amount of C element [87]

### 5.3.4 Formation of Supersaturated Matrix

The rapid solidification rates of the AM processes can extend the solid solubility limit of alloying elements in the matrix and cause some distortions in its lattice structure. These distortions can be identified by the shift occurred in the position of XRD peaks corresponding to the matrix phases [81, 82]. The supersaturation of matrix from alloying elements may contribute to the enhancement of mechanical properties through solid solution strengthening mechanism [48, 158–161]. As reported by Prashanth et al. [29], in addition to the in-situ synthesized high strength $Al_6MoTi$ reinforcement phase formed in Al12Si-TNM system, development of an Al matrix supersaturated with both Ti and Si elements was responsible for further improvement in the hardness of SLM fabricated composites. The same phenomenon has been observed by Gopagoni et al. [87] when using LENS process to fabricate in-situ synthesized TiC reinforced Ni matrix composites from mixed elemental powder system of Ni-10Ti-10C. Despite the negligible solid solubility of C in Ni, the rapid solidification rates led to diffusion-limited trapping of C in Ni to form saturated solid solution matrix, as confirmed by high resolution Auger electron spectroscopy (AES) analysis of the fabricated MMCs (Fig. 30).

### 5.3.5 Formation of Dislocations in the Matrix

In order to accommodate the misfit in CTE as well as EM induced between the reinforcements and the matrix during rapid melting and cooling, a large amount of thermal stresses are formed in the matrix of AM processed MMCs. These stresses may be high enough to cause plastic deformation which is inclined to be portioned more toward the substantially softer phase (the metallic side of interface), especially adjacent to the reinforcements/matrix interface. For instance, when cooled down from laser processing to room temperature, the difference between average CTEs of TiC reinforcement and Inconel 625 matrix can generate a thermal stress over 500 MPa, which is beyond the yield strength of Inconel 625 in the annealed form [86]. The strains induced by these stresses are partially relieved by the formation of misfit dislocation network substructures at the interface which are known as geometrically necessary dislocations (GNDs). High resolution TEM (HRTEM) observation of the TiC/Ni interface in AM processed TiC reinforced Ni matrix composites provided in Fig. 31 indicates the formation of dislocations which separate coherent

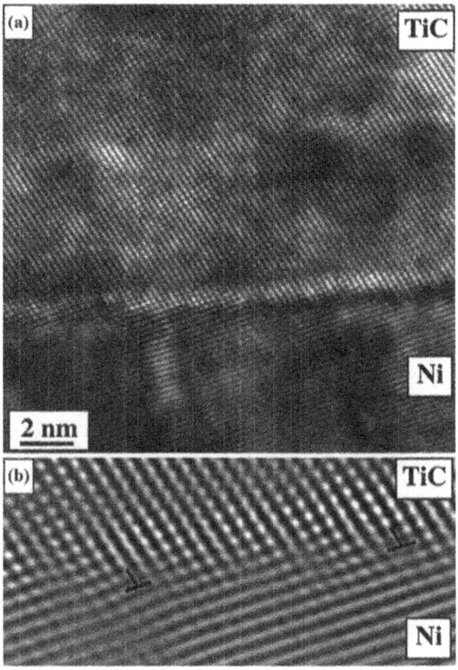


**Fig. 31** (**a**) HRTEM image of the TiC/Ni interface in LENS processed composites and (**b**) higher magnification image indicating a semi-coherent interface containing dislocations [87]

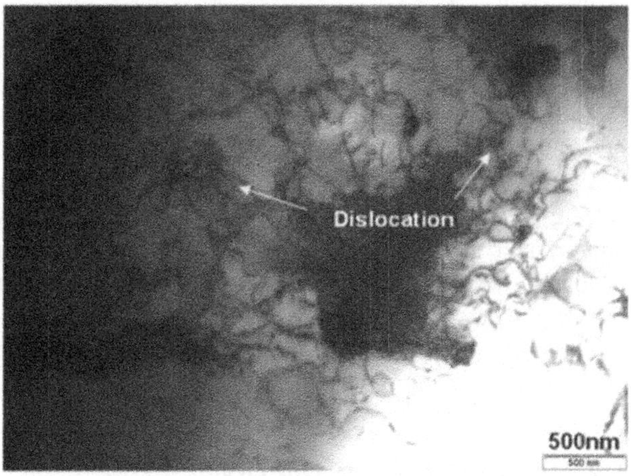


**Fig. 32** TEM image showing high density of dislocations in the matrix of LENS processed TiC reinforced Inconel 625 matrix composites [86]

regions of excellent lattice matching at the TiC/Ni interface. The intensity of the thermal stress and consequently the density of GNDs are reduced by moving away from the reinforcement/matrix interface.

In addition to the reinforcement/matrix interfacial regions, misfit dislocation substructures can be also generated at the interface of different phases existing in the matrix. For instance, as revealed by TEM analysis, both the β/TiB and α/β interfaces in AM processed TiB reinforced Ti6Al4V matrix composites contain misfit dislocation substructures [51]. These dislocations could be beneficial to the enhancement of strength based on dislocation strengthening mechanism discussed.

It is worth noting that the fast cooling rates (in the order of ~$10^6$–$10^8$ K/s) as well as the multiple reheating cycles associated with the AM processes can also lead to the generation of dislocations in the matrix of MMCs, as shown in Fig. 32.

## 6 Part Quality and Surface Integrity of AM Processed MMCs

Generally, the most important defects in AM processed parts limiting the densification level are porosities and cracks. The porosity in AM processes is originated mainly from the melt pool behavior as well as the quality of the starting powder particles. Several resources are responsible for the formation of porosities in AM processed parts including gas voids, solidification shrinkage, lack of melting and local evaporation of some elements when subjected to temperatures exceeding their boiling point [125, 127].

The bonding performance between the adjacent tracks and subsequent layers play significant roles in the densification level of parts [82]. In fact, a good bonding between the tracks and layers can guarantee the fabrication of parts with sound densification levels. This can be achieved by favorable wetting and effective contribution of previously solidified materials in the freshly deposited molten material. Moreover, the wetting behavior of solid reinforcements with the surrounding molten material affects the porosity level in AM processing of MMCs.

If the process parameters are not correctly chosen, the instability of laser track can generate high surface roughness or even cause balling effect. To the same, the incorrect overlapping of adjacent tracks and/or layers caused by inappropriate selection of hatch spacing and/or layer thickness may also leave the inter-track and/or inter-layer regions unmelted, leading to the formation of porosities. The following describes the effects of applied energy density and the characteristics of mixed powder system as two major factors influencing the part quality and surface integrity of AM processed MMCs.

## *6.1 Applied Energy Density*

The applied processing parameters (energy density) play significant roles in part quality and surface integrity of AM processed parts. Figure 33 presents the surface morphology for SLM fabricated TiC reinforced TMCs processed with varying energy densities by changing the scanning speeds. At relatively high energy density, the fabricated part showed a dense surface having no pores and spherical balls. However, the decrease in the applied energy density (increasing the scan speed) enhanced the surface roughness by the formation of larger balls as well as porosities among neighbor balls or adjacent tracks. The formation of balls on the surface is due to the occurrence of balling effect, where the fresh molten material is not able to wet the underlying substrate. Under this condition, the continuous melt track transforms to several spherical-shaped agglomerates to obtain the equilibrium state.

The poor surface integrity and quality of AM processed MMCs parts obtained at relatively low energy densities could be discussed in terms of the melt pool features under this condition which favor the formation of less dense parts. These features involve:

(i) The low working temperatures, decreased volume of melt pool and limited amount of molten material [79, 127]
(ii) The shortened lifetime of the molten material induced by higher solidification rates [79, 89]
(iii) The enhanced viscosity, limited convective flow and reduced liquid-solid rheological performance in the melt pool [120, 125]
(iv) The decreased wettability of solid reinforcements with the surrounding melt [78, 83, 125]

The reduced wettability and limited spreading of molten material to the outward combined with the poor dilution between the fresh depositing track and the previously deposited track/layer leads to a weak interfacial bonding and consequently results in the formation of porosities in inter-layer/inter-track locations [78, 82, 83, 89], as shown in Fig. 34. However, the increased volume of melt and the decreased

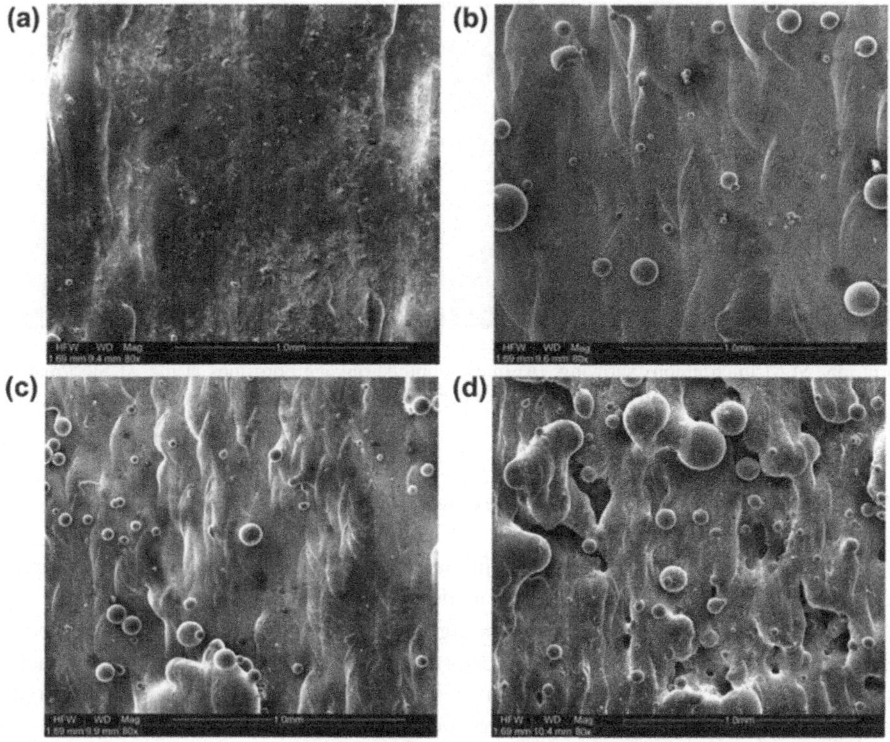


**Fig. 33** SEM images of the surface morphology for SLM fabricated TiC reinforced TMCs processed with energy density of: (**a**) 360, (**b**) 180, (**c**) 120 and (**d**) 90 J/mm$^3$ [43]

**Fig. 34** Cross-sectional optical micrograph of SLM processed SiC/AlSi10Mg mixed powder system showing the formation of inter-layer porosities [18]

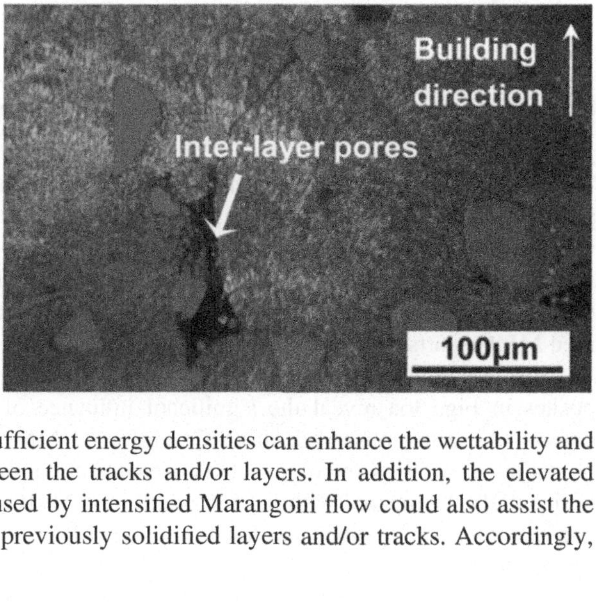


melt viscosity obtained at sufficient energy densities can enhance the wettability and improve the bonding between the tracks and/or layers. In addition, the elevated stirring in the melt pool caused by intensified Marangoni flow could also assist the melt with better wetting of previously solidified layers and/or tracks. Accordingly,

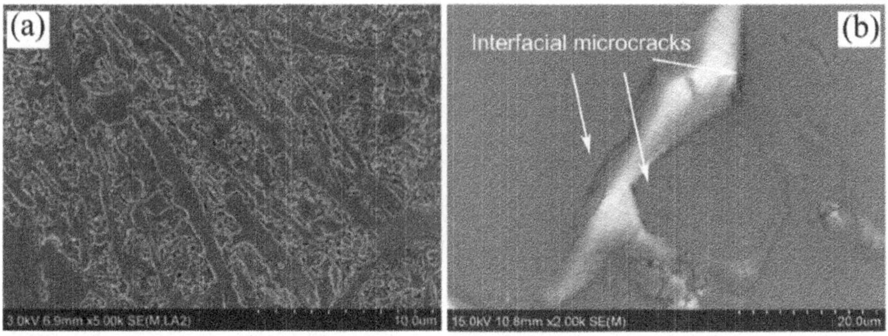


**Fig. 35** SEM micrographs showing: (**a**) microscopic shrinkage porosities at the $Al_4SiC_4$/matrix interface of SLM processed AMCs, and (**b**) interfacial micro-cracks at the $Al_2O_3$/Al interface of SLM fabricated AMCs [25]

MMCs with better surface integrities and higher densification levels could be achieved.

The high temperature gradients induced at excessive energy densities may enhance the risk of balling and lead to the accumulation of stress in parts of the previously deposited material experiencing remelting. These stresses may be responsible for the delamination of layers, distortion of part or even appearance of cracks [79, 196]. Moreover, excessive energy density may result in the formation of interfacial microscopic shrinkage porosities and thermal micro-cracks at the reinforcement/matrix interface in AM processed MMCs [19, 25]. Figure 35a shows such porosities at the reinforcement/matrix interfaces in SLM processed SiC/AlSi10Mg powder mixture while Fig. 35b indicates thermally induced interfacial micro-cracks formed in AM processed $Al_2O_3$ reinforced AMCs.

## 6.2 *Characteristics of Mixed Powder System*

The characteristics of mixed powder system affecting the densification level of AM processed MMCs include the type, size and volume fraction of reinforcements as well as the category of mixed powder system.

For a given volume fraction of reinforcements, the limited wettability of larger reinforcing particles leads to higher amounts of larger-sized porosities in the solidified MMCs parts.

Cross-sectional optical micrographs of AM processed Inconel 625 matrix composites in Fig. 36a reveal the significant influence of reinforcement type on the defects. As being observed, the SiC reinforced MMCs contain large volume of porosities as well as interconnected macro-cracks propagated both vertically and horizontally. Compared with the SiC reinforced MMCs, the $Al_2O_3$ reinforced ones had much lower porosities and fewer numbers of cracks. Formation of porosities in

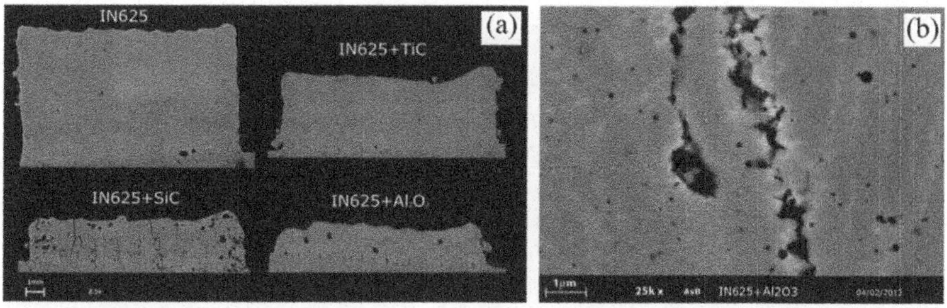


**Fig. 36** (**a**) Cross-sectional micrographs of additively layer manufactured Inconel 625 matrix composites with different types of reinforcements, and (**b**) SEM micrograph illustrating cracking in $Al_2O_3$ reinforced composites shown in (**a**) [76]

this type of MMCs was attributed to the thermal destruction of $Al_2O_3$ during the process, leading to the release of oxygen and their subsequent capturing to form bubbles. This is while the enrichment of edges of cracks from Nb and Mo elements gave evidence of liquation cracking (Fig. 36b). Despite the SiC and $Al_2O_3$ reinforced MMCs, those reinforced with TiC were almost free of defects.

The enhanced volume fraction of reinforcements is generally associated with worsening of the densification level mainly due to: (i) the increased melt pool viscosity which results in decreased wettability and favored spheroidization of melt pool, (ii) the elevated melt pool instability, and (iii) the occurrence of balling effect [121, 123].

Due to the higher brittleness of ceramic reinforcements compared to the metallic matrix, the enhanced volume fraction of reinforcements and their probable agglomeration may favor the formation of cracks in the AM processed MMCs. As an example, aggregation of Ca and P elements (formed by dissolution of HA) along the boundaries of laser tracks in SLM processed HA reinforced stainless steel matrix composites introduced the boundaries as potential locations for crack nucleation and propagation when subjected to the residual stresses of the process. While the enhanced volume fraction of HA promoted this mechanism and led to the formation of larger and higher amounts of cracks, the elevated scanning speeds declined the amount of aggregation and reduced the risk of cracking (Fig. 37). The decrease in the melt pool size caused by the elevated scanning speeds may also lead to insufficient overlapping of tracks and consequently increase the porosity level (Fig. 37) [119]. However, it should be noted that the local accumulation of residual stress caused by application of relatively high scan speeds (high cooling rates) can enhance the crack formation susceptibility in some systems [79, 162].

The category of mixed powder system can also affect both the type and amount of defects. Using two different powder categories; i.e. directly mixed and ball milled $TiB_2$/316 L stainless steel mixed powder systems, AlMangour et al. [124] indicated that while the porosities were more evident in the directly mixed powder system,

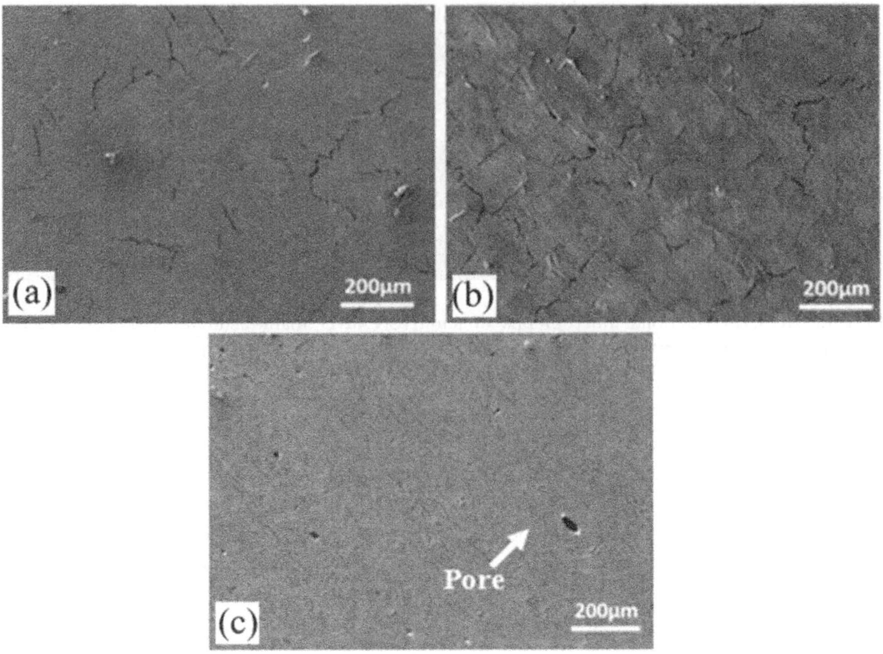


**Fig. 37** SEM micrographs of SLM fabricated HA reinforced stainless steel matrix composites with HA content and scanning speed of: (**a**) $HA = 5$ vol. %, $\upsilon = 250$ mm/s, (**b**) $HA = 15$ vol. %, $\upsilon = 250$ mm/s and (**c**) $HA = 5$ vol. %, $\upsilon = 400$ mm/s [119]

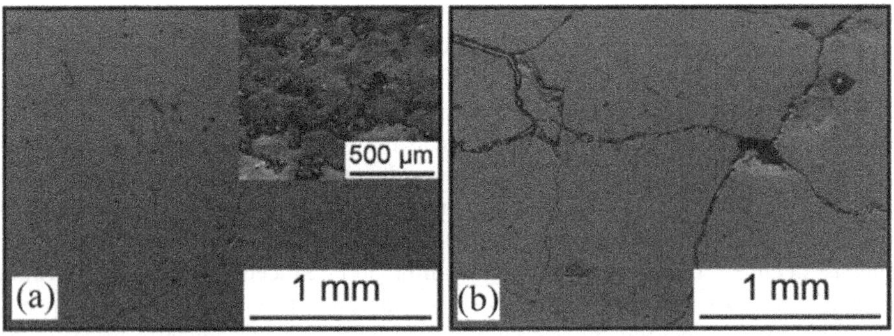


**Fig. 38** SEM micrographs indicating $TiB_2$ reinforced 316L stainless steel matrix composites fabricated by SLM processing of: (**a**) directly mixed and (**b**) ball milled powder systems [124]

larger and more severe cracks were characterized for the ball milled powder category (Fig. 38). The formation of cracks were related to the liquation cracking at the boundaries, solidification shrinkage as well as the limited wettability caused by application of inappropriate processing parameters.

# 7 Mechanical Properties of AM Processed MMCs

As far as the AM technology has been considered, it has been revealed that it provides the fabrication of parts with superior mechanical properties including higher hardness, improved strength and comparable ductility than those manufactured by conventional processing techniques. While the addition of reinforcements to the matrix to form MMCs structures helps to enhance the hardness, strength and work hardening compared to the non-reinforced parts, they may render the fabricated MMCs parts extremely brittle (Fig. 39) [29, 49, 50]. As shown in compressive stress-strain curves in Fig. 39a, Al-12Si-TNM AMCs fabricated by SLM showed yield and ultimate strengths of ~420 and ~636 MPa, respectively, which were higher than those for the non-reinforced Al-12Si samples having yield and ultimate strengths of ~283 and ~475 MP, respectively. However, the AMCs broke at ~6% total strain while the non-reinforced parts did not experience fracture until 60% of deformation. Toughness measurements of the CP-Ti and TiB reinforced TMCs fabricated with the same SLM processing parameters also shows a ~50% decrease for the composites than the non-reinforced part [50]. The decreased ductility and toughness of MMCs compared to the non-reinforced structures can be ascribed to the insufficient ductility, the high brittleness and limited ability of the hard and brittle reinforcements to accommodate the plastic deformation [48, 50]. The increase in the content of these reinforcements can make the fabricated MMCs less ductile [45].

## 7.1 Strengthening Mechanisms

Generally, the strengthening of MMCs is determined by the contribution of direct and indirect strengthening mechanisms. The direct strengthening is also known as "classical" strengthening and is concerned with the load transfer from the softer

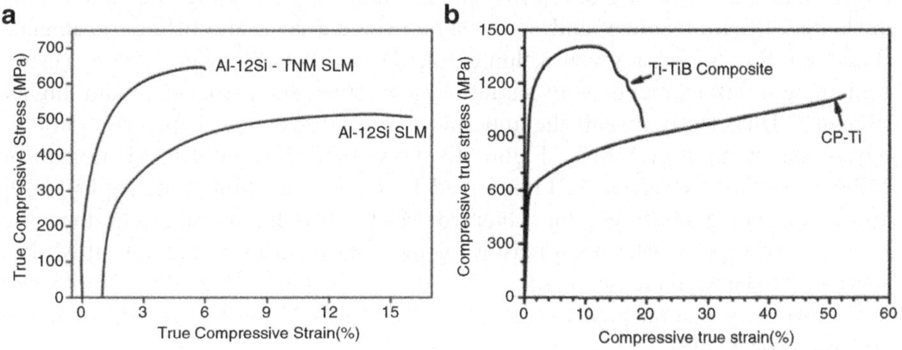


**Fig. 39** Compressive stress-strain curves of SLM processed: (**a**) non-reinforced Al-12Si and Al-12Si-TNM composite; and (**b**) non-reinforced CP-Ti and TiB reinforced TMC [49]

matrix to the harder reinforcements. However, the indirect strengthening is based on the matrix strengthening induced by the presence of reinforcements.

### 7.1.1 Direct Strengthening

Based on the simple rule of mixture, incorporation of reinforcements into the matrix to form MMC structures is almost associated with the improved hardness and strength since a fraction of the matrix is substituted by harder phase(s) [5, 48, 86, 123, 161, 163, 164]. The rapid solidification rates associated with the AM processes result in the formation of inherently strong bonding coherence between the reinforcements and the matrix. Based on the interfacial strengthening mechanism, such kind of interface aids to the effective load transfer from the matrix to the reinforcement and consequently improves the flow stress and strength of AM processed MMCs [30, 45, 48, 128, 129]. The formation of graded interfacial layers at the interface of ceramic reinforcements and the metallic matrix in some systems can also enhance the interfacial bonding strength and contribute to the increase in mechanical properties [89]. By hindering the movement of dislocations, the reinforcements distributed in the matrix can restrain its local micro-deformation and consequently improve the strength as well [82, 128, 129, 165].

The amount of increment in strength of discontinuously reinforced MMCs is affected by some variables including the type, size, shape, distribution state and volume fraction of reinforcements as well as their orientation with the applied loading direction. For the MMCs containing short whisker-shaped reinforcements aligned in the tensile loading direction, the shear-lag model developed by Nardone and Prewo predicts the increase in yield strength as [166, 167]:

$$\Delta\sigma = \sigma_{ym} V_r \left[\frac{(l+t)A}{4l}\right] \tag{5}$$

where $\sigma_{ym}$ is the yield strength of the matrix, $V_r$ represents the volume fraction of reinforcements, $l$ is the size of reinforcement parallel to the loading direction, $t$ is the thickness of reinforcement, and $A = l/t$ signifies the aspect ratio of reinforcements. Based on Eq. 5, for a given volume fraction of reinforcements, those MMCs containing reinforcements with larger aspect ratios are associated with higher strengths. This clearly reveals the importance of reinforcement shape and morphology on the strength of MMCs. Figure 40 shows the variation in yield strength of LENS processed CP-Ti and TiB reinforced TMCs as a function of the applied laser power. As being observed, the obtained MMCs had higher strengths than the non-reinforced parts. The main reason for the improvement in strength of MMCs was found to be the increase in aspect ratio of TiB precipitates caused by the elevated laser power (e.g. change in aspect ratio from 8.35 to 13.32 for laser powers of 125 to 200 W, respectively).

The following describes the effects some other reinforcement features have on the mechanical properties of AM processed MMCs.

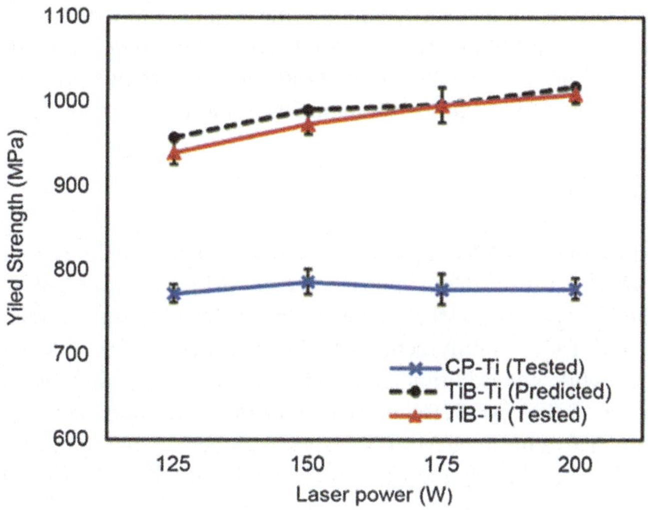


**Fig. 40** Variation of yield strength for LENS processed CP-Ti and TiB reinforced TMCs as a function of applied laser power [50]

## Reinforcement Volume Fraction

When the MMCs contain equiaxed particulate shaped reinforcements with aspect ratio of 1, Eq. 5 is simplified as:

$$\Delta\sigma = \frac{1}{2}\sigma_{ym}V_r \tag{6}$$

Accordingly, the increase in yield strength of MMCs could be linearly correlated with the volume fraction of reinforcements distributed in the matrix. The significant improvement in mechanical properties of AM processed MMCs with the enhancement of reinforcement volume fraction has been reported in several researches [31, 45, 56, 59, 77, 121, 168]. For instance, the increase in volume fraction of in-situ synthesized TiC reinforcements from 20 to 60 vol.% led to the significant enhancement of hardness from 365 to 1898 HV in TiC reinforced Ni matrix composites [77]. In another research, the increase in $TiB_2$ content from 5 to 10 wt. % resulted in noticeably high hardnesses of 560 and 750 HV, respectively for the obtained in-situ TiB reinforced Ti6Al4V matrix composites. This is while the non-reinforced Ti6Al4V parts processed with the same parameters had a hardness of 340 HV [37]. The research performed by Balla et al. [59] to fabricate TiN reinforced Ti6Al4V composite coatings using LENS process revealed an increasing trend in hardness from 527 to 1138 HV with the enhancement of TiN content from 10 to 40 wt.%. However, the LENS processed pure Ti6Al4V in their research had a hardness of 394 HV.

Recently, the concept of improvement in mechanical properties due to enhancement of reinforcement volume fraction has been applied to develop AM processed FGCMs. These materials benefit from varying content of reinforcements and consequently gradient mechanical properties along the building direction [5, 45, 169, 170].

Reinforcement Type

Due to the differences in characteristic of reinforcements (melting point, hardness, EM, CTE and etc.), the mechanical properties of AM processed MMCs are strongly dependent on the type of reinforcement. Using three different ceramic particles of $Al_2O_3$, SiC and TiC as reinforcement to Inconel 625 matrix, Cooper et al. [76] indicated that while the addition of $Al_2O_3$ did not have a notable influence on hardness, major ~130% and 32% improvements in the mean hardness were obtained for the SiC and TiC reinforced MMCs, respectively, compared to the non-reinforced part. In another research, Oh and Lee [171] found that while the Ti6Al4V substrate had a hardness of 320 HV in their research, different levels of hardening (520, 715 and 728 HV for TiC, SiC and hybrid SiC+TiC reinforced MMCs, respectively) were achieved for the deposited coatings containing different reinforcements.

When it comes to the high temperature applications, the reinforcement type plays a crucial role in MMC structure. Using various ceramic reinforcements (TiC, TiN and TiC+TiN) added to the Ti6Al6V matrix, Yun et al. [161] found the highest hardness of 937 HV for TiN reinforced composite and the lowest hardness for the TiC reinforced one (554 HV). The hardness of the hybrid reinforced composite was 726 HV, located between those having either TiC or TiN as reinforcement. The high temperature hardness measurements of the fabricated MMCs revealed a gradual decreasing from room temperature to 450 °C, followed by a rather rapid decrease to 900 °C. To the same feature as the room temperature hardness, the TiN and TiC reinforced MMCs had the highest and lowest hardness values, respectively. Even at the temperature as high as 450 °C, the TiN reinforced specimen showed a hardness 3.5 times greater than that of the substrate, introducing TiN as an appropriate reinforcement for Ti matrix at high temperature applications. In cases where there exists no intermediate phase between the reinforcement and the matrix (e.g. TiB and Ti), the reinforcements are stable and in equilibrium with the matrix, leading to the extended high temperature application of the MMC structure.

Reinforcement Size

One of the major concerns with conventionally processed particle reinforced MMCs is the limited interfacial wetting and bonding coherence between the reinforcements and the matrix, making the large-sized ceramic reinforcements prone to cracking when subjected to external loading. This consequently results in poor ductility as well as premature failure of MMCs parts [172]. Both the strength and ductility of MMCs are noticeably influenced by the size of reinforcements [173]. Generally, the MMCs containing finer reinforcements are associated with higher hardness and strength as well as improved ductility. The improvement in strength due to the reinforcement refinement can be described based on Orowan mechanism. This mechanism accounts for the interaction of dislocations with reinforcements and is active when the non-shearable reinforcements act as pinning points for dislocations. It should be noted that this mechanism is effective when the size of reinforcements is smaller than 1 μm [174]. The Orowan-Ashby equation defines the Orowan stress as [175]:

$$\Delta\sigma_{or} = \frac{0.13Gb}{\lambda} \ln\frac{d}{2b} \quad (7)$$

in which $b$ and $G$ are the Burger's vector and shear modulus of the matrix, respectively. $d$ represents the equivalent diameter of reinforcement and $\lambda$ is the inter-particle spacing.

For a given volume fraction of reinforcements, the decrease in size of reinforcements ($d$) is associated with the declined inter-particle spacing ($\lambda$). Accordingly, the pinning effect of particles for mobile dislocations becomes more pronounced, leading to higher Orowan stresses required for further deformation and consequently enhanced strength. It is worth noting that unpinning of dislocations caused by bowing leads to the formation of dislocation loops around the reinforcements. This also decreases the effective inter-particle spacing and elevates the required bypassing stress for the next dislocations encountering the reinforcements, leading to higher work hardening rates and strengthening of the MMCs structure [176, 177].

In addition to the enhanced hardness, strength and even ductility, the refining of reinforcements can significantly affect the toughness of MMCs. The decreased size of TiB precipitates combined with their improved distribution pattern along the grain boundaries of the Ti matrix (3D quasi-continuous network microstructure (3DCQN)) led to the significant increase in toughness of TMCs from 201 to 320 J/mm$^3$ for parts processed with laser powers of 125 and 200 W, respectively [50]. As schematically illustrated in Fig. 41, this finding was ascribed to: (i) the refinement of TiB precipitates, (ii) crack propagation along both the TiB-rich regions (grain boundaries) and TiB-lean areas (within grains) of the 3DQCN structure, and (iii) the crack tip blunting and deflection by the ductile matrix of the 3DQCN structure.

The effect of reinforcement refining on the improvement of mechanical properties could be even more effective when the size of reinforcements is further reduced from micrometer to nanometer scale. This can further intensify the interaction of dislocations with the reinforcements and result in more remarkable improvement in mechanical properties [178, 179].

### 7.1.2 Indirect Strengthening

As thoroughly described in Sect. 5.3, incorporation of reinforcement into the system to fabricate MMCs structures is associated with some microstructural changes in the matrix. These changes can affect the mechanical properties of the matrix and

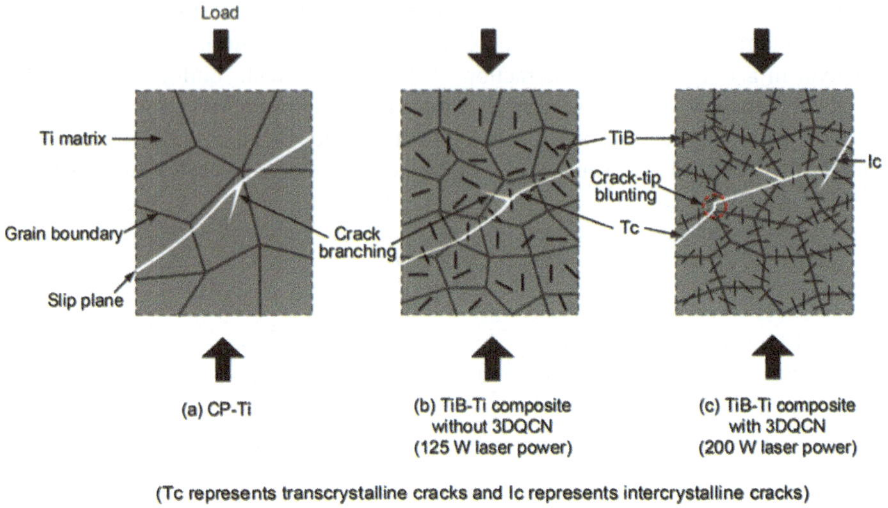


**Fig. 41** Schematic illustrations of failure modes in compression test for SLM fabricated CP-Ti and TiB reinforced TMCs [50]

**Table 6** A comparison between the hardness of AM processed Ti6Al4V alloy in non-reinforced state and in MMCs containing different types of reinforcements

| Mixed powder system | Matrix hardness in MMC | Hardness of non-reinforced Ti6Al4V | References |
|---|---|---|---|
| TiC/Ti6Al4V | 340 | 320 | [161, 171] |
| SiC/Ti6Al4V | 415 | | [171] |
| (TiC+SiC)/Ti6Al4V | 460 | | [171] |
| TiN/Ti6Al4V | 412 | | [161] |
| (TiC+TiN)/Ti6Al4V | 434 | | [161] |

consequently the MMCs parts. To gain a better insight into the effect of reinforcements on matrix strengthening, Table 6 provides some comparison between the hardness of Ti6Al4V alloy in non-reinforced state and the Ti6Al4V matrix of AM processed TMCs reinforced with various types of reinforcements. As being observed, the Ti6Al4V matrix in TMCs shows considerably higher hardness values than that of the non-reinforced Ti6Al4V alloy. However, the amount of improvement in hardness varies depending on the employed reinforcement.

The mechanisms involved in matrix hardening of AM fabricated MMCs are known as indirect strengthening mechanisms and include:

#### Grain Refinement of the Matrix

As discussed in Sect. 5.3.1, the reinforcements incorporated into the matrix of AM processed MMCs refine the microstructure by acting as preferential grain nucleation sites as well as grain growth inhibitors. Because of the significant role grain boundaries play in hindering the movement of dislocations, the increased volume of grain boundaries caused by grain refinement enhances the microstructure resistance to plastic deformation during loading and consequently improves the strength of material [192, 194]. Hall-Petch effect is a well-known strengthening mechanism which considers the effect of grain size on the strength of material. Based on this mechanism, the effect of reinforcement-induced grain refinement on the enhancement of strength can be expressed as [180]:

$$\Delta\sigma_{HP} = k_{HP}\left(\frac{1}{\sqrt{d_2}} - \frac{1}{\sqrt{d_1}}\right) \tag{8}$$

in which $k_{HP}$ is the Hall-Petch constant for the matrix while $d_2$ and $d_1$ represent the average grain sizes of the matrix in MMC structure and non-reinforced specimen, respectively. As discussed previously, incorporation of reinforcements into the system led to microstructural refinement of the matrix than that of the non-reinforced state. This refining was more severe for MMCs containing higher volume fractions and/or finer reinforcements. The relationship among matrix grain size, reinforcement size and reinforcement content has been demonstrated by the Zener equation as [181]:

$$d_m = \frac{4\alpha d_p}{3v_p} \tag{9}$$

where $d_m$ represents the matrix grain size, $\alpha$ is a proportionality constant, $d_p$ signifies the reinforcement size, and $v_p$ is the reinforcement volume fraction.

Increased Density of Dislocations in the Matrix

Based on discussions in Sect. 5.3.5, noticeable density of dislocations can be generated in the matrix of AM processed MMCs. These dislocations are believed to have various resources including multiple reheating cycles induced in parts due to the nature of AM process as well as the difference between the CTE and EM of reinforcements and the matrix. While the generation of dislocation by repeated reheating cycle is common in all AM processes, those induced by CTE and EM differences are limited to the composite structures. In order to accommodate these mismatches, GNDs are formed adjacent to the reinforcement/matrix interface (in the matrix). This phenomenon increases the work hardening rate and strengthens the MMCs. The increase in the yield strength of MMCs caused by both the CTE and EM mismatches can be expressed as [182]:

$$\Delta\sigma_{dis} = \sqrt{(\Delta\sigma_{CTE})^2 + (\Delta\sigma_{EM})^2} \tag{10}$$

where $\Delta\sigma_{CTE}$ and $\Delta\sigma_{EM}$ represent the stress increment caused by CTE and EM mismatch between the reinforcement and the matrix, respectively. The effects of CTE and EM mismatches on the strength of MMCs can be expressed by Taylor relationship as Eqs. 11 and 12, respectively [167, 183]:

$$\sigma_{CTE} = M\beta Gb\sqrt{\rho_{CTE}} \tag{11}$$

$$\sigma_{EM} = \sqrt{3}\alpha Gb\sqrt{\rho_{EM}} \tag{12}$$

in which $M$ is the Taylor factor, $\beta$ and $\alpha$ are constants in the order of 1.25 and 0.5 respectively. $\rho_{CTE}$ and $\rho_{EM}$ represent the density of dislocations induced by CTE and EM mismatches, respectively, and are defined as [178, 184]:

$$\rho_{CTE} = \frac{A\Delta\alpha\Delta T V_r}{bd_p(1 - V_r)} \tag{13}$$

$$\rho_{EM} = \frac{6V_r}{\pi d^3}\varepsilon \tag{14}$$

where $A$ is a geometric constant varying between 10–12 based on the geometry of reinforcement, $\Delta\alpha$ represents the difference in CTE, $\Delta T$ is the temperature change from the processing to testing temperature, and $\varepsilon$ signifies the plastic strain of the matrix.

It is noteworthy that as the deformation proceeds, the increase in dislocation density results in the reduced space between them which consequently elevates the load required for further deformation.

#### The Matrix Strengthening Caused by Microstructural Modifications

Referring to different microstructural variations discussed in Sect. 5.3.3, microstructural modifications of the matrix induced by the presence of reinforcements include:

(i) The transition in the solidification mode
(ii) The change in the morphology of phases(s)
(iii) The change in the volume fraction of phase(s)
(iv) The phase transformation

Each of these microstructural modifications can lead to the improvement in mechanical properties of fabricated MMCs. For instance, phase transformation of the iron matrix from ferrite in pure state to pearlite and/or martensite in SiC reinforced iron matrix MMCs has been reported as one of the mechanisms contributing to the enhanced strength of SiC reinforced iron matrix MMCs compared to the non-reinforced parts [128, 129]. Moreover, the formation of martensitic $\alpha'$ phase in TMCs caused by dissolution of reinforcements in the matrix well assists to the improved mechanical properties of AM processed parts [161, 171]. While the

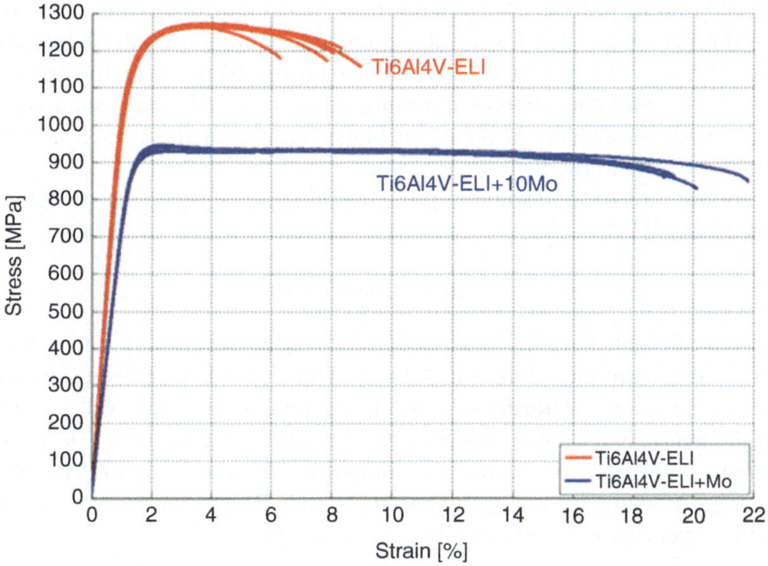


**Fig. 42** Stress-strain curves of SLM produced Ti6Al4V and Ti6Al4V + 10Mo parts [63]

general concept behind the fabrication of most MMCs is the improvement in hardness and strength, the outstanding properties of AM technology combined with the microstructural engineering of the matrix has recently led to the development of MMCs with excellent ductility [62, 63].

Figure 42 illustrates the stress-strain curves of SLM processed non-reinforced Ti6Al4V alloy and 10 wt.% Mo/Ti6Al4V powder mixture. In case of MMC containing Mo as metallic reinforcement, the partial dissolution of Mo particles into the matrix changed the fully $\alpha'$ martensitic microstructure of the matrix to a $\beta$ phase one. Because of the lower strength and higher ductility of $\beta$ phase compared to $\alpha'$, fabricated composites showed lower Young's modulus, lower yield and ultimate tensile strengths, and noticeably higher elongation compared with the non-reinforced Ti6Al4V parts made by the same processing parameters.

## Solid Solution Strengthening of the Matrix

Solid solutions can be formed in both equilibrium and non-equilibrium states. For any rational concentration of solutes, the spacing between the solute atoms will be noticeably less than even the grain sizes in nanoscale. The non-equilibrium alloys fabricated by non-equilibrium processing routes can contain solid solutions with much larger size misfit than the equilibrium alloys. AM processing techniques, due to their extremely rapid cooling rates as well as non-equilibrium nature, can significantly extend the solution limit of alloying elements into the matrix and consequently favor the solid solution strengthening mechanism. It is worth mentioning that the ball milling process used as a promising technique to mix the powders prior to the AM processing of MMCs can also extend the equilibrium solid solubility limit of solute elements into the matrix [131]. In fact, the mechanical alloying provides the synthesis of metastable (non-equilibrium) supersaturated solid solutions from blended elemental powders in binary or higher order systems as well as ceramic-metallic powder mixtures.

The contribution of solid solution strengthening mechanism to the strength can be expressed as [175]:

$$\Delta\sigma_{ss} = \beta G \delta^{p} c^{q} \tag{15}$$

where $\beta$ is a constant, $\delta = |(r_{matrix} - r_{solute})/r_{matrix}|$ is the misfit strain, and $c$ represents the atomic fraction of the solute. $p$ and $q$ values are dependent on solute spacing and the dislocation-solute interactions [185]. The solute atoms entrapped in the lattice sites of the matrix make some distortions to the lattice structure [190, 191]. Accordingly, higher amounts of stress will be required for the movement of dislocations in the distorted structure, leading to enhanced strength. The larger the misfit strain, the higher the supersaturated solid solution strengthening.

## 7.2 *Weakening Mechanisms*

While several strengthening mechanisms can contribute to the enhancement of mechanical properties in AM processed MMCs, there are some other factors inversely affecting these properties. The weakening mechanisms can lead to the fabrication of MMCs having even inferior mechanical properties than the non-reinforced state [20, 90]. Moreover, these mechanisms may intensify the degrees of fluctuations in the mechanical properties of fabricated MMCs [123]. The most important weakening mechanisms which may be active in AM processed MMCs include:

### 7.2.1 Decreased Densification Level

As discussed in Sect. 6, the decreased densification level in AM processed MMCs mostly involve the formation of porosities as well as micro-cracks. When existing in the structure, these defects may act as stress concentrating regions, favoring the initiation and propagation of cracks during loading. Moreover, their growth or coalescence by mechanical or thermal stresses may result in catastrophic failure of engineering MMCs components [186].

### 7.2.2 Microstructure Coarsening

The microstructure coarsening in the AM processed MMCs may include the coarsening of matrix structure and/or the reinforcements. The major reason for such a phenomenon is the excessive energy density applied to the system during processing [19, 25, 127]. The coarsening of reinforcements may also lead to the microstructural inhomogeneity in MMCs and deteriorate the mechanical properties [29].

### 7.2.3 Microstructural Inhomogeneity

The inhomogeneity in distribution state of reinforcements in AM processed MMCs is in fact due to the application of inappropriate processing parameters. Compared with the MMCs possessing homogeneous distribution pattern of reinforcements, those containing non-homogeneous distribution states are generally associated with decreased strength and ductility. In fact, when having non-homogenous distribution of reinforcements, some regions of the matrix are left non-reinforced, making them weaker than the reinforced areas. Moreover, if the MMC microstructures with inhomogeneous distribution pattern of reinforcements contain clusters or agglomerates of hard and brittle reinforcements, the parts may be much more prone to cracking when subjected to external loading.

## 8 Wear Behavior

Wear is defined as the progressive loss of material caused by the relative motion of contacting substance and the surface. Wear is regarded as a complex phenomenon during which the real contact area between contacting surfaces is consistently smaller than the apparent contact area, leading to relatively high localized load transfer through the contact points. Abrasive, erosive, adhesive, surface fatigue, cavitation and tribo-chemical are the classifications presented for wear. Pin-on-Disc is one of the most common tests used for measuring the sliding friction and wear characteristics of materials. Coefficient of friction (COF) and wear rate obtained from the wear test are employed to study the wear behavior of materials. The lower the COF and wear rate, the higher the wear resistance of the material. Microstructural observation of the wear surface is a helpful tool for characterizing the wear properties.

Besides the extrinsic factors (e.g. applied load, sliding velocity and distance, surface finish, environment and temperature), the intrinsic properties of MMCs (e.g. characteristics of reinforcements and the matrix as well as the reinforcement/matrix interface) play significant roles in wear performance of MMCs. The following discusses the effects of reinforcement feature and applied energy density on wear behavior of AM processed MMCs.

### *8.1 Effect of Size and Volume Fraction of Reinforcements*

For a given MMC system, the volume fraction of reinforcements is known as the parameter playing the most important role in wear resistance [187, 188]. However, when investigating the change in wear resistance as a function of reinforcement volume fraction, the size and morphology of reinforcements need to be considered. The effect of size and volume fraction of reinforcements on wear resistance of MMCs could be discussed in terms of inter-particle spacing (λ) which is defined as [112]:

$$\lambda \propto \frac{d}{\sqrt{f}} \tag{16}$$

where $d$ represents the size, and $f$ signifies the volume fraction of reinforcements. The decrease in size and increase in volume fraction of reinforcements is believed to improve the wear resistance of composites. In fact, the enhanced content of reinforcements in MMCs provides the matrix with higher amounts of reinforcements protecting the plastic flow of ductile matrix. This enhances the hardness and strength of composite. In addition, it may lead to the formation of larger areas of stain-hardened tribolayer on wear surface during sliding [123]. Moreover, in some cases where the reinforcements have lubricating behavior (e.g. BN ceramic reinforcements in Inconel 718 matrix composites [90]), MMCs containing higher amounts of them may show improved wear behavior. It should be noted that the improving trend in wear resistance achieved with the increased content of reinforcements is valid as long as the spalling of reinforcements is avoided, and they are remained in the matrix during wear sliding [112]. This clearly reveals the supporting ability of the matrix to the reinforcements as an importance point in wear resistance of MMCs.

In addition to the improvement in wear resistance, the increase in reinforcement volume fraction may change the wear mechanism. For instance, as shown in Fig. 43, the increase in TiC content from 2.5 to 15 vol.% in SLM processed TiC reinforced 316L stainless steel matrix nanocomposites changed the wear mechanism from abrasive to adhesive [121]. In another study on DLF processing of TiC reinforced Ni matrix composites, the increase in volume fraction of in-situ synthesized TiC

**Fig. 43** SEM micrographs of the worn surface for SLM processed TiC reinforced 316L stainless steel matrix nano-composites having: (**a**) 2.5 and (**b**) 15 vol.% TiC [121]

precipitates from 20 to 60 vol.% changed the morphology of worn surface from gross grooves to fine scratches [77].

The improved wear resistance of AM processed MMCs obtained by the decrease in size of reinforcements could be attributed to the following reasons:

(i) The increased densification level and improved microstructural homogeneity which both decrease the preferential locations for crack nucleation [18]
(ii) Compared with the coarser reinforcements, the stronger interfacial bonding of finer ones with the matrix lowers their protrusion and separation when subjected to the shear stress during sliding
(iii) The enhanced hardness and strength of MMCs
(iv) The restricting actions that fine reinforcements play in plastic deformation and flow of surface material during sliding. Compared with the rough surface containing deep and irregular grooves showing severe surface plowing during sliding for MMCs with relatively coarse reinforcements, those with finer ones may lead to narrower and shallower grooves
(v) The elevated grain refinement of the matrix which leads to much more severe plastic deformation of the wear surface and paves the way for the formation of strain-hardened tribolayer [18].

## *8.2 Effect of Applied Energy Density*

The amount of energy density subjected to the system during AM processing has a significant influence on the wear resistance of fabricated MMCs parts. Several parameters have been reported in the literature for the weak wear resistance of AM processed MMCs fabricated by relatively low energy densities. The appropriate selection of process parameters can provide the powder system with sufficient energy density [195]. This consequently results in the production of MMCs with sound wear resistance due to the following phenomena:

(i) The improvement in densification level [25, 80, 83]
(ii) The formation of microstructures with much uniform and homogenous distribution pattern of reinforcements [15, 25, 80, 83]
(iii) The enhanced microhardness [25, 78, 83]
(iv) The improvement of reinforcement/matrix interfacial bonding or the formation of more effective graded interfacial reaction layer between the reinforcements and the matrix in some systems. These layers can effectively transfer the load from softer matrix to the harder reinforcements and provide the MMCs parts with less wear.

When employing excessive energy density, the coarsening of microstructure (both the reinforcement ad the matrix) as well as the microscopic shrinkage porosities and micro-cracks formed especially at the reinforcement/matrix interface may weaken the wear performance of MMCs [14, 25, 82].

## References

1. A.C.F.o.A.M. Technologies, A.C.F.o.A.M.T.S.F.o. Terminology, Standard Terminology for Additive Manufacturing Technologies, ASTM International 2012.
2. D. Gu, W. Meiners, K. Wissenbach, R. Poprawe, Laser additive manufacturing of metallic components: materials, processes and mechanisms, International materials reviews 57 (3) (2012) 133–164.
3. S.M. Thompson, L. Bian, N. Shamsaei, A. Yadollahi, An overview of Direct Laser Deposition for additive manufacturing; Part I: Transport phenomena, modeling and diagnostics, Additive Manufacturing 8 (2015) 36–62.
4. F. Wang, J. Mei, X. Wu, Compositionally graded Ti6Al4V+ TiC made by direct laser fabrication using powder and wire, Materials & design 28(7) (2007) 2040–2046.
5. J.M. Wilson, Y.C. Shin, Microstructure and wear properties of laser-deposited functionally graded Inconel 690 reinforced with TiC, Surface and Coatings Technology 207 (2012) 517–522.
6. S.-w. Lim, T. Imai, Y. Nishida, T. Choh, High strain rate superplasticity of TiC particulate reinforced magnesium alloy composite by vortex method, Scripta metallurgica et materialia 32 (11) (1995) 1713–1717.
7. M.O. Shabani, A. Mazahery, Computational modeling of cast aluminum 2024 alloy matrix composites: Adapting the classical algorithms for optimal results in finding multiple optima, Powder technology 249 (2013) 77–81.
8. T. Rajmohan, K. Palanikumar, S. Arumugam, Synthesis and characterization of sintered hybrid aluminium matrix composites reinforced with nanocopper oxide particles and microsilicon carbide particles, Composites Part B: Engineering 59 (2014) 43–49.
9. V. Umasankar, M.A. Xavior, S. Karthikeyan, Experimental evaluation of the influence of processing parameters on the mechanical properties of SiC particle reinforced AA6061 aluminium alloy matrix composite by powder processing, Journal of Alloys and Compounds 582 (2014) 380–386.
10. S. Ozden, R. Ekici, F. Nair, Investigation of impact behaviour of aluminium based SiC particle reinforced metal–matrix composites, Composites Part A: Applied Science and Manufacturing 38(2) (2007) 484–494.
11. D. Buchbinder, H. Schleifenbaum, S. Heidrich, W. Meiners, J. Bültmann, High power selective laser melting (HP SLM) of aluminum parts, Physics Procedia 12 (2011) 271–278.
12. E. Louvis, P. Fox, C.J. Sutcliffe, Selective laser melting of aluminium components, Journal of Materials Processing Technology 211(2) (2011) 275–284.
13. D. Gu, H. Wang, D. Dai, P. Yuan, W. Meiners, R. Poprawe, Rapid fabrication of Al-based bulk-form nanocomposites with novel reinforcement and enhanced performance by selective laser melting, Scripta Materialia 96 (2015) 25–28.
14. D. Gu, H. Wang, D. Dai, Laser additive manufacturing of novel aluminum based nanocomposite parts: tailored forming of multiple materials, Journal of Manufacturing Science and Engineering 138(2) (2016) 021004.
15. D. Gu, H. Wang, D. Dai, F. Chang, W. Meiners, Y.-C. Hagedorn, K. Wissenbach, I. Kelbassa, R. Poprawe, Densification behavior, microstructure evolution, and wear property of TiC nanoparticle reinforced AlSi10Mg bulk-form nanocomposites prepared by selective laser melting, Journal of Laser Applications 27(S1) (2015) S17003.
16. P. Yuan, D. Gu, D. Dai, Particulate migration behavior and its mechanism during selective laser melting of TiC reinforced Al matrix nanocomposites, Materials & Design 82 (2015) 46–55.
17. D. Gu, P. Yuan, Thermal evolution behavior and fluid dynamics during laser additive manufacturing of Al-based nanocomposites: Underlying role of reinforcement weight fraction, Journal of Applied Physics 118(23) (2015) 233109.

18. F. Chang, D. Gu, D. Dai, P. Yuan, Selective laser melting of in-situ Al4SiC4+ SiC hybrid reinforced Al matrix composites: Influence of starting SiC particle size, Surface and Coatings Technology 272 (2015) 15–24.
19. D. Gu, F. Chang, D. Dai, Selective laser melting additive manufacturing of novel aluminum based composites with multiple reinforcing phases, Journal of Manufacturing Science and Engineering 137(2) (2015) 021010.
20. D. Manfredi, F. Calignano, M. Krishnan, R. Canali, E.P. Ambrosio, S. Biamino, D. Ugues, M. Pavese, P. Fino, Additive manufacturing of Al alloys and aluminium matrix composites (AMCs), Light metal alloys applications, InTech 2014.
21. A. Simchi, D. Godlinski, Effect of SiC particles on the laser sintering of Al–7Si–0.3 Mg alloy, Scripta Materialia 59(2) (2008) 199–202.
22. A. Simchi, D. Godlinski, Densification and microstructural evolution during laser sintering of A356/SiC composite powders, Journal of materials science 46(5) (2011) 1446–1454.
23. S.K. Ghosh, P. Saha, Crack and wear behavior of SiC particulate reinforced aluminium based metal matrix composite fabricated by direct metal laser sintering process, Materials & Design 32(1) (2011) 139–145.
24. S.K. Ghosh, P. Saha, S. Kishore, Influence of size and volume fraction of SiC particulates on properties of ex situ reinforced Al–4.5 Cu–3Mg metal matrix composite prepared by direct metal laser sintering process, Materials Science and Engineering: A 527(18) (2010) 4694–4701.
25. J. Jue, D. Gu, K. Chang, D. Dai, Microstructure evolution and mechanical properties of Al-Al 2 O 3 composites fabricated by selective laser melting, Powder Technology 310 (2017) 80–91.
26. L. Wang, J. Jue, M. Xia, L. Guo, B. Yan, D. Gu, Effect of the Thermodynamic Behavior of Selective Laser Melting on the Formation of In situ Oxide Dispersion-Strengthened Aluminum-Based Composites, Metals 6(11) (2016) 286.
27. D. Dai, D. Gu, Influence of thermodynamics within molten pool on migration and distribution state of reinforcement during selective laser melting of AlN/AlSi10Mg composites, International Journal of Machine Tools and Manufacture 100 (2016) 14–24.
28. P. Ma, Y. Jia, K.G. Prashanth, S. Scudino, Z. Yu, J. Eckert, Microstructure and phase formation in Al–20Si–5Fe–3Cu–1Mg synthesized by selective laser melting, Journal of Alloys and Compounds 657 (2016) 430–435.
29. K.G. Prashanth, S. Scudino, A.K. Chaubey, L. Löber, P. Wang, H. Attar, F.P. Schimansky, F. Pyczak, J. Eckert, Processing of Al–12Si–TNM composites by selective laser melting and evaluation of compressive and wear properties, Journal of Materials Research 31(1) (2016) 55–65.
30. K. Prashanth, H.S. Shahabi, H. Attar, V. Srivastava, N. Ellendt, V. Uhlenwinkel, J. Eckert, S. Scudino, Production of high strength Al85Nd8Ni5Co2 alloy by selective laser melting, Additive Manufacturing 6 (2015) 1–5.
31. S. Dadbakhsh, L. Hao, Effect of Al alloys on selective laser melting behaviour and microstructure of in situ formed particle reinforced composites, Journal of alloys and compounds 541 (2012) 328–334.
32. L. Xiao, W. Lu, J. Qin, Y. Chen, D. Zhang, M. Wang, F. Zhu, B. Ji, Creep behaviors and stress regions of hybrid reinforced high temperature titanium matrix composite, Composites Science and Technology 69(11) (2009) 1925–1931.
33. A.A. da Silva, J.F. dos Santos, T.R. Strohaecker, An investigation of the fracture behaviour of diffusion-bonded Ti6Al4V/TiC/10p, Composites science and technology 66(13) (2006) 2063–2068.
34. J. Vreeling, V. Ocelik, J.T.M. De Hosson, Ti–6Al–4V strengthened by laser melt injection of WC p particles, Acta Materialia 50(19) (2002) 4913–4924.
35. Y. Pei, V. Ocelik, J.T.M. De Hosson, SiC p/Ti6Al4V functionally graded materials produced by laser melt injection, Acta Materialia 50(8) (2002) 2035–2051.
36. J. Candel, V. Amigó, J. Ramos, D. Busquets, Sliding wear resistance of TiC p reinforced titanium composite coating produced by laser cladding, Surface and Coatings Technology 204 (20) (2010) 3161–3166.

37. F. Wang, J. Mei, X. Wu, Direct laser fabrication of Ti6Al4V/TiB, Journal of materials processing technology 195(1) (2008) 321–326.
38. J. DuttaMajumdar, L. Li, Development of titanium boride (TiB) dispersed titanium (Ti) matrix composite by direct laser cladding, Materials Letters 64(9) (2010) 1010–1012.
39. W. Pang, H. Man, T. Yue, Laser surface coating of Mo–WC metal matrix composite on Ti6Al4V alloy, Materials Science and Engineering: A 390(1) (2005) 144–153.
40. T. Yamamoto, A. Otsuki, K. Ishihara, P. Shingu, Synthesis of near net shape high density TiB/Ti composite, Materials Science and Engineering: A 239 (1997) 647–651.
41. D. Gu, W. Meiners, Y.-C. Hagedorn, K. Wissenbach, R. Poprawe, Bulk-form TiCx/Ti nanocomposites with controlled nanostructure prepared by a new method: selective laser melting, Journal of Physics D: Applied Physics 43(29) (2010) 295402.
42. D. Gu, H. Wang, G. Zhang, Selective laser melting additive manufacturing of Ti-based nanocomposites: the role of nanopowder, Metallurgical and Materials Transactions A 45 (1) (2014) 464–476.
43. D. Gu, Y.-C. Hagedorn, W. Meiners, K. Wissenbach, R. Poprawe, Nanocrystalline TiC reinforced Ti matrix bulk-form nanocomposites by Selective Laser Melting (SLM): Densification, growth mechanism and wear behavior, Composites Science and Technology 71 (13) (2011) 1612–1620.
44. D. Gu, G. Meng, C. Li, W. Meiners, R. Poprawe, Selective laser melting of TiC/Ti bulk nanocomposites: Influence of nanoscale reinforcement, Scripta Materialia 67(2) (2012) 185–188.
45. Y. Zhang, Z. Wei, L. Shi, M. Xi, Characterization of laser powder deposited Ti–TiC composites and functional gradient materials, journal of materials processing technology 206 (1) (2008) 438–444.
46. D. Gu, Z. Wang, Y. Shen, Q. Li, Y. Li, In-situ TiC particle reinforced Ti–Al matrix composites: powder preparation by mechanical alloying and selective laser melting behavior, Applied Surface Science 255(22) (2009) 9230–9240.
47. D. Gu, D. Dai, G. Zhang, H. Wang, Growth mechanisms of in situ TiC in laser melted Ti–Si–C ternary system, Applied Physics Letters 101(17) (2012) 171603.
48. H. Attar, L. Löber, A. Funk, M. Calin, L. Zhang, K. Prashanth, S. Scudino, Y. Zhang, J. Eckert, Mechanical behavior of porous commercially pure Ti and Ti–TiB composite materials manufactured by selective laser melting, Materials Science and Engineering: A 625 (2015) 350–356.
49. H. Attar, M. Bönisch, M. Calin, L.-C. Zhang, S. Scudino, J. Eckert, Selective laser melting of in situ titanium–titanium boride composites: processing, microstructure and mechanical properties, Acta Materialia 76 (2014) 13–22.
50. Y. Hu, W. Cong, X. Wang, Y. Li, F. Ning, H. Wang, Laser deposition-additive manufacturing of TiB-Ti composites with novel three-dimensional quasi-continuous network microstructure: Effects on strengthening and toughening, Composites Part B: Engineering 133 (2018) 91–100.
51. R. Banerjee, P. Collins, A. Genc, H. Fraser, Direct laser deposition of in situ Ti–6Al–4V–TiB composites, Materials Science and Engineering: A 358(1) (2003) 343–349.
52. A. Genç, R. Banerjee, D. Hill, H. Fraser, Structure of TiB precipitates in laser deposited in situ, Ti-6Al-4V–TiB composites, Materials Letters 60(7) (2006) 859–863.
53. S. Nag, S. Samuel, A. Puthucode, R. Banerjee, Characterization of novel borides in Ti–Nb–Zr–Ta+ 2B metal-matrix composites, Materials characterization 60(2) (2009) 106–113.
54. Y. Zhang, J. Sun, R. Vilar, Characterization of (TiB+ TiC)/TC4 in situ titanium matrix composites prepared by laser direct deposition, Journal of Materials Processing Technology 211(4) (2011) 597–601.
55. M. Das, V.K. Balla, D. Basu, S. Bose, A. Bandyopadhyay, Laser processing of SiC-particle-reinforced coating on titanium, Scripta Materialia 63(4) (2010) 438–441.
56. M. Das, S. Bysakh, D. Basu, T.S. Kumar, V.K. Balla, S. Bose, A. Bandyopadhyay, Microstructure, mechanical and wear properties of laser processed SiC particle reinforced coatings on titanium, Surface and Coatings Technology 205(19) (2011) 4366–4373.

57. D. Gu, C. Hong, G. Meng, Densification, microstructure, and wear property of in situ titanium nitride-reinforced titanium silicide matrix composites prepared by a novel selective laser melting process, Metallurgical and Materials Transactions A 43(2) (2012) 697–708.
58. H. Attar, K.G. Prashanth, L.-C. Zhang, M. Calin, I.V. Okulov, S. Scudino, C. Yang, J. Eckert, Effect of powder particle shape on the properties of in situ Ti–TiB composite materials produced by selective laser melting, Journal of Materials Science & Technology 31 (10) (2015) 1001–1005.
59. V.K. Balla, A. Bhat, S. Bose, A. Bandyopadhyay, Laser processed TiN reinforced Ti6Al4V composite coatings, Journal of the mechanical behavior of biomedical materials 6 (2012) 9–20.
60. M. Das, K. Bhattacharya, S.A. Dittrick, C. Mandal, V.K. Balla, T.S. Kumar, A. Bandyopadhyay, I. Manna, In situ synthesized TiB–TiN reinforced Ti6Al4V alloy composite coatings: microstructure, tribological and in-vitro biocompatibility, Journal of the mechanical behavior of biomedical materials 29 (2014) 259–271.
61. M. Das, V.K. Balla, D. Basu, I. Manna, T.S. Kumar, A. Bandyopadhyay, Laser processing of in situ synthesized TiB–TiN-reinforced Ti6Al4V alloy coatings, Scripta Materialia 66 (8) (2012) 578–581.
62. S.L. Sing, W.Y. Yeong, F.E. Wiria, Selective laser melting of titanium alloy with 50 wt% tantalum: Microstructure and mechanical properties, Journal of Alloys and Compounds 660 (2016) 461–470.
63. B. Vrancken, L. Thijs, J.-P. Kruth, J. Van Humbeeck, Microstructure and mechanical properties of a novel β titanium metallic composite by selective laser melting, Acta Materialia 68 (2014) 150–158.
64. H.N. Moosavy, M.-R. Aboutalebi, S.H. Seyedein, M. Goodarzi, M. Khodabakhshi, C. Mapelli, S. Barella, Modern fiber laser beam welding of the newly-designed precipitation-strengthened nickel-base superalloys, Optics & Laser Technology 57 (2014) 12–20.
65. T.-Y. Kuo, Effects of pulsed and continuous Nd–YAG laser beam waves on welding of Inconel alloy, Science and Technology of Welding and Joining 10(5) (2005) 557–565.
66. X. Xing, X. Di, B. Wang, The effect of post-weld heat treatment temperature on the microstructure of Inconel 625 deposited metal, Journal of Alloys and Compounds 593 (2014) 110–116.
67. Y. Chen, K. Zhang, J. Huang, S.R.E. Hosseini, Z. Li, Characterization of heat affected zone liquation cracking in laser additive manufacturing of Inconel 718, Materials & Design 90 (2016) 586–594.
68. C.-M. Kuo, Y.-T. Yang, H.-Y. Bor, C.-N. Wei, C.-C. Tai, Aging effects on the microstructure and creep behavior of Inconel 718 superalloy, Materials Science and Engineering: A 510 (2009) 289–294.
69. D. Thakur, B. Ramamoorthy, L. Vijayaraghavan, Study on the machinability characteristics of superalloy Inconel 718 during high speed turning, Materials & Design 30(5) (2009) 1718–1725.
70. R.C. Reed, The superalloys: fundamentals and applications, Cambridge university press 2008.
71. M. Henderson, D. Arrell, R. Larsson, M. Heobel, G. Marchant, Nickel based superalloy welding practices for industrial gas turbine applications, Science and Technology of Welding and Joining 9(1) (2004) 13–21.
72. J.C. Lippold, S.D. Kiser, J.N. DuPont, Welding metallurgy and weldability of nickel-base alloys, John Wiley & Sons 2011.
73. K.A. Mumtaz, P. Erasenthiran, N. Hopkinson, High density selective laser melting of Waspaloy®, Journal of materials processing technology 195(1) (2008) 77–87.
74. T. Lei, J. Ouyang, Y. Pei, Y. Zhou, Microstructure and wear resistance of laser clad TiC particle reinforced coating, Materials science and technology 11(5) (1995) 520–525.
75. B. Mei, R. Yuan, X. Duan, Investigation of Ni 3 Al-matrix composites strengthened by TiC, Journal of materials research 8(11) (1993) 2830–2834.

76. D.E. Cooper, N. Blundell, S. Maggs, G.J. Gibbons, Additive layer manufacture of Inconel 625 metal matrix composites, reinforcement material evaluation, Journal of Materials Processing Technology 213(12) (2013) 2191–2200.
77. Y. Li, P. Bai, Y. Wang, J. Hu, Z. Guo, Effect of TiC content on Ni/TiC composites by direct laser fabrication, Materials & Design 30(4) (2009) 1409–1412.
78. Q. Jia, D. Gu, Selective laser melting additive manufacturing of TiC/Inconel 718 bulk-form nanocomposites: Densification, microstructure, and performance, Journal of Materials Research 29(17) (2014) 1960–1969.
79. Q. Shi, D. Gu, M. Xia, S. Cao, T. Rong, Effects of laser processing parameters on thermal behavior and melting/solidification mechanism during selective laser melting of TiC/Inconel 718 composites, Optics & Laser Technology 84 (2016) 9–22.
80. D. Gu, C. Hong, Q. Jia, D. Dai, A. Gasser, A. Weisheit, I. Kelbassa, M. Zhong, R. Poprawe, Combined strengthening of multi-phase and graded interface in laser additive manufactured TiC/Inconel 718 composites, Journal of Physics D: Applied Physics 47(4) (2013) 045309.
81. C. Hong, D. Gu, D. Dai, A. Gasser, A. Weisheit, I. Kelbassa, M. Zhong, R. Poprawe, Laser metal deposition of TiC/Inconel 718 composites with tailored interfacial microstructures, Optics & Laser Technology 54 (2013) 98–109.
82. C. Hong, D. Gu, D. Dai, M. Alkhayat, W. Urban, P. Yuan, S. Cao, A. Gasser, A. Weisheit, I. Kelbassa, Laser additive manufacturing of ultrafine TiC particle reinforced Inconel 625 based composite parts: tailored microstructures and enhanced performance, Materials Science and Engineering: A 635 (2015) 118–128.
83. S. Cao, D. Gu, Laser metal deposition additive manufacturing of TiC/Inconel 625 nanocomposites: Relation of densification, microstructures and performance, Journal of Materials Research 30(23) (2015) 3616–3628.
84. S. Cao, D. Gu, Q. Shi, Relation of microstructure, microhardness and underlying thermodynamics in molten pools of laser melting deposition processed TiC/Inconel 625 composites, Journal of Alloys and Compounds 692 (2017) 758–769.
85. C. Hong, D. Gu, D. Dai, S. Cao, M. Alkhayat, Q. Jia, A. Gasser, A. Weisheit, I. Kelbassa, M. Zhong, High-temperature oxidation performance and its mechanism of TiC/Inconel 625 composites prepared by laser metal deposition additive manufacturing, Journal of Laser Applications 27(S1) (2015) S17005.
86. B. Zheng, T. Topping, J.E. Smugeresky, Y. Zhou, A. Biswas, D. Baker, E.J. Lavernia, The influence of Ni-coated TiC on laser-deposited IN625 metal matrix composites, Metallurgical and Materials Transactions A 41(3) (2010) 568–573.
87. S. Gopagoni, J. Hwang, A. Singh, B. Mensah, N. Bunce, J. Tiley, T. Scharf, R. Banerjee, Microstructural evolution in laser deposited nickel–titanium–carbon in situ metal matrix composites, Journal of Alloys and Compounds 509(4) (2011) 1255–1260.
88. J. Hwang, A. Neira, T. Scharf, J. Tiley, R. Banerjee, Laser-deposited carbon nanotube reinforced nickel matrix composites, Scripta Materialia 59(5) (2008) 487–490.
89. T. Rong, D. Gu, Formation of novel graded interface and its function on mechanical properties of WC 1− x reinforced Inconel 718 composites processed by selective laser melting, Journal of Alloys and Compounds 680 (2016) 333–342.
90. S.H. Kim, G.-H. Shin, B.-K. Kim, K.T. Kim, D.-Y. Yang, C. Aranas, J.-P. Choi, J.-H. Yu, Thermo-mechanical improvement of Inconel 718 using ex situ boron nitride-reinforced composites processed by laser powder bed fusion, Scientific reports 7(1) (2017) 14359.
91. D. Gu, Y. Shen, Influence of phosphorus element on direct laser sintering of multicomponent Cu-based metal powder, Metallurgical and Materials Transactions B 37(6) (2006) 967–977.
92. J. Zhang, Y. Zhou, Microstructure, mechanical, and electrical properties of Cu–Ti 3 AlC 2 and in situ Cu–TiC x composites, Journal of Materials Research 23(4) (2008) 924–932.
93. P. Deshpande, J. Li, R. Lin, Infrared processed Cu composites reinforced with WC particles, Materials Science and Engineering: A 429(1) (2006) 58–65.
94. P. Deshpande, R. Lin, Wear resistance of WC particle reinforced copper matrix composites and the effect of porosity, Materials Science and Engineering: A 418(1) (2006) 137–145.

95. G. Gusmano, A. Bianco, R. Polini, P. Magistris, G. Marcheselli, Chemical synthesis and sintering behaviour of highly dispersed W/Cu composite powders, Journal of materials science 36(4) (2001) 901–907.
96. H. Xing, X. Cao, W. Hu, L. Zhao, J. Zhang, Interfacial reactions in 3D-SiC network reinforced Cu-matrix composites prepared by squeeze casting, Materials Letters 59(12) (2005) 1563–1566.
97. S. Rathod, O. Modi, B. Prasad, A. Chrysanthou, D. Vallauri, V. Deshmukh, A. Shah, Cast in situ Cu–TiC composites: synthesis by SHS route and characterization, Materials Science and Engineering: A 502(1) (2009) 91–98.
98. D. Gu, Y. Shen, Influence of Cu-liquid content on densification and microstructure of direct laser sintered submicron W–Cu/micron Cu powder mixture, Materials Science and Engineering: A 489(1) (2008) 169–177.
99. E. Breval, J. Cheng, D. Agrawal, P. Gigl, M. Dennis, R. Roy, A. Papworth, Comparison between microwave and conventional sintering of WC/Co composites, Materials Science and Engineering: A 391(1) (2005) 285–295.
100. D. Gu, Y. Shen, L. Zhao, J. Xiao, P. Wu, Y. Zhu, Effect of rare earth oxide addition on microstructures of ultra-fine WC–Co particulate reinforced Cu matrix composites prepared by direct laser sintering, Materials Science and Engineering: A 445 (2007) 316–322.
101. D. Gu, Y. Shen, J. Xiao, Influence of processing parameters on particulate dispersion in direct laser sintered WC–Co p/Cu MMCs, International Journal of Refractory Metals and Hard Materials 26(5) (2008) 411–422.
102. D. Gu, Y. Shen, Influence of reinforcement weight fraction on microstructure and properties of submicron WC–Co p/Cu bulk MMCs prepared by direct laser sintering, Journal of alloys and compounds 431(1) (2007) 112–120.
103. D. Gu, Y. Shen, Processing and microstructure of submicron WC–Co particulate reinforced Cu matrix composites prepared by direct laser sintering, Materials Science and Engineering: A 435 (2006) 54–61.
104. D. Gu, Y. Shen, The role of La2O3 in direct laser sintering of submicrometre WC–Cop/Cu MMCs, Journal of Physics D: Applied Physics 41(9) (2008) 095308.
105. D. Gu, Y. Shen, Microstructures and properties of direct laser sintered tungsten carbide (WC) particle reinforced Cu matrix composites with RE–Si–Fe addition: A comparative study, Journal of Materials Research 24(11) (2009) 3397–3406.
106. L. Lu, J. Fuh, Z. Chen, C. Leong, Y. Wong, In situ formation of TiC composite using selective laser melting, Materials Research Bulletin 35(9) (2000) 1555–1561.
107. D. Gu, Y. Shen, Effects of processing parameters on consolidation and microstructure of W–Cu components by DMLS, Journal of Alloys and Compounds 473(1) (2009) 107–115.
108. D. Gu, Y. Shen, WC–Co particulate reinforcing Cu matrix composites produced by direct laser sintering, Materials Letters 60(29) (2006) 3664–3668.
109. A. Bhat, V.K. Balla, S. Bysakh, D. Basu, S. Bose, A. Bandyopadhyay, Carbon nanotube reinforced Cu–10Sn alloy composites: Mechanical and thermal properties, Materials Science and Engineering: A 528(22) (2011) 6727–6732.
110. J.F. Shackelford, Y.-H. Han, S. Kim, S.-H. Kwon, CRC materials science and engineering handbook, CRC press 2016.
111. I. Ibrahim, F. Mohamed, E. Lavernia, Particulate reinforced metal matrix composites—a review, Journal of materials science 26(5) (1991) 1137–1156.
112. E. Pagounis, V. Lindroos, M. Talvitie, Influence of reinforcement volume fraction and size on the microstructure and abrasion wear resistance of hot isostatic pressed white iron matrix composites, Metallurgical and Materials Transactions A 27(12) (1996) 4171–4181.
113. F. Akhtar, Microstructure evolution and wear properties of in situ synthesized TiB 2 and TiC reinforced steel matrix composites, Journal of Alloys and Compounds 459(1) (2008) 491–497.
114. T. Shen, C. Koch, K. Wang, M. Quan, J. Wang, Solid-state reaction in nanocrystalline Fe/SiC composites prepared by mechanical alloying, Journal of materials science 32(14) (1997) 3835–3839.

115. J. Abenojar, F. Velasco, J. Torralba, J. Bas, J. Calero, R. Marce, Reinforcing 316L stainless steel with intermetallic and carbide particles, Materials Science and Engineering: A 335 (1) (2002) 1–5.
116. K.K. Chawla, Metal matrix composites, Wiley Online Library 2006.
117. I. Nadra, A.R. Boccaccini, P. Philippidis, L.C. Whelan, G.M. McCarthy, D.O. Haskard, R.C. Landis, Effect of particle size on hydroxyapatite crystal-induced tumor necrosis factor alpha secretion by macrophages, Atherosclerosis 196(1) (2008) 98–105.
118. M. Fathi, A. Hanifi, V. Mortazavi, Preparation and bioactivity evaluation of bone-like hydroxyapatite nanopowder, Journal of materials processing technology 202(1) (2008) 536–542.
119. Q. Wei, S. Li, C. Han, W. Li, L. Cheng, L. Hao, Y. Shi, Selective laser melting of stainless-steel/nano-hydroxyapatite composites for medical applications: Microstructure, element distribution, crack and mechanical properties, Journal of Materials Processing Technology 222 (2015) 444–453.
120. B. AlMangour, D. Grzesiak, J.-M. Yang, Nanocrystalline TiC-reinforced H13 steel matrix nanocomposites fabricated by selective laser melting, Materials & Design 96 (2016) 150–161.
121. B. AlMangour, D. Grzesiak, Selective laser melting of TiC reinforced 316L stainless steel matrix nanocomposites: Influence of starting TiC particle size and volume content, Materials & Design 104 (2016) 141–151.
122. B. AlMangour, D. Grzesiak, J.-M. Yang, Scanning strategies for texture and anisotropy tailoring during selective laser melting of TiC/316L stainless steel nanocomposites, Journal of Alloys and Compounds 728 (2017) 424–435.
123. B. AlMangour, D. Grzesiak, J.-M. Yang, Rapid fabrication of bulk-form TiB 2/316L stainless steel nanocomposites with novel reinforcement architecture and improved performance by selective laser melting, Journal of Alloys and Compounds 680 (2016) 480–493.
124. B. AlMangour, D. Grzesiak, J.-M. Yang, Selective laser melting of TiB 2/316L stainless steel composites: The roles of powder preparation and hot isostatic pressing post-treatment, Powder Technology 309 (2017) 37–48.
125. B. AlMangour, D. Grzesiak, J.-M. Yang, Selective laser melting of TiB 2/H13 steel nanocomposites: Influence of hot isostatic pressing post-treatment, Journal of Materials Processing Technology 244 (2017) 344–353.
126. A. Gåård, P. Krakhmalev, J. Bergström, Microstructural characterization and wear behavior of (Fe, Ni)–TiC MMC prepared by DMLS, Journal of Alloys and Compounds 421(1) (2006) 166–171.
127. L. Hao, S. Dadbakhsh, O. Seaman, M. Felstead, Selective laser melting of a stainless steel and hydroxyapatite composite for load-bearing implant development, Journal of Materials Processing Technology 209(17) (2009) 5793–5801.
128. B. Song, S. Dong, C. Coddet, Rapid in situ fabrication of Fe/SiC bulk nanocomposites by selective laser melting directly from a mixed powder of microsized Fe and SiC, Scripta Materialia 75 (2014) 90–93.
129. B. Song, S. Dong, P. Coddet, G. Zhou, S. Ouyang, H. Liao, C. Coddet, Microstructure and tensile behavior of hybrid nano-micro SiC reinforced iron matrix composites produced by selective laser melting, Journal of Alloys and Compounds 579 (2013) 415–421.
130. H. Niu, I. Chang, Selective laser sintering of gas and water atomized high speed steel powders, Scripta Materialia 41(1) (1999) 25–30.
131. C. Suryanarayana, Mechanical alloying and milling, Progress in materials science 46 (1) (2001) 1–184.
132. J. Xie, P. Fox, W. O'Neill, C. Sutcliffe, Effect of direct laser re-melting processing parameters and scanning strategies on the densification of tool steels, Journal of materials processing technology 170(3) (2005) 516–523.
133. L. Thijs, F. Verhaeghe, T. Craeghs, J. Van Humbeeck, J.-P. Kruth, A study of the microstructural evolution during selective laser melting of Ti–6Al–4V, Acta Materialia 58(9) (2010) 3303–3312.
134. H. De Vries, G. Den Ouden, Influence of welding on structure of Al–Mg–Si matrix composites, Materials science and technology 15(2) (1999) 202–206.

135. D. Stefanescu, B. Dhindaw, S. Kacar, A. Moitra, Behavior of ceramic particles at the solid-liquid metal interface in metal matrix composites, Metallurgical Transactions A 19(11) (1988) 2847–2855.
136. B. Dutta, M. Surappa, Directional dendritic solidification of a composite slurry: Part II. Particle distribution, Metallurgical and Materials Transactions A 29(4) (1998) 1329–1339.
137. B. Zheng, Y. Zhou, J. Smugeresky, J. Schoenung, E. Lavernia, Thermal behavior and microstructure evolution during laser deposition with laser-engineered net shaping: part II. Experimental investigation and discussion, Metallurgical and Materials Transactions A 39(9) (2008) 2237–2245.
138. B. Zheng, J. Smugeresky, Y. Zhou, D. Baker, E. Lavernia, Microstructure and properties of laser-deposited Ti6Al4V metal matrix composites using Ni-coated powder, Metallurgical and Materials Transactions A 39(5) (2008) 1196–1205.
139. I. Gotman, E. Gutmanas, P. Mogilevsky, Interaction between SiC and Ti powder, Journal of materials research 8(10) (1993) 2725–2733.
140. J.E. Smugeresky, B. Zheng, Y. Zhou, D. Baker, A. Biswas, E.J. Lavernia, SYNTHESIS AND CHARACTERIZATION OF HYBRID AL/AL3NI FOAM MATERIALS VIA LENS (R) DEPOSITION OF NI-COATED AL POWDER, Sandia National Laboratories (SNL-CA), Livermore, CA (United States), 2008.
141. Z. Liu, Y. Gong, W. Zhou, L. Ma, J. Yu, J.C. Idrobo, J. Jung, A.H. MacDonald, R. Vajtai, J. Lou, Ultrathin high-temperature oxidation-resistant coatings of hexagonal boron nitride, Nature communications 4 (2013) 2541.
142. J. Bolton, M. Youseffi, Fracture toughness of sintered metal matrix composites based upon high speed steels enriched with hard ceramic carbides, Powder metallurgy 36(2) (1993) 142–152.
143. H. Attar, M. Calin, L. Zhang, S. Scudino, J. Eckert, Manufacture by selective laser melting and mechanical behavior of commercially pure titanium, Materials Science and Engineering: A 593 (2014) 170–177.
144. P. Ma, K.G. Prashanth, S. Scudino, Y. Jia, H. Wang, C. Zou, Z. Wei, J. Eckert, Influence of annealing on mechanical properties of Al-20Si processed by selective laser melting, Metals 4 (1) (2014) 28–36.
145. B.E. Carroll, T.A. Palmer, A.M. Beese, Anisotropic tensile behavior of Ti–6Al–4V components fabricated with directed energy deposition additive manufacturing, Acta Materialia 87 (2015) 309–320.
146. X. Tan, Y. Kok, Y.J. Tan, M. Descoins, D. Mangelinck, S.B. Tor, K.F. Leong, C.K. Chua, Graded microstructure and mechanical properties of additive manufactured Ti–6Al–4V via electron beam melting, Acta Materialia 97 (2015) 1–16.
147. L. Thijs, M.L.M. Sistiaga, R. Wauthle, Q. Xie, J.-P. Kruth, J. Van Humbeeck, Strong morphological and crystallographic texture and resulting yield strength anisotropy in selective laser melted tantalum, Acta Materialia 61(12) (2013) 4657–4668.
148. Z. Wang, T.A. Palmer, A.M. Beese, Effect of processing parameters on microstructure and tensile properties of austenitic stainless steel 304L made by directed energy deposition additive manufacturing, Acta Materialia 110 (2016) 226–235.
149. K. Kunze, T. Etter, J. Grässlin, V. Shklover, Texture, anisotropy in microstructure and mechanical properties of IN738LC alloy processed by selective laser melting (SLM), Materials Science and Engineering: A 620 (2015) 213–222.
150. A. Hussein, L. Hao, C. Yan, R. Everson, Finite element simulation of the temperature and stress fields in single layers built without-support in selective laser melting, Materials & Design 52 (2013) 638–647.
151. I. Sulima, P. Klimczyk, P. Malczewski, Effect of TiB2 particles on the tribological properties of stainless steel matrix composites, Acta Metallurgica Sinica (English Letters) 27(1) (2014) 12–18.
152. B. AlMangour, M.-S. Baek, D. Grzesiak, K.-A. Lee, Strengthening of stainless steel by titanium carbide addition and grain refinement during selective laser melting, Materials Science and Engineering: A 712 (2018) 812–818.

153. K. Zhang, G. Chen, Effect of SiC particles on crystal growth of Al–Si alloy during laser rapid solidification, Materials Science and Engineering: A 292(2) (2000) 229–231.
154. B. Dutta, M. Surappa, Microstructure evolution during multidirectional solidification of Al-Cu-SiC composites, Composites Part A: Applied Science and Manufacturing 29(5–6) (1998) 565–573.
155. J. Sekhar, R. Trivedi, Solidification microstructure evolution in the presence of inert particles, Materials Science and Engineering: A 147(1) (1991) 9–21.
156. P. Rohatgi, K. Pasciak, C. Narendranath, S. Ray, A. Sachdev, Evolution of microstructure and local thermal conditions during directional solidification of A356-SiC particle composites, Journal of materials science 29(20) (1994) 5357–5366.
157. S. Sathish, M. Geetha, N. Pandey, C. Richard, R. Asokamani, Studies on the corrosion and wear behavior of the laser nitrided biomedical titanium and its alloys, Materials Science and Engineering: C 30(3) (2010) 376–382.
158. E.J. Lavernia, T.S. Srivatsan, The rapid solidification processing of materials: science, principles, technology, advances, and applications, Journal of Materials Science 45(2) (2010) 287.
159. M. Salehi, K. Dehghani, Structure and properties of nanostructured aluminum A413. 1 produced by melt spinning compared with ingot microstructure, Journal of Alloys and Compounds 457(1) (2008) 357–361.
160. E. Huskins, B. Cao, K. Ramesh, Strengthening mechanisms in an Al–Mg alloy, Materials Science and Engineering: A 527(6) (2010) 1292–1298.
161. E. Yun, K. Lee, S. Lee, Correlation of microstructure with high-temperature hardness of (TiC, TiN)/Ti–6Al–4V surface composites fabricated by high-energy electron-beam irradiation, Surface and Coatings Technology 191(1) (2005) 83–89.
162. C. Ma, D. Gu, D. Dai, G. Yu, M. Xia, H. Chen, Thermodynamic behaviour and formation mechanism of novel titanium carbide dendritic crystals within a molten pool of selective laser melting TiC/Ti–Ni composites, CrystEngComm 19(7) (2017) 1089–1099.
163. E. Fereiduni, S.G. Banadkouki, Reliability/unreliability of mixture rule in a low alloy ferrite–martensite dual phase steel, Journal of alloys and compounds 577 (2013) 351–359.
164. E. Fereiduni, S.G. Banadkouki, Improvement of mechanical properties in a dual-phase ferrite–martensite AISI4140 steel under tough-strong ferrite formation, Materials & Design 56 (2014) 232–240.
165. C. Ma, D. Gu, D. Dai, W. Chen, F. Chang, P. Yuan, Y. Shen, Aluminum-based nanocomposites with hybrid reinforcements prepared by mechanical alloying and selective laser melting consolidation, Journal of Materials Research 30(18) (2015) 2816–2828.
166. R. Aikin, L. Christodoulou, The role of equiaxed particles on the yield stress of composites, Scripta metallurgica et materialia 25(1) (1991) 9–14.
167. Q.B. Nguyen, M. Gupta, Enhancing compressive response of AZ31B using nano-Al 2 O 3 and copper additions, Journal of Alloys and Compounds 490(1) (2010) 382–387.
168. S. Dadbakhsh, L. Hao, P. Jerrard, D. Zhang, Experimental investigation on selective laser melting behaviour and processing windows of in situ reacted Al/Fe 2 O 3 powder mixture, Powder technology 231 (2012) 112–121.
169. W. Liu, J. DuPont, Fabrication of functionally graded TiC/Ti composites by laser engineered net shaping, Scripta Materialia 48(9) (2003) 1337–1342.
170. R. Mahamood, E.T. Akinlabi, Laser metal deposition of functionally graded Ti6Al4V/TiC, Materials & Design 84 (2015) 402–410.
171. J.C. Oh, S. Lee, Correlation of microstructure with hardness and fracture properties of (TiC, SiC)/Ti–6Al–4V surface composites fabricated by high-energy electron-beam irradiation, Surface and Coatings Technology 179(2) (2004) 340–348.
172. J.S. Moya, S. Lopez-Esteban, C. Pecharroman, The challenge of ceramic/metal microcomposites and nanocomposites, Progress in materials science 52(7) (2007) 1017–1090.
173. S.C. Tjong, Novel nanoparticle-reinforced metal matrix composites with enhanced mechanical properties, Advanced engineering materials 9(8) (2007) 639–652.
174. W. Miller, F. Humphreys, Strengthening mechanisms in particulate metal matrix composites, Scripta metallurgica et materialia 25(1) (1991) 33–38.
175. D. Hull, D.J. Bacon, Introduction to dislocations, Butterworth-Heinemann 2001.

176. W. Wong, M. Gupta, Development of Mg/Cu nanocomposites using microwave assisted rapid sintering, Composites Science and Technology 67(7) (2007) 1541–1552.
177. Z. Szaraz, Z. Trojanova, M. Cabbibo, E. Evangelista, Strengthening in a WE54 magnesium alloy containing SiC particles, Materials Science and Engineering: A 462(1) (2007) 225–229.
178. A. Sanaty-Zadeh, Comparison between current models for the strength of particulate-reinforced metal matrix nanocomposites with emphasis on consideration of Hall–Petch effect, Materials Science and Engineering: A 531 (2012) 112–118.
179. Z. Zhang, D. Chen, Contribution of Orowan strengthening effect in particulate-reinforced metal matrix nanocomposites, Materials Science and Engineering: A 483 (2008) 148–152.
180. E. Hall, The deformation and ageing of mild steel: III discussion of results, Proceedings of the Physical Society. Section B 64(9) (1951) 747.
181. C. Goh, J. Wei, L. Lee, M. Gupta, Properties and deformation behaviour of Mg–Y 2 O 3 nanocomposites, Acta Materialia 55(15) (2007) 5115–5121.
182. T. Clyne, P. Withers, An introduction to metal matrix composites, Cambridge University Press 1995.
183. R.E. Smallman, A. Ngan, Physical metallurgy and advanced materials, Butterworth-Heinemann 2011.
184. Q. Nguyen, M. Gupta, Enhancing compressive response of AZ31B magnesium alloy using alumina nanoparticulates, Composites Science and Technology 68(10) (2008) 2185–2192.
185. R. Labusch, A statistical theory of solid solution hardening, physica status solidi (b) 41 (2) (1970) 659–669.
186. W. Soboyejo, Mechanical properties of engineered materials, CRC press 2002.
187. P. Bindumadhavan, H.K. Wah, O. Prabhakar, Dual particle size (DPS) composites: effect on wear and mechanical properties of particulate metal matrix composites, Wear 248(1) (2001) 112–120.
188. G.V. Kumar, C. Rao, N. Selvaraj, Mechanical and tribological behavior of particulate reinforced aluminum metal matrix composites–a review, Journal of minerals and materials characterization and engineering 10(01) (2011) 59.
189. E. Fereiduni, M. Movahedi, A. Baghdadchi, (2017) Ultrahigh-strength friction stir spot welds of aluminium alloy obtained by Fe O nanoparticles. Science and Technology of Welding and Joining 23 (1):63–70.
190. Fereiduni, E., & Banadkouki, S. G. (2014). Ferrite hardening response in a low alloy ferrite–martensite dual phase steel. Journal of Alloys and Compounds, 589, 288–294.
191. Banadkouki, S. G., & Fereiduni, E. (2014). Effect of prior austenite carbon partitioning on martensite hardening variation in a low alloy ferrite–martensite dual phase steel. Materials Science and Engineering: A, 619, 129–136.
192. E. Fereiduni, M. Movahedi, A.H. Kokabi, (2015) Aluminum/steel joints made by an alternative friction stir spot welding process. Journal of Materials Processing Technology 224:1–10.
193. E. Fereiduni, M. Movahedi, A. H. Kokabi, (2016) Dissimilar Al/steel friction stir spot welding: To penetrate into the lower steel sheet or not?. Science and Technology of Welding and Joining 21 (6):466–472.
194. Eskandar Fereiduni, Mojtaba Movahedi, Amir Hossein Kokabi, Hossein Najafi, (2017) Effect of Dwell Time on Joint Interface Microstructure and Strength of Dissimilar Friction Stir Spot-Welded Al-5083 and St-12 Alloy Sheets. Metallurgical and Materials Transactions A 48 (4):1744–1758.
195. Mostafa Yakout, Andrea Cadamuro, M. A. Elbestawi, Stephen C. Veldhuis, (2017) The selection of process parameters in additivemanufacturing for aerospace alloys. The International Journal of Advanced Manufacturing Technology 92 (5-8):2081–2098.
196. Mostafa Yakout, M. A. Elbestawi, Stephen C. Veldhuis, (2018) On the characterization of stainless steel 316L parts produced by selective laser melting. The International Journal of Advanced Manufacturing Technology 95 (5-8):1953–1974.

# Additive Manufacturing of Titanium Alloys for Biomedical Applications

**Lai-Chang Zhang and Yujing Liu**

## 1 Introduction

More than 90% of patients over the age of 40 suffer from varying degrees of joint disease. For patients with advanced arthritis, artificial implants made from biomedical materials help release pain and increase people's life [1]. All of these require orthopaedic surgery, resulting in an increasing number of replacements. Biomedical applications are designed and fabricated primarily in accordance with the requirements of the implant and are commonly used in different parts of the body. The ultimate goal of the researchers is to complete the implantation of the human body without failure or modification of the operation. Therefore, the choice of materials is significant and need to be given priority consideration. Such a material should have the following advantages in human body fluid environment including great corrosion resistance, high strength, low Young's modulus, good wear resistance and no cytotoxicity. So far, three common metals have been used for implants, i.e. stainless steel, Co based alloys and titanium alloys [2, 3]. Titanium alloys have been extensively studied on the basis of excellent mechanical properties such as the light weight, high strength, corrosion resistance, good biocompatibility and low modulus [4–7]. Conventional titanium alloy manufacturing methods such as casting and powder metallurgy need subsequent mechanical processing, which consumes more time and energy.

Titanium alloys are usually manufactured through casting technology, powder metallurgy technology and foaming technology [8, 9]. However, these traditional techniques involve multiple processing steps, requiring longer periods of time, as well as more material resource and energy consumption [10]. In addition, the high reactivity of titanium and oxygen and its high melting point bring some challenges to

L.-C. Zhang (✉) · Y. Liu
Edith Cowan University, Perth, WA, Australia
e-mail: l.zhang@ecu.edu.au

 
B. AlMangour (ed.), *Additive Manufacturing of Emerging Materials*,
https://doi.org/10.1007/978-3-319-91713-9_5

these typical technologies. Compared with many other metal alloys, the complexity of the extraction process, the difficulty of melting, the manufacturing and mechanical problems make the titanium alloys expensive. The material removal from the traditional multistep manufacturing process is difficult. High cost and difficult processing are the two main reasons that titanium alloy is difficult to be widely used. The porous structure method is very necessary for the manufacture of the parts with complex shapes.

In recent decades, the development of additive manufacturing (AM) technology has stimulated and improved the development of the implants [11–14]. The porous implants can be produced directly from the additive manufacturing system without any subsequent processing program. These implants have obvious advantages, such as low Young's modulus and light weight, which can promote the ingrowth of bone cells. Compared with the conventional methods, additive manufacturing can realize the pore structure of complex unit structure by precision machining, which has aroused great concerns. Until now, selective laser melting (SLM) and electron beam melting (EBM) are the two most common additive manufacturing technologies used for manufacturing metal components [7]. The parts produced could have complex structures and have high mechanical properties. The combination of additive manufacturing technology and high performance biomedical titanium materials will help to succeed in the field of implants. This paper mainly focuses on the development of biomedical titanium alloys and their AM as-produced parts as a medical implant.

## 2 Additive Manufacturing for Biomedical Application

Additive Manufacturing, commonly referred to as 3D printing, is a method of fabricating objects from three-dimensional model data based on the principle, by melting a layer of powder using a laser or electron beam, Through the computer control to produce the structure of complex components [7]. Due to its rapid, efficient and accurate control of the internal pore structure and the complex shape of the produced components, additive manufacturing has been widely concerned in the fields of aerospace and biomedicine. Generally, there are two representative additive manufacturing techniques, i.e. selective laser melting and electron beam melting [7]. In this part, the additive manufacturing working principle and production process (including selective laser melting and electron beam melting), microstructure, mechanical properties of as-produced products and their applications in biomedical field are briefly discussed.

## 2.1 Selective Laser Melting (SLM)

Selective laser melting system was first reported by Fraunhofer Institute ILT in Aachen in 1995 in Germany. Differently, selective laser melting systems use a laser spot as an input energy to completely melt the powder bed, which was deposited on the metal plate in advance. The computer controls the laser beam through two mirrors and then the laser beam is focused on the powder bed. The input laser beam with a high energy of up to 1 kW and the mechanical movements of the scanning mirror permit accurate laser beam scanning up to ~15 m/s scan rate to make sure the powder melted completely [15]. The processing chamber is filled with argon gas atmosphere during manufacturing process to avoid occurrence of oxidation [16]. So far, a wide range of materials including metals, polymers and ceramic [8, 17–21] have been used for industry applications. The SLM-produced components including the solid parts and complex porous structures exhibit excellent mechanical properties without subsequent treatment [22, 23].

The main goal of the selective laser melting is to obtain completely dense and defect-free parts. However, achieving this goal is not easy because of the absence of mechanical stress in the selective laser melting, which is dominated by gravity, recoil forces, melt pool surface tension and thermal effects [7]. The selective laser melting process involves a large number of process parameters, so proper control of the relevant parameters will result in a high quality product. Some parameters, such as laser wavelength and laser operating mode, cannot be changed for a specific selective laser melting system. In addition, certain properties of the powder, such as surface tension and thermal conductivity, which defines the boundary conditions for the selective laser melting process, are fixed. In contrast, other parameters, called manufacturing or process parameters, can be calculated and optimized [11]. Many previous articles have studied and optimized in details for processing parameters of some specific materials.

In general, for a given material, the laser energy density, E, applied to a certain volume of powder material during selective laser melting is defined by

$$\mathrm{E} = \frac{P}{vts} \tag{1}$$

where $P$ is the input power, $v$ is the scanning speed, $t$ is the layer thickness and $s$ is the hatch space. A near-full density component can be obtained by a group of optimized parameters which balance all factors above during selective laser melting process. As can be seen in Eq. 1, increasing the input power and/or reducing the scan speed or layer thickness or scan hatch space can rise the laser energy density, thereby increasing the temperature of the melt pool. A higher laser energy density will result in a larger amount of melting area and therefore higher final density [10]. Since the formation of a complete melting is a necessary condition for the manufacture of dense parts, obtaining high-density parts requires sufficient laser energy density. Laser energy density is a key factor that affects the densification and quality of SLM-made parts. For a given material, the value of the energy density can be

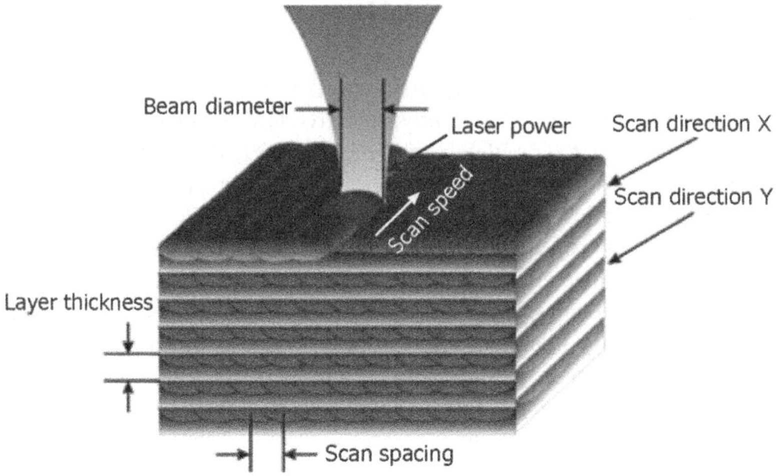


**Fig. 1** Schematic of selective laser melting parameters [24]

precisely calculated from the known processing parameters. The optimized energy density usually has a range rather than a fixed value. In general, the minimum critical laser energy density produces the densest part. For example, the critical laser energy densities (CP-Ti, Ti-6Al-4V and Ti2448 of pure titanium with selective laser melting are around 120 and 40 J/mm$^3$, respectively [25]. A schematic of these selective laser melting processing parameters is shown in Fig. 1. In selective laser melting, the laser beam travels to the powder bed at a constant rate, called scanning speed (v), which controls the selective laser melting production time. In other words, if you need a short production time, you need a higher scanning speed. However, the maximum laser power for a particular selective laser melting device must be taken into account when increasing the scan speed [25]. Layer thickness defines the energy required to melt/solidify a layer of powder and the production time. The thickness of the layer is very important, as good inter-layer connectivity is only possible when previously processed layers are re-melted. If a larger layer thickness is used in the SLM process, the production time of manufacturing a component can be reduced. However, higher energy inputs also require full melting of the thick layer, which can result in increasing surface roughness and reducing dimensional accuracy. The scanning distance (s) is usually chosen parallel to the selective laser melting line, so it is also named as "hatch space" [25]. Scanning strategy is laser scanning track referred to the scanning length and pattern, the pattern can be a straight line or a circle. There are several common scanning strategies including cross-hatching zigzag scanning pattern, unidirectional scanning, inter-layer scanning and interlayer rotation scanning (Fig. 2). For a give material, different strategies will result in different relative density and mechanical properties; for example, the relative density of SLM-produced Ti-6Al-4V samples with cross-hatching zigzag scanning pattern can reach 99.9% [25].

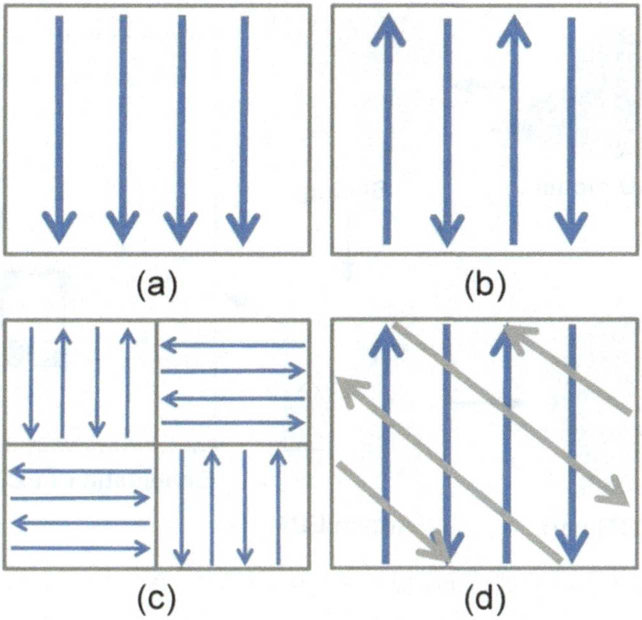


**Fig. 2** Different selective laser melting processing patterns: (**a**) Uni-directional scanning, (**b**) Bi-directional scanning, (**c**) Inter-layer scanning, and (**d**) Interlayer rotation scanning [26]

## 2.2 *Electron Beam Melting (EBM)*

Electron beam melting is another AM system equipped with an electron beam launching device, which can produce nearly full density parts in a vacuum environment [29, 30]. The working principle and process of electron beam melting are similar to that of selective laser melting (Fig. 3) [27, 28]. The main difference with the selective laser melting is that the electron beam melting uses different heat source and chamber atmosphere. The heat source of electron beam melting is electron beam with a voltage of 60 kV, which preheats the substrate plate to a pre-setting temperature before dropping the powder [6, 7]. The electron beam will prescan the powder for sintering the powder and then scan the powder bed based on the sample CAD geometric shape. These differences in beam energy input and chamber environment of electron beam melting and selective laser melting result in the different microstructure (referred to melt pool size, phase type and grain size) and mechanical properties (referred to hardness, compressive, tensile and fatigue properties) of selective laser melting and electron beam melting products. The densification rate and microstructural homogeneity of EBM-produced part with an optimized parameter results in an improvement of relative density and mechanical properties [6, 7]. Some studies have been conducted to study the performance of EBM as-fabricated components and improve the properties of those samples. Some previous works have reported that plenty of implants such as knee, hip joint, and jaw replacements have been produced successfully by electron beam melting.

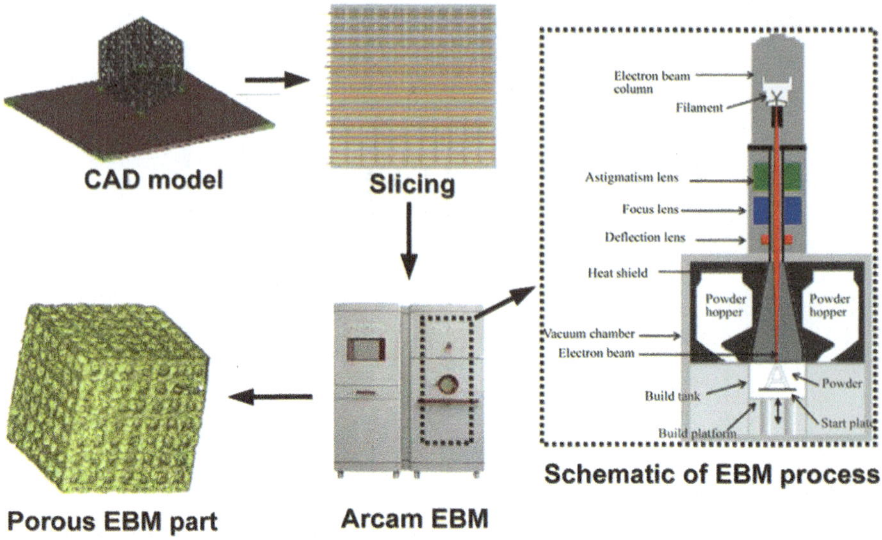


**Fig. 3** Schematic diagram of electron beam melting system [27, 28]

## 3 Development of AM Biomedical Titanium Alloy

The ideal goal of researchers is to get a material that can be implanted into the human body for a long time without reoperation. Therefore, the performance of materials is considered a priority. Such a material should have the advantages of strong corrosion resistance, high strength, low modulus of elasticity, good wear resistance and no cytotoxicity in human environment [25, 29]. Among the three common metal materials used in medical field, titanium alloys have excellent mechanical properties in terms of low density, high strength, excellent corrosion resistance, good biocompatibility and low modulus [25, 29]. As a result, they prefer biomedical applications because they have good performance.

It is reported that the first titanium and its alloys were developed in the United States in 1940s [25]. Until now, titanium alloys have been widely used in aerospace, chemical and medical industries. Different titanium alloys also have different properties. For example, commercially pure titanium has good biocompatibility, but its low strength (~500MPa) limits its wide application [9]. In the past few decades, a lot of efforts have been made to develop titanium based alloys with excellent mechanical properties and excellent biocompatibility. At present, the most commonly used titanium alloy is an $\alpha + \beta$ type Ti-6Al-4V, originally developed for aerospace industry. Subsequent research had found that it has excellent corrosion resistance and high strength making it an ideal choice for biomaterials. Zhang et al. [31] reported that the cell had a very high survival rate on the thin disk fabricated by Ti-6Al-4V alloy. Johansson et al. [32] showed that Ti-6Al-4V also applied to the soft tissue of rats after implantation. However, recent studies [16, 33] shows the adverse properties, including some cytotoxic components such as Al and V, which are associated with neurotoxicity and neurodegenerative diseases [34]. Furthermore, Ti-6Al-4V presents much higher elastic modulus (~110 GPa) compared to human bone (less than 30 GPa), this can cause stress shielding, resulting in bone loss and premature failure. Some researchers reported that β-type titanium alloys composing of non-toxic elements with low modulus are good materials for potential applications as implant [35].

The β phase of titanium alloys, which is metastable under the β/α transition temperature, can be reserved at room temperature through introducing some β stabilizer elements, for example, the elements of Nb, Ta, Mo, Zr and Sn can be used for producing βtitanium alloys. Due to their non-toxic property, these alloys can be manufactured as implants to pursue an improvement of long-term performance [1, 36]. Synthes et al. [37] made a comparison for the performance of Ti-7.5Mo, Ti-15Mo, Ti-10Mo and Ti-13Nb-13Zr, and considered Ti-15Mo alloy as potential of medical material with fine microstructure, high strength and low modulus. After systematical study, Ho et al. [38] pointed out that the Ti-15Mo comprised the low modulus with a value of ~77 GPa. Furthermore, some β type TiNb-based alloys presented the super elasticity; for example, Kim et al. [39] observed the super elasticity in Ti-(15–35) (at.%) Nb alloys with a recoverable strain of 3.3%, this excellent performance expands the scope of application of TiNb-based alloys. Hao et al. found that Ti-24Nb-4Zr-7.5Sn alloy had the lowest Young's modulus of 52 GPa and great super elasticity at room temperature [40]. The following studies reported that Ti2448 alloy in the as hot-rolled state exhibited peculiar non-linear super-elastic behaviour with the greatest recoverable strain up to 3.3% and incipient Young's modulus of 42 GPa. Until now, many titanium alloys have been processed by AM technologies. The materials, manufacturing methods and their mechanical properties are listed in Table 1.

### *3.1 Selective Laser Melting (SLM) of Titanium Alloys*

Selective laser melting is able to successfully produce a variety of titanium alloys, such as commercially pure titanium (CP-Ti), Ti-6Al-4V and Ti-24Nb-4Zr-8Sn (Ti2448), and some of the them have been well applied in medical field [50]. As a traditional implant material, CP-Ti is one of the most commonly used titanium alloys in AM printing medical applications. SLM as-produced CP–Ti samples mostly demonstrate the importance of manufacturing parameters, microstructures and mechanical properties [11]. It was reported that the input energy density (E) of 120 J/mm$^3$ is suitable to melt the powders sufficiently and build almost fully dense CP-Ti parts with the relative density is high than 99.5% [51]. However, the input power and scan speed should be adjusted at this energy density for achieving high-density parts (Fig. 4).

Laser processing parameters affect the microstructures of the SLM-produced CP-Ti samples, variation from plate-like α to acicular martensitic α' phase (Fig. 5)

**Table 1** The mechanical properties of EBM-manufactured titanium materials

| Material | Method | UTS (MPa) | $\sigma_{0.2}$ (MPa) | Elongation (%) | Hardness (GPa) | E (GPa) | References |
|---|---|---|---|---|---|---|---|
| Cp-Ti | SLM | 757 | 555 | 19.5 | 2.61 | 106 ± 3 | [9] |
| Cp-Ti | SLM | 650 | 500 | 17 | | | [41] |
| Ti–6Al–4V | SLM | 1267 | 1110 | 7.28 | 4.09 | 109 | [42] |
| Ti-24Nb-4Zr-7.9Sn (Ti2448) | SLM | 665 ± 18 | 563 ± 38 | 13.8 ± 4.1 | 2.2 | 53 | [16] |
| Ti–6Al–4V | EBM | 950–990 | 910–940 | 14–16 | ... | 120 | [27] |
| Ti–6Al–4V | EBM | 910 ± 10 | 830 ± 5 | ... | 3.21 ± 0.02 | 118 ± 5 | [43] |
| Ti–6Al–4V | EBM | 1100–1400 | ... | 12–25 | 4.3 | ... | [44] |
| Ti–6Al–4V | EBM | 1150–1200 | 1100–1150 | 16–25 | 3.6–3.9 | ... | [45] |
| Ti–6Al–4V | EBM | 775 | 735 | 2.3 | 3.69 | 93 | [46] |
| Ti–6Al–4V | EBM | 994–1029 | 883–938 | 11.6–13.6 | ... | ... | [41] |
| Ti–6Al–4V | EBM | 930 | 865 | 12–17 | ... | ... | [47] |
| Ti–6Al–4V | EBM | 944.5–964.5 | 823.4–851.8 | 13.2–16.3 | 3.19–3.27 | ... | [12] |
| Ti–6Al–4V | EBM | ... | 832–1049 | ... | ... | ... | [5] |
| Ti–6Al–4V | EBM | ~990–1180 | ~900–1100 | ~18–23 | ... | ... | [13] |
| Ti-48Al-2Nb-0.7Cr-0.3Si | EBM | 336 ± 26 | 253 ± 13 | 0.27 ± 0.1 | ... | 166 ± 2 | [48] |
| γ-TiAl | EBM | ... | 1400 | ... | 4.1 | ... | [29] |
| Ti-24Nb-4Zr-7.9Sn (Ti2448) | EBM | ... | ... | ... | 2.5 | ... | [49] |

UTS is the ultimate tensile strength; $\sigma_{0.2}$ is yield strength; and E is Young's modulus

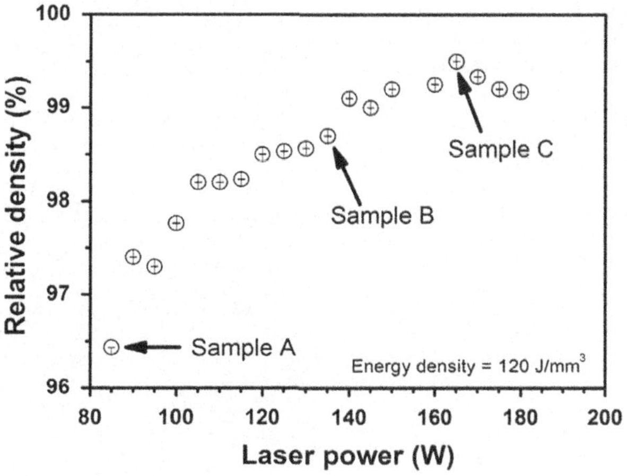


**Fig. 4** Relationship between relative density and laser power for the SLM-fabricated CP–Ti parts at fixed energy density of 120 J/mm$^3$. Samples A, B, and C show different examples of relative densities [9]

which can be determined by different laser scan speed (v). With a constant input energy density (E) of 120 J/mm$^3$, once the laser scan speed is lower than 100 mm/s complete allotropic transformation of β to a takes place in the process of solidification because of energy thermalization in the melt pool (Fig. 5a). On the other hand, with increasing the laser scan speed over 100 mm/s, both kinetic and thermal characteristics below cooling go up, resulting in the growth in temperature gradients in the melt pool [60], and thereby causing the generation of α' in the final SLM-produced sample (Fig. 5).

As mentioned earlier, Ti–6Al–4V is another commonly used titanium alloy with an (α + β)-type phase in medical field. Literature showed that SLM as-produced Ti–6Al–4V samples are composed of dominant fine acicular α' martensite and some prior β grains [26, 52]. This microstructure is different with the typical α + β morphology of Grade 5 sample (Fig. 6). SLM-produced sample comprising acicular α' martensite as a result of fast cooling rate exhibit a high tensile strength with a value of 1267 MPa [42]. This is because cooling rate is faster during selective laser melting solidification process ($10^3$–$10^8$ K/s) [53] than the critical cooling rate of Ti–6Al–4V martensitic transformation with the value of 410 K/s from β to α'. The near fully-dense SLM as-produced Ti–6Al–4V samples comprise microhardness of 409 Hv, which is higher than that of the samples manufactured by superplastic forming with a value of 346 Hv [54]. Furthermore, other mechanical properties, such as tensile UTS and yield properties, of the SLM-produced Ti–6Al–4V samples (Table 1) are higher than that of cast counterparts [26]. Corrosion resistance of SLM-produced sample is another important property. However, most the studies on selective laser melting process have mainly focused on the manufacturing process, densification and mechanical properties for as-produced parts. There is a lack of research into the corrosion behaviour of titanium parts (and other metallic materials) produced by selective laser melting. Recent studies have demonstrated the corrosion behaviour of SLM-produced Ti-6Al-4V in 3.5% NaCl solution compared to the commercial Grade 5 alloy [4, 52]. Potentiodynamic measurements pointed out that the corrosion

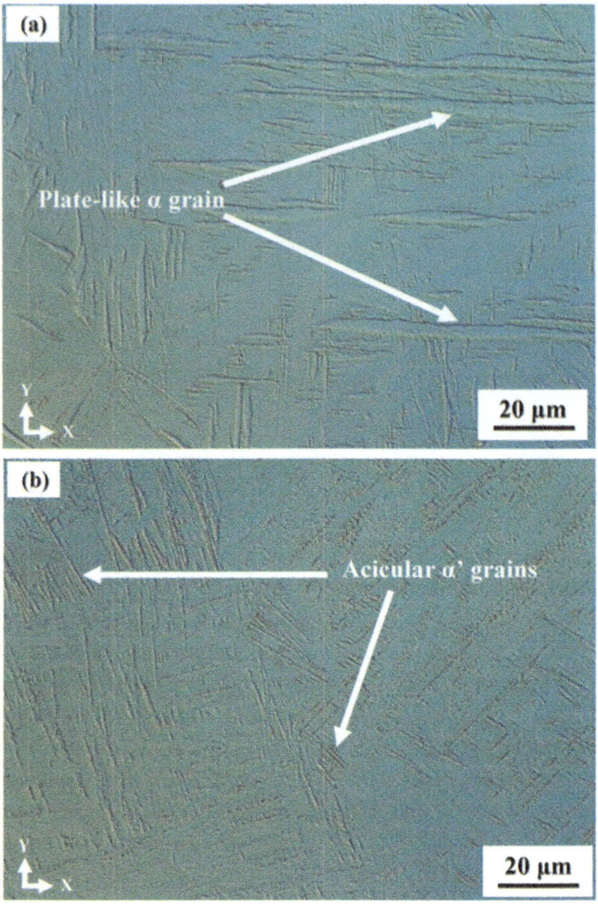


**Fig. 5** Cross-sectional (Y-X) optical microstructure of the SLM-manufactured CP–Ti Samples, (**a**) Plate-like a and (**b**) acicular [9]

current density in the passive range of the SLM-produced Ti–6Al–4V alloy is twice that of the commercial Grade 5 alloy. This illustrates that the SLM-processed sample has worse corrosion resistance than the commercial Grade 5 alloy. The corrosion behavior of selective laser melted samples can also be affected by the building directions. Dai et al. revelled that XY-plane of SLM-produced sample had a greater corrosion resistance in comparison with XZ-plane in 1 M HCl solution, although having slight difference in 3.5 wt.% NaCl solution, pointing out that the different building planes presented more pronounced difference in corrosion resistance in harsher solution system [4]. Such a phenomenon was also found in SLM-produced Al-12Si alloy [61].

Recently, some β type titanium alloys produced by selective laser melting have been reported. Zhang et al. manufactured β type Ti2448 solid bulk parts by selective laser melting [16]. The relative density and micro-hardness generally increased with decreasing the laser scan speed, which corresponds to a higher laser energy density. Nearly full density samples (>99%) were obtained with a laser power of 200 W and a

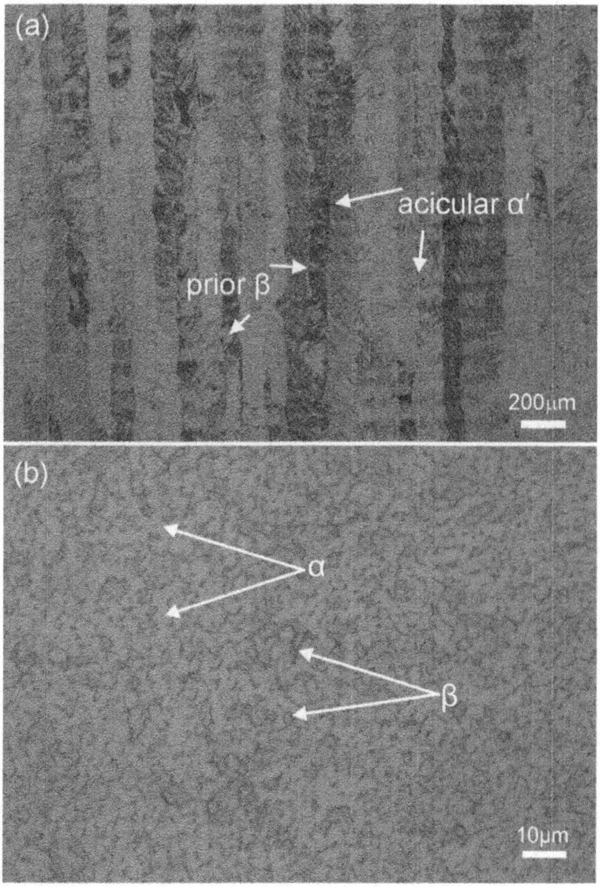


**Fig. 6** Optical microstructure of the (**a**) SLM-produced Ti–6Al–4V sample and (**b**) Grade 5 alloy [52]

scan speed range of 300–600 mm/s. Compared with material prepared by conventional processing methods, SLM-produced samples with similar tensile property but without pronounced super-elastic properties as a result of the high oxygen content of the pre alloy powder. An example of a SLM-produced acetabular hip cup sample was completed with complex outer scaffold (Fig. 7). Liu et al. [10] reported an topology optimized porous structure with 85% porosity manufactured by Ti2448 material [10]. The relative density was influenced by laser scan speed and input energy. A high relative density specimen (>99%) was obtained under a scan speed of 750 mm/s and an input power of 175 W. The compressive strength reached ~51 MPa with a high ductility of ~14%. The results above showed that SLM-produced Ti2448 meets the requirement of implant in terms of mechanical properties. Very recently, the author also found that the topology optimized structure exhibits the excellent balance of bending and buckling stress with a high elastic energy absorption, a low Young's modulus (~2.3 GPa) and a high compressive strength (~58 MPa) [63].

**Fig. 7** An example of the SLM-produced precise Ti–24Nb–4Zr–8Sn acetabular cup [16]

## 3.2 *Electron Beam Melting (EBM) of Titanium Alloys*

Electron beam melting technology could process titanium components under a vacuum environment and obtain lower costs and more efficient production than conventional methods. Moreover, the advantages of electron beam melting include the ability to produce complex shapes to meet specific industry needs, as well as other advantages such as short processing cycles and efficient material utilization, making it an excellent alternative to titanium products. So far, Ti-6Al-4V alloy has been widely used as electron beam melting production. In order to improve the mechanical properties of the resulting components, extensive efforts have been made to optimize the process parameters to manipulate the microstructure of EBM-produced samples. Researches have also been conducted on another type of titanium alloy such as β-type Ti2448) [14, 30].

The microstructure of EBM as-produced Ti-6Al-4V is very complicated. In the electron beam melting process, the electron beam scans at high speed the powder bed and melts to generate a melt pool, and the liquid metal rapidly solidifies. As the building layers accumulate, the solidification of the solid part of the building below the building level is affected by the multiple thermal cycling effects in the vacuum environment due to lower cooling rates and heat build-up in the electron beam melting process. This results in the overall temperature of the sample being maintained at a high level, and this plays a key role in stress annealing elimination. Such a process facilitates good matching of sample strength and plasticity and uniformity of part performance [7, 14, 55–58]. The solid part of EBM as-produced Ti-6Al-4V contains columnar prior β grains which are delineated by α grain boundary and a transformed α/β structure, and Widmanstätten pattern and lamellar colony within the prior β grains [12, 41]. The generation of an α grain boundary along the grain boundaries of the prior β grains indicates the diffusive nature of the β-α transformation. In this way, the microstructure obtained in EBM-produced sample differs with the one observed in other AM-produced parts. Such as in selective laser

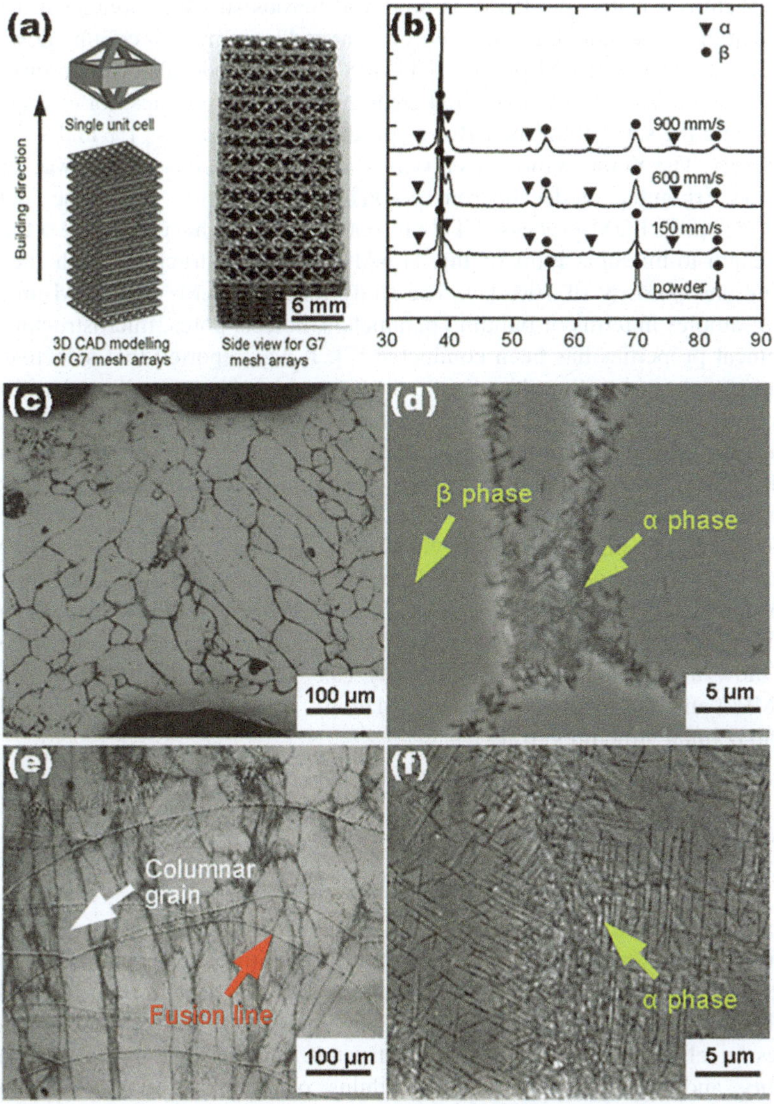


**Fig. 8** (**a**) Porous structure model used and side view of EBM-processed Ti2448 component, (**b**) XRD profiles of the starting powder and EBM-processed components, (**c**) and (**e**) OM, and (**d**) and (**f**) SEM images of Group A. (**c**, **d**) are at horizontal plane and (**e**, **f**) are vertical views [6]

melting process with extremely fast cooling rates lead to a diffusion less martensitic β-α′ transformation. The texture of the parent β grain in EBM-produced sample exhibits a strong <001> β pole along the build direction (z axis) [5].

The EBM as-produced β titanium porous sample exhibits excellent properties. Fig. 8 shows the side view of EBM-processed, X-ray diffraction (XRD) patterns and microstructure. Clearly, the EBM as-produced porous structure contained α phase due to a high temperature building environment and long time for cooling down. Liu et al. reported that the EBM-produced Ti2448 rhombic dodecahedron porous structures with a porosity of 70% exhibited great mechanical properties, for example, the ductility and the strength of the EBM as-produced sample are ~11% and ~37 MPa, respectively. The Young's modulus is only ~0.86 GPa, which is much lower than the Ti-6Al-4V rhombic dodecahedron porous samples with the same porosity (~1.4 GPa) [59]. EBM-processed Ti2448 porous samples have about two times of the strength-to-modulus ratio of the Ti-6Al-4V porous structures with the same structure and porosity of 70% [7]. The study for both SLM- and EBM-produced Ti2448 samples in terms of building parameters, defects pores, microstructures and mechanical properties has been conducted [7]. It was reported that selective laser melting sample with a finer laser beam spot created a deeper melt pool with more defect pores than that of EBM-produced sample. The formation of defect pore was significantly affected by the tin vaporization during the manufacturing process. The defects with a size smaller than 50 μm had limited effect on the compressive properties of the porous samples, but could decrease the fatigue properties of the sample under the higher stress levels [7]. More results on the processing and mechanical performance of the EBM-produced porous titanium structures have been summarized in our recent review article [30]. Very recently, it was reported that EBM-produced functionally graded Ti-6Al-4V alloy interconnected mesh structures show a combination of low density (0.5–2 g/cm3), high fatigue strength (~70 MPa) and energy absorption (~50 MJ/mg), which is superior to the ordinary uniform cellular structures [62].

## 4 Conclusion

This chapter summarizes the development of biomedical titanium for selective laser melting and electron beam melting. Biomedical titanium alloys such as Ti-6Al-4V are generally preferred materials for medical implants because of their low Young's modulus, superior biocompatibility, and high corrosion resistance compared to stainless steels and CoCr alloys. Porous structure can further improve the biocompatibility and reduce the Young's modulus of titanium alloy. Ti-24Nb-4Zr-8Sn (Ti2448) and other new biomedical titanium alloys with low modulus and non-toxic components will be the promisng material of choice for the next generation of biomedical implants.

Additive manufacturing (AM) technology based on titanium biomaterials has great potential in the medical industry. The complex structure made by AM provides enhanced mechanical properties and improved bone in-growth in the manufacture of artificial implants. Further research needs to focus on improving the roughness of AM implants by surface treatment, and reducing defects by optimizing building process. More studies are needed to prepare and fabricate for gradient porous structures with gradient Young's modulus to reduce the stress concentration of the implants, and to design and manufacture more reliable implants that are more suitable for the human body and to bring better clinic results for patients.

## References

1. Long M, Rack H (1998) Titanium alloys in total joint replacement—a materials science perspective. Biomaterials 19 (18):1621-1639
2. Gu DD, Meiners W, Wissenbach K, Poprawe R (2012) Laser additive manufacturing of metallic components: materials, processes and mechanisms. International Materials Review 57 (3):133-164.
3. Wang XJ, Xu SQ, Zhou SW, Xu W, Leary M, Choong P, Qian M, Brandt M, Xie YM (2016) Topological design and additive manufacturing of porous metals for bone scaffolds and orthopaedic implants: a review. Biomaterials 83:127-141
4. Dai N, Zhang LC, Zhang J, Zhang X, Ni Q, Chen Y, Wu M, Yang C (2016) Distinction in Corrosion Resistance of Selective Laser Melted Ti-6Al-4V Alloy on Different Planes. Corrosion Science 111:703-710
5. de Formanoir C, Michotte S, Rigo O, Germain L, Godet S (2016) Electron beam melted Ti–6Al–4V: Microstructure, texture and mechanical behavior of the as-built and heat-treated material. Materials Science and Engineering: A 652:105-119
6. Liu YJ, Li SJ, Hou WT, Wang SG, Hao YL, Yang R, Sercombe TB, Zhang LC (2016) Electron beam melted beta-type Ti-24Nb-4Zr-8Sn porous structures with high strength-to-modulus ratio. Journal of Materials Science & Technology 32 (6):505-508
7. Liu YJ, Li SJ, Wang HL, Hou WT, Hao YL, Yang R, Sercombe TB, Zhang LC (2016) Microstructure, defects and mechanical behavior of beta-type titanium porous structures manufactured by electron beam melting and selective laser melting. Acta Materialia 113:56-67
8. Attar H, Bönisch M, Calin M, Zhang LC, Scudino S, Eckert J (2014) Selective laser melting of in situ titanium–titanium boride composites: Processing, microstructure and mechanical properties. Acta Materialia 76 (9):13–22
9. Attar H, Calin M, Zhang LC, Scudino S, Eckert J (2014) Manufacture by selective laser melting and mechanical behavior of commercially pure titanium. Materials Science and Engineering: A 593:170-177
10. Liu YJ, Li X, Zhang LC, Sercombe T (2015) Processing and properties of topologically optimised biomedical Ti–24Nb–4Zr–8Sn scaffolds manufactured by selective laser melting. Materials Science and Engineering: A 642:268-278
11. Attar H, Löber L, Funk A, Calin M, Zhang LC, Prashanth KG, Scudino S, Zhang YS, Eckert J (2015) Mechanical behavior of porous commercially pure Ti and Ti-TiB composite materials manufactured by selective laser melting. Materials Science Engineering: A 625:350-356
12. Tan XP, Kok YH, Tan YJ, Descoins M, Mangelinck D, Tor SB, Leong KF, Chua CK (2015) Graded microstructure and mechanical properties of additive manufactured Ti–6Al–4V via electron beam melting. Acta Materialia 97:1-16
13. Zhao XL, Li SJ, Zhang M, Liu YD, Sercombe TB, Wang SG, Hao YL, Yang R, Murr LE (2016) Comparison of the microstructures and mechanical properties of Ti–6Al–4V fabricated by selective laser melting and electron beam melting. Materials & Design 95:21-31
14. Liu YJ, Wang HL, Li SJ, Wang SG, Wang WJ, Hou WT, Hao YL, Yang R, Zhang LC (2017) Compressive and fatigue behavior of beta-type titanium porous structures fabricated by electron beam melting. Acta Materialia 126:58-66
15. Chua CK, Leong KF (2015) 3D printing and additive manufacturing: principles and applications

16. Zhang LC, Klemm D, Eckert J, Hao YL, Sercombe TB (2011) Manufacture by selective laser melting and mechanical behavior of a biomedical Ti–24Nb–4Zr–8Sn alloy. Scripta Materialia 65 (1):21-24
17. Liu ZH, Zhang DQ, Chua CK, Leong KF (2013) Crystal structure analysis of M2 high speed steel parts produced by selective laser melting. Materials Characterization 84 (10):72–80
18. Ramirez DA, Murr LE, Martinez E, Hernandez DH, Martinez JL, Machado BI, Medina F, Frigola P, Wicker RB (2011) Novel precipitate-microstructural architecture developed in the fabrication of solid copper components by additive manufacturing using electron beam melting. Acta Materialia 59 (10):4088-4099
19. Sun SH, Koizumi Y, Kurosu S, Li YP, Chiba A (2015) Phase and grain size inhomogeneity and their influences on creep behavior of Co–Cr–Mo alloy additive manufactured by electron beam melting. Acta Materialia 86:305–318
20. Riedlbauer D, Drexler M, Drummer D, Steinmann P, Mergheim J (2014) Modelling, simulation and experimental validation of heat transfer in selective laser melting of the polymeric material PA12. Computational Materials Science 93:239-248
21. Wilkes J, Hagedorn Y-C, Meiners W, Wissenbach K (2013) Additive manufacturing of ZrO2-Al2O3 ceramic components by selective laser melting. Rapid Prototyping Journal 19 (1):51-57
22. Prashanth K, Scudino S, Klauss H, Surreddi KB, Löber L, Wang Z, Chaubey A, Kühn U, Eckert J (2014) Microstructure and mechanical properties of Al–12Si produced by selective laser melting: Effect of heat treatment. Materials Science and Engineering: A 590:153-160
23. Hrabe NW, Heinl P, Flinn B, Körner C, Bordia RK (2011) Compression-compression fatigue of selective electron beam melted cellular titanium (Ti-6Al-4V). Journal of Biomedical Materials Research Part B: Applied Biomaterials 99 (2):313-320
24. Lin CY, Wirtz T, LaMarca F, Hollister SJ (2007) Structural and mechanical evaluations of a topology optimized titanium interbody fusion cage fabricated by selective laser melting process. Journal of Biomedical Materials Research Part A 83 (2):272-279
25. Zhang LC, Attar H (2016) Selective laser melting of titanium alloys and titanium matrix composites for biomedical applications: a review. Advanced Engineering Materials 18 (4):463-475
26. Thijs L, Verhaeghe F, Craeghs T, Van Humbeeck J, Kruth JP (2010) A study of the microstructural evolution during selective laser melting of Ti–6Al–4V. Acta Materialia 58 (9):3303-3312
27. Parthasarathy J, Starly B, Raman S, Christensen A (2010) Mechanical evaluation of porous titanium (Ti6Al4V) structures with electron beam melting (EBM). Journal of the Mechanical Behavior of Biomedical Materials 3 (3):249-259
28. Mohammadhosseini A, Masood SH, Fraser D, Jahedi M (2015) Dynamic compressive behaviour of Ti-6Al-4V alloy processed by electron beam melting under high strain rate loading. Advanced Manufacturing 3 (3):232-243
29. Murr LE, Gaytan SM, Ceylan A, Martinez E, Martinez JL, Hernandez DH, Machado BI, Ramirez DA, Medina F, Collins S (2010) Characterization of titanium aluminide alloy components fabricated by additive manufacturing using electron beam melting. Acta Materialia 58 (5):1887-1894
30. Zhang LC, Liu YJ, Li SJ, Hao YL (2018) Additive manufacturing of titanium alloys by electron beam melting: a review. Advanced Engineering Materials 20 (5):1700842
31. Zhang H, Lewis CG, Aronow MS, Gronowicz GA (2004) The effects of patient age on human osteoblasts' response to Ti–6Al–4V implants in vitro. Journal of Orthopaedic Research 22 (1):30-38
32. Johansson CB, Albrektsson T, Ericson LE, Thomsen P (1992) A quantitative comparison of the cell response to commercially pure titanium and Ti-6Al-4V implants in the abdominal wall of rats. Journal of Materials Science Materials in Medicine 3 (2):126-136
33. Haghighi SE, Lu H, Jian G, Cao G, Habibi D, Zhang LC (2015) Effect of $\alpha''$martensite on the microstructure and mechanical properties of beta-type Ti–Fe–Ta alloys. Materials & Design 76:47-54

34. Rao S, Okazaki Y, Tateishi T, Ushida T, Ito Y (1997) Cytocompatibility of new Ti alloy without Al and V by evaluating the relative growth ratios of fibroblasts L929 and osteoblasts MC3T3-E1 cells. Materials Science and Engineering: C 4 (4):311-314
35. Niinomi M, Nakai M, Hieda J (2012) Development of new metallic alloys for biomedical applications. Acta Biomaterialia 8 (11):3888-3903
36. Eisenbarth E, Velten D, Müller M, Thull R, Breme J (2004) Biocompatibility of β-stabilizing elements of titanium alloys. Biomaterials 25 (26):5705-5713
37. Zardiackas LD, Mitchell DW, Disegi JA (1996) Characterization of ti-15Mo Beta Titanium Alloy for Orthopaedic. Medical Applications of Titanium and Its Alloys: The Material and Biological Issues (1272):60
38. Ho W, Ju C, Lin JC (1999) Structure and properties of cast binary Ti–Mo alloys. Biomaterials 20 (22):2115-2122
39. Kim H, Ikehara Y, Kim J, Hosoda H, Miyazaki S (2006) Martensitic transformation, shape memory effect and superelasticity of Ti–Nb binary alloys. Acta Materialia 54 (9):2419-2429
40. Hao Y, Li S, Sun S, Yang R (2006) Effect of Zr and Sn on Young's modulus and superelasticity of Ti–Nb-based alloys. Materials Science and Engineering: A 441 (1):112-118
41. Al-Bermani SS, Blackmore ML, Zhang W, Todd I (2010) The origin of microstructural diversity, texture, and mechanical properties in electron beam melted Ti-6Al-4V. Metallurgical and Materials Transaction A 41 (13):3422-3434
42. Vrancken B, Thijs L, Kruth J-P, Van Humbeeck J (2012) Heat treatment of Ti6Al4V produced by Selective Laser Melting: Microstructure and mechanical properties. Journal of Alloys and Compounds 541:177-185
43. Facchini L, Magalini E, Robotti P, Molinari A (2009) Microstructure and mechanical properties of Ti-6Al-4V produced by electron beam melting of pre-alloyed powders. Rapid Prototyping Journal 15 (3):171-178
44. Gaytan SM, Murr LE, Medina F, Martinez E, Lopez MI, Wicker RB (2009) Advanced metal powder based manufacturing of complex components by electron beam melting. Materials Technololgy 24:180-190
45. Murr LE, Esquivel EV, Quinones SA, Gaytan SM, Lopez MI, Martinez EY, Medina F, Hernandez DH, Martinez E, Martinez JL (2009) Microstructures and mechanical properties of electron beam-rapid manufactured Ti-6Al-4V biomedical prototypes compared to wrought Ti–6Al–4V. Materials Characterization 60 (2):96-105
46. Koike M, Martinez K, Guo L, Chahine G, Kovacevic R, Okabe T (2011) Evaluation of titanium alloy fabricated using electron beam melting system for dental applications. Journal of Materials Processing Technology 211 (8):1400-1408
47. Scharowsky T, Juechter V, Singer RF, Körner C (2015) Influence of the scanning strategy on the microstructure and mechanical properties in selective electron beam melting of Ti–6Al–4V. Advanced Engineering Materials 17 (11):1573-1578
48. Baudana G, Biamino S, Klöden B, Kirchner A, Weißgärber T, Kieback B, Pavese M, Ugues D, Fino P, Badini C (2016) Electron beam melting of Ti-48Al-2Nb-0.7 Cr-0.3 Si: feasibility investigation. Intermetallics 73:43-49
49. Hernandez J, Li SJ, Martinez E, Murr LE, Pan XM, Amato KN, Cheng XY, Yang F, Terrazas CA, Gaytan SM (2013) Microstructures and Hardness Properties for β-Phase Ti-24Nb-4Zr-7.9 Sn Alloy Fabricated by Electron Beam Melting. Journal of Materials Science & Technology 29 (11):1011-1017
50. Imanishi J, Choong PF (2015) Three-dimensional printed calcaneal prosthesis following total calcanectomy. International Journal of Surgery Case Reports 10:83-87
51. Gu D, Hagedorn Y-C, Meiners W, Meng G, Batista RJS, Wissenbach K, Poprawe R (2012) Densification behavior, microstructure evolution, and wear performance of selective laser melting processed commercially pure titanium. Acta Materialia 60 (9):3849-3860
52. Dai N, Zhang LC, Zhang J, Chen Q, Wu M (2016) Corrosion Behaviour of Selective Laser Melted Ti-6Al-4V Alloy in NaCl Solution. Corrosion Science 102:484-489

53. Das M, Balla VK, Basu D, Bose S, Bandyopadhyay A (2010) Laser processing of SiC-particle-reinforced coating on titanium. Scripta Materialia 63 (4):438-441
54. Niinomi M (1998) Mechanical properties of biomedical titanium alloys. Materials Science and Engineering: A 243 (1):231-236
55. Suo J, Chen H, Li Z (2009) Mechanical Properties of Ti-6Al-4V Alloys by Electron Beam Melting (EBM)[J]. Aerospace Manufacturing Techology 6:6
56. Malinauskas M, Rekštytė S, Lukoševičius L, Butkus S, Balčiūnas E, Pečiukaitytė M, Baltriukienė D, Bukelskienė V, Butkevičius A, Kucevičius P (2014) 3D Microporous Scaffolds Manufactured via Combination of Fused Filament Fabrication and Direct Laser Writing Ablation. Micromachines 5 (4):839-858
57. Bai Y, Gai X, Li SJ, Zhang LC, Liu YJ, Hao YL, Zhang X, Yang R, Gao YB (2017) Improved corrosion behaviour of electron beam melted Ti-6Al–4 V alloy in phosphate buffered saline. Corrosion Science 123:289-296
58. Zhao S, Li SJ, Hou WT, Hao YL, Yang R, Misra RDK (2016) The influence of cell morphology on the compressive fatigue behavior of Ti-6Al-4V meshes fabricated by electron beam melting. Journal of the Mechanical Behavior of Biomedical Materials 59:251-264
59. Kufelt O, Eltamer A, Sehring C, Schliewolter S, Chichkov BN (2014) Hyaluronic acid based materials for scaffolding via two-photon polymerization. Biomacromol 15 (2):650-659
60. Liu YJ, Liu Z, Jiang Y, Wang GW, Yang Y, Zhang LC (2018) Gradient in microstructure and mechanical property of selective laser melted AlSi10Mg. Journal of Alloys and Compounds 735:1414-1421
61. Chen Y, Zhang J, Gu X, Dai N, Qin P, Zhang LC (2018) Distinction of corrosion resistance of selective laser melted Al-12Si alloy on different planes. Journal of Alloys and Compounds 747:648-658
62. Zhao S, Li SJ, Wang SG, Hou WT, Li Y, Zhang LC, Hao YL, Yang R, Misra RDK, Murr LE (2018) Compressive and fatigue behavior of functionally graded Ti-6Al-4V meshes fabricated by electron beam melting. Acta Materialia 150:1-15
63. Liu YJ, Li SJ, Zhang LC, Hao YL, Sercombe TB (2018) Early plastic deformation behaviour and energy absorption in porous β-type biomedical titanium produced by selective laser melting. Scripta Materialia 153:99-103

# Corrosion Behaviors of Additive Manufactured Titanium Alloys

**Lai-Chang Zhang and Peng Qin**

## 1 Introduction

In recent decades, thanks to their outstanding properties such as high strength, lower Young's modulus, high wear resistance properties, excellent biocompatibility and corrosion resistance, titanium and titanium-based alloys have met the needs of various industries, in particular in the field of bone implants [1–3]. Coupled with the recent series of bone diseases caused by the frequent traffic accidents and aging, the demand for biological materials has dramatically increased, especially for titanium alloys [4–6]. However, the production of these biomaterials (especially titanium alloys) often requires very complicated steps, resulting in high cost of a product. The typical production methods often include solidification, casting, powder metallurgy and foaming [7]. Unfortunately, these methods usually provide large amounts of material waste, energy consumption, complicated manufacturing processes and relatively long production cycles including the subsequent material removal steps to achieve the desired shape [7]. The recent emerging metal additive manufacturing (AM) technologies that can manufacture metal components with complex shapes. For example, selective laser melting (SLM) and electron beam melting (EBM) are the almost perfect solutions to the above problems. These technologies mainly utilize a direct selective linear scanning of high-energy sources like laser or electron beam to heat a layer of metal powder homogeneous laid on the powder bed, so that these metal powder can be instantly melt and solidified, which can continue to be accumulated layers to construct complex parts [8]. This complex component just meets the basic needs of the shape of the bone implants.

In the previous works, the SLM-produced Ti-6Al-4V reveals a different biphasic microstructure that contains acicular α' phase and columnar prior β phase [9–11].

L.-C. Zhang (✉) · P. Qin
Edith Cowan University, Perth, Australia
e-mail: l.zhang@ecu.edu.au



B. AlMangour (ed.), *Additive Manufacturing of Emerging Materials*,
https://doi.org/10.1007/978-3-319-91713-9_6

This biphasic structure has demonstrated a significant improvement in mechanical properties due to the rapid solidification from β phase to α phase. Compared with Ti-6Al-4V alloy made by the traditional method, the work [9, 12] has proved that SLM-produced Ti-6Al-4V has higher hardness, wear resistance, improvement Young's modulus and UTS. This enhanced Ti-6Al-4V alloy can significantly reduce the metal debris generated by friction and wear in human applications. In the other work [4, 6, 13], the SLM- and EBM-produced porous Ti-24Nb-4Zr-8Sn also reflect an idea mechanical properties (close to bones) in the application of biomedical, in which the compressive strength reached ~50 MPa and Young's modulus arrived at ~1GPa. However, the α' phase of Ti-6Al-4V produced by SLM is not optimistic in the corrosion test, and the corrosion occurs preferentially on the supersaturating solid solution - needle like α' phase [9]. However, not all of the alloys produced by AM technology have the same result. For example, the EBM-produced Ti-6Al-4V has caught the traditional wrought of Ti-6Al-4V regarding corrosion resistance [14]. Nevertheless, the AM technology is not only limited to Ti-6Al-4V alloy, like the other (such as CP-Ti, Ti-6Al-4V and TiB [9, 14, 15]), CoCr-based alloys (like CoCrW, CoCrMo and CoCrWMo [16–18]), stainless steel [19, 20]. Al-based alloys [21, 22], and even metallic glass [23] are also produced by AM techniques. Therefore, this chapter mainly reviews the corrosion behavior of the alloys (Ti alloy, CoCr alloy and stainless steel) recently fabricated by AM method.

## 2 SLM-Produced Ti-6Al-4V

SLM as one of the AM techniques can produce the Ti-6Al-4V alloy by using the laser selectively melting the metal powder directly from the computer-aided design (CAD) model. Because the laser can rapidly generate a significant amount of heat to melt the metal powder and the molten metal powder solidifies instantaneously due to the temperature difference of the melt pool, it is favored to make titanium alloy. The production of finished products also has a comparable or even better mechanical properties than the alloy manufactured by conventional methods. In this section, the difference in corrosion performance between Ti-6Al-4V alloy made by SLM and commercial Grade 5, the corrosion behavior of Ti-6Al-4V alloy between the different build planes, and the SLM-produced Ti-6Al-4V alloy after heat treatment at various temperatures were reviewed.

As the SLM manufacturing strategy is scanning line by line through the laser, and then layers stack together to form a composite. It is the unique scanning trajectory of SLM that makes the titanium alloys have different microstructure compared with the conventional manufacturing methods. Figure 1 [9, 10] summarizes the microstructure after heat treatment at different temperatures of the Ti-6Al-4V manufactured by SLM and commercial Grade 5 alloy. Figure 1a shows a classical titanium alloy [24] with biphasic microstructure made by conventional methods: α phase (bright region) and β phase (dark area). In Fig. 1b, the SLM-produced Ti-6Al-4V have clear and visible columnar prior β phase grows along the direction of manufacturing (normal

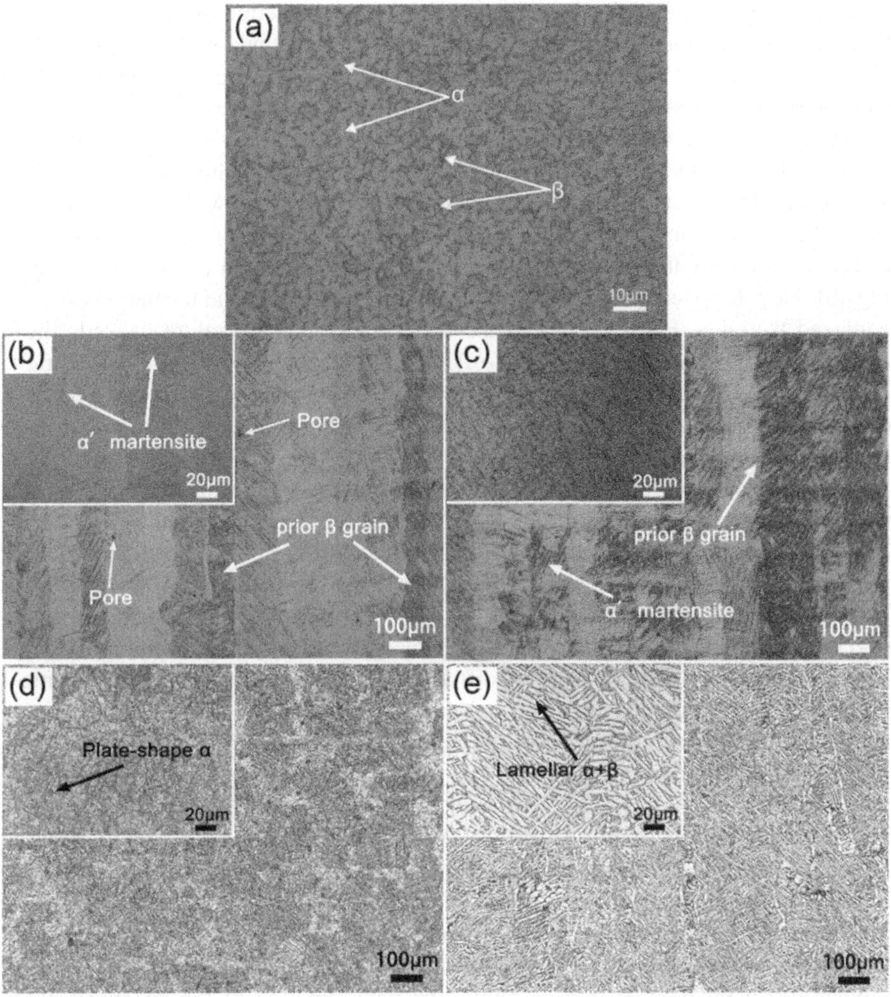


**Fig. 1** Optical microstructure of: (**a**) commercial Grade 5, (**b**) SLM-produced Ti-6Al-4V alloy, and heat-treated samples at (**c**) 500 °C, (**d**) 850 °C and (**e**) 1000 °C. The insets are the magnified microstructures [9, 11]

to the paper), and the plate-like α phase presented in commercial Grade 5 alloy entirely replaced by the overwhelming acicular α' phase. The formation of acicular α' phase is associated with SLM manufacturing process: the kinetic energy and thermal characteristics of the melted powder has been increased due to the faster laser scanning speed, thus increased the temperature gradient in the molten pool will result in the formation of the acicular α' phase [7]. The forms α' phase due to the rapid cooling usually causing an un-equilibrium supersaturated solid solution in the microstructure, which is considered to be a "high energy state" that is detrimental to corrosion [9]. In addition, during the SLM process, the defect pores shown in Fig. 1b are formed due to the surface roughness, and the accumulation of the metal powder surround the molten pool [25]. The presence of this kind of defect pores also increases the pitting potential of the alloy. When the SLM-produced Ti-6Al-4V is subject to heat treatment, the microstructure will also be changed. However, there is no apparent change of the original acicular α' phase after a 500 °C heat treatment, as shown in Fig. 1c, the combination of acicular α' phase and columnar prior β phase still visible. However, after the heat treatment at 850 °C (below the temperature of the β phase transition), the acicular α' phase and columnar prior β phase completely disappeared and replaced by the plate-like α phase and the β phase, as shown in Fig. 1d. This disappearance of the columnar β phase indicates the formation of semi-equiaxed β grains. However, when the heat treatment temperature exceeds the β phase transition temperature of the Ti-6Al-4V alloy (~995 °C), the microstructure is further changed, and the previously formed plate-like α phase evolves into a mixture of lamellar α + β phase. The microstructures of these heat treated alloys are similar to the literature in [26, 27].

The calculated volume fraction from XRD patterns in the literature [9] reveals that the distribution of different volume fraction of α, α' and β phase in SLM-produced Ti-6Al-4V alloy and the commercial Grade 5 alloy, which alters the microstructure of the alloy and ultimately leads to the differences in corrosion behavior. The SLM-produced Ti-6Al-4V alloy contains 95% metastable structure α'-Ti phase, and the α'-Ti phase has a high energy state in corrosion and it dissolves easily in corrosive media. This is different from the results in the mechanical properties, that is grain refinement and the α'-Ti can improve the wear properties of the alloy [7]. Correspondingly, the content of β-phase in the microstructre is relatively low. At the same time, more V element in the β phase can resist the dissolution of the alloy, and the $Ti_2O$ oxide film formed on the β phase is more stable than the α phase [28]. Therefore, the commercial Grade 5 alloy with a relatively large β-phase content (13.3%) shows a higher corrosion resistance than the SLM-produced counterparts with lower β-phase content (5%) [9].

During the electrochemical tests, the open circuit potential not only can predict the changes of the electrode surface but also can know the relative corrosion probability of the sample. For example, the determination of the produced oxide film surface of the sample is observed through the rise of open circuit potential (OCP) curve. Such a rise in OCP indicates the improvement of the difficulty of the electron transfer and thus the reduction of the kinetic energy of the anode reaction [9, 11, 29]. Once the oxide film is dissolved, the electrolyte will reach the surface of the electrode, therefore a decreasing trend of the OCP curve can be observed. Figure 2 [9–11] compares the OCP curves of SLM-produced Ti-6Al-4V under different states and different electrolyte environments. Figure 2a is the OCP of commercial Grade 5 alloy and SLM-produced Ti-6Al-4V alloy tested in 3.5 wt.% NaCl solution and Fig. 2a inset shows the schematic diagram of test plane schematic diagram. The OCP of the two alloys is continuously positive shift, indicating that the two alloys have the same cathode process, and the anode process is inhibited. However, there are two processes, i.e. the initial formation process of the passivation film and the stabilization and thickening process of the passivation film, can be

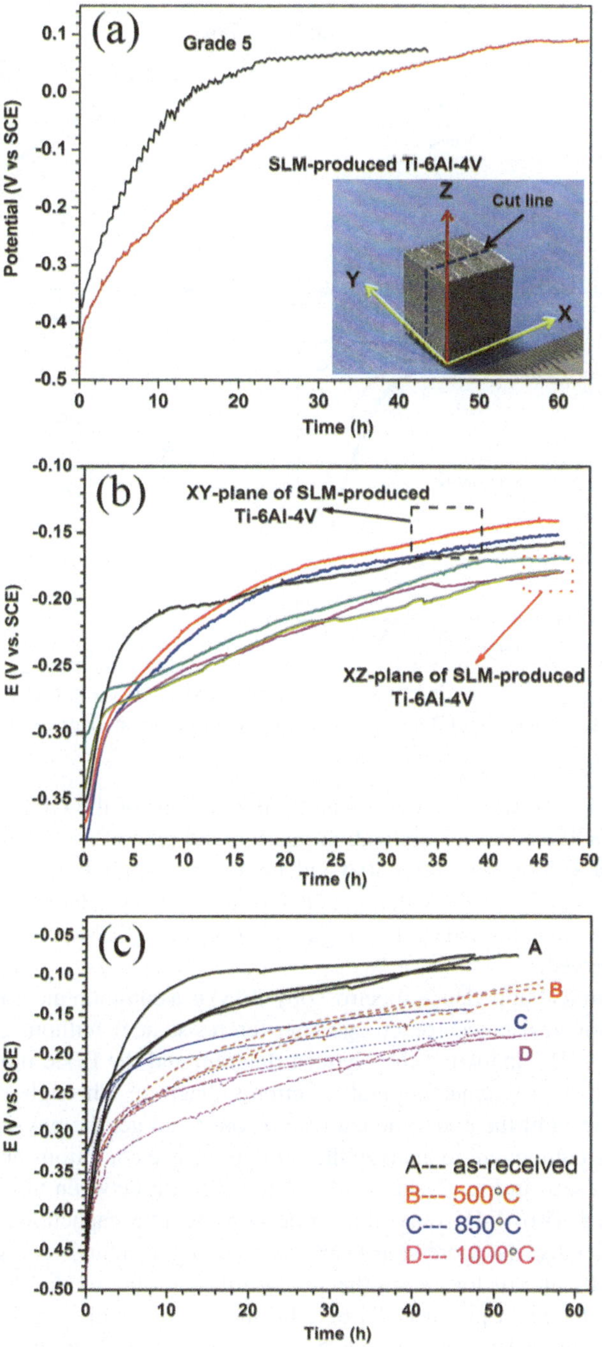


**Fig. 2** OCP of: (**a**) commercial Grade 5 and SLM-produced Ti-6Al-4V in 3.5 wt.% NaCl, and SLM-produced Ti-6Al-4V of (**b**) XY-plane and XZ-plane in 1 M HCl, and (**c**) at different heat treatment temperatures in 3.5 wt.% NaCl [9–11]

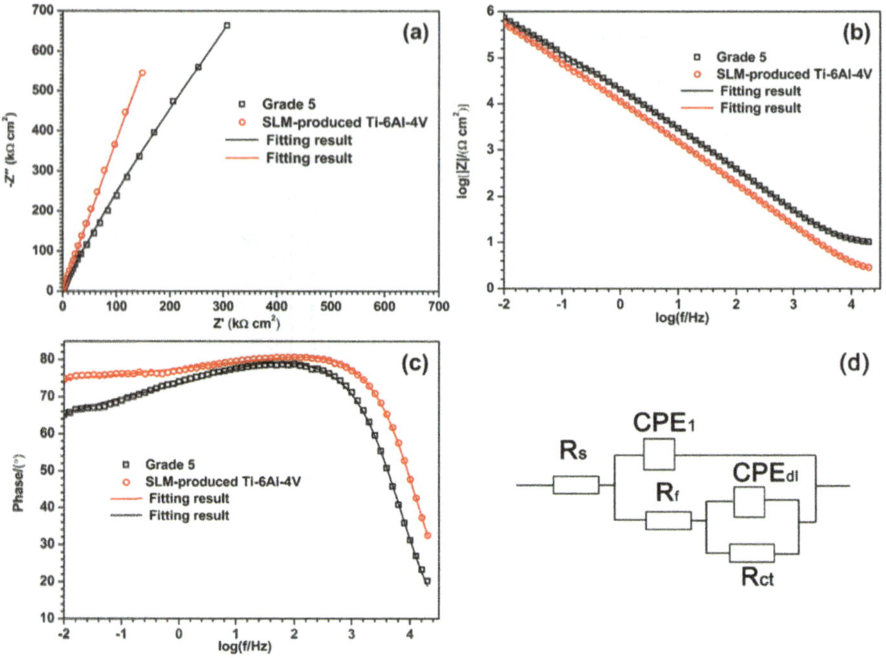


**Fig. 3** The EIS test of commercial Grade 5 alloy and the SLM-produced Ti-6Al-4V alloy in 3.5 wt. % NaCl: (**a**) Nyquist plot, (**b**, **c**) Bode plot, and (**d**) equivalent circuit diagram [9]

observed in Fig. 2a, which is evidenced by fluctuations of the OCP. For the Grade 5 alloy, the OCP reaches a stabilization phase at the time of 25 h, but the relatively stable stage of SLM reaches after 50 h with continued rise OCP in a small amplitude. The results of this figure show that the OCP of the alloys produced by two different methods is the same, but the slightly higher OCP value of SLM-produced Ti-6Al-4V has been observed.

Electrochemical impedance spectroscopy (EIS) is a common method that predicts the polarization resistance (or charge transfer resistance) without destroying the sample surface. By analyzing the Nyquist diagram and the Bode plot diagram can determine the approximate equivalent circuit diagram. Through the equivalent circuit diagram to fit the experimental results, once can get the physical parameters of each fitting component to analyze the sample surface conditions and polarization resistance. As seen in Fig. 3, the results of the EIS test between Grade 5 alloy and SLM-produced Ti-6Al-4V only show one considerable capacitor arc in Nyquist diagram. Generally, the more significant arc (i.e. larger radius) represents the more substantial difficulty of ions from the solution through the oxide film [14, 30]. As seen in Fig. 3, the SLM-produced Ti-6Al-4V shows a relatively larger arc, indicating that the corrosion resistance of Grade 5 are inferior to SLM-produced alloy. However, this conclusion is not always inevitable, and its ultimate resistance is determined by the polarization impedance. In EIS test, the polarization resistance is

**Table 1** Summary of the results of EIS fitting for recent SLM-produced Ti-6Al-4V

| Sample | State | Test solution | $R_s$ | $R_f$ | $R_{ct}$ | Ref. |
|---|---|---|---|---|---|---|
| | | | ($\Omega \cdot cm^2$) | ($k\Omega \cdot cm^2$) | ($M\Omega \cdot cm^2$) | |
| Grade 5 | – | 3.5 wt.% NaCl | – | ~94.12 | ~58.99 | [9] |
| SLM-produced Ti-6Al-4V | – | 3.5 wt.% NaCl | – | ~40.26 | ~17.41 | [9] |
| | XY-plane | 3.5 wt.% NaCl | – | ~26.84 | ~0.82 | [11] |
| | XZ-plane | 3.5 wt.% NaCl | – | ~22.69 | ~0.80 | [11] |
| | XY-plane | 1 M HCl | – | ~5.63 | ~0.29 | [11] |
| | XZ-plane | 1 M HCl | – | ~1.67 | ~0.13 | [11] |
| | Original | 3.5 wt.% NaCl | ~4.7 | – | ~4.7 | [10] |
| | Heat treated 500 °C | 3.5 wt.% NaCl | ~3.8 | – | ~2.4 | [10] |
| | Heat treated 850 °C | 3.5 wt.% NaCl | ~5.3 | – | ~1.8 | [10] |
| | Heat treated 1000 °C | 3.5 wt.% NaCl | ~4.9 | – | ~0.6 | [10] |

usually estimated by the summation of the oxide film resistance and charge transfer resistance [30], which are summarized in Table 1. As seen from Table 1, the $R_s$, $R_f$ and $R_{ct}$ represent the solution resistance, oxide film resistance and charge transfer resistance.

The equivalent circuit diagram also needs the analysis of bode plot. In Bode phase angle diagram (Fig. 3), a relatively broad plateau has been observed from the middle to low frequency. Such a broad plateau can be considered as the coupling of two relatively large plateaus. Generally, the closer of the plateau to 90°, the denser of the oxide film formed on the electrode surface [29]. So the two-time constants equivalent circuit diagram can be the preliminary estimated to fit the EIS results in Fig. 3. Also the thickness of the oxide film formed on the sample surface is not uniform due to the electron irregular accumulated on the electrode surface when the sample immersed in the electrolyte. This phenomenon will lead to the different ability to generate the oxide film on sample surface that results in the ununiformed thickness of the oxide film as well as the roughness of the sample surface [9]. Besides, the formed oxide film can be considered as the combination of compact inner layer and loose outer layer, which can also cause the nonuniform surface roughness of sample [9, 29]. So the capacitor component of the fitting circuit diagram can be replaced by the constant phase element (CPE), whereby the final two-time constant equivalent circuit diagram can be determined to simulate the EIS data in Fig. 3.

Nonetheless, the fitting circuit diagram does not have a unified standard as long as it can reasonably explain the EIS plot. It is apparent from Table 1 (fitting parameters of the component) that the oxide film resistance of all tested alloys does not have significant effect on the charge transfer resistance; that is since the impedance value obtained is not on a same magnitude order, the polarization resistance is affected only by the charge transfer resistance. In the comparison of Grade 5 alloy and SLM-produced Ti-6Al-4V alloy, the charge transfer resistance of the Grade 5 (~58.99 $M\Omega \cdot cm^2$) is 5 times higher than that of the SLM-produced Ti-6Al-4V alloy (~17.41 $M\Omega \cdot cm^2$). So, the Grade 5 alloy shows a better corrosion resistance than SLM-produced Ti-6Al-4V alloy. Similar results are obtained in the test of different manufacturing planes of SLM-produced Ti-6Al-4V. The fitting results in Table 1 are consistent with the subsequent polarization curves.

The potentiodynamic polarization curves of SLM-produced Ti-6Al-4V were tested and studied in 3.5 wt.% NaCl solution, 1 M HCl solution and human simulated solution (Hanks' solution), respectively. Figure 4a shows the polarization curves of commercial Grade 5 alloy and SLM-produced Ti-6Al-4V alloy tested in 3.5 wt.% NaCl solution. Firstly, these two alloys have a pronounced passivation area, which can be determined from the curve perpendicular to the X-axis. In the passivation zone, the current density remains substantially constant as the potential rises. The passivation current density in this figure is presented by $i_{P,A}$ (0.39 $\mu A/cm^2$) and $i_{P,B}$ (0.84 $\mu A/cm^2$), for Grade 5 and SLM-produced Ti-6Al-4V, individually. The smaller the passivation current density represents the ability of the alloy with advance access to the passivation zone. From the value of the passivation current density of these two alloys, the Grade 5 alloy is twice smaller than that the SLM-produced Ti-6Al-4V, which indicates that the passivation film is formed on the Grade 5 alloy only with a small current density, and the passivation ability is better than SLM-produced Ti-6Al-4V. There is another set of comparable values in this figure, $E_{b,A}$ and $E_{b,B}$, which represent the breakdown potential of the passivation film for Grade 5 and SLM-produced Ti-6Al-4V, respectively. This potential can be determined from the sudden increase in the passivation current density from the end of the passivation zone. It is apparent that the passivation current density of the commercial grade 5 alloy is the higher one. Also, the higher the passivation current density represents a tremendous ability of the alloy to maintain the passivation zone, and it is difficult to breakdown or dissolves by halide ions. Meanwhile, the size of the passivation region also can determine the corrosion resistance, such as the $\Delta E_1$ and $\Delta E_2$ in Fig. 4b.

Figure 4b shows the polarization curves of the different manufacturing plane (XY plane and XZ plane) for SLM-produced Ti-6Al-4V. Firstly, the corrosion behaviors of the two samples in 3.5 wt.% NaCl solution are similar due to the almost overlapped curves. Then the passivation region starts at the potential about 0.2 V for all tests. However, the oxide film ruptures at 1.2 V in 3.5 wt.% NaCl solution and 1.5 V in 1 M HCl solution. The results show that the passivation interval of SLM-produced Ti-6Al-4V in hydrochloric acid solution is larger than that in sodium chloride solution. Additionally, the breakdown potential of XZ plane is slightly more enormous than that of the XY plane, but they have different starting passivation

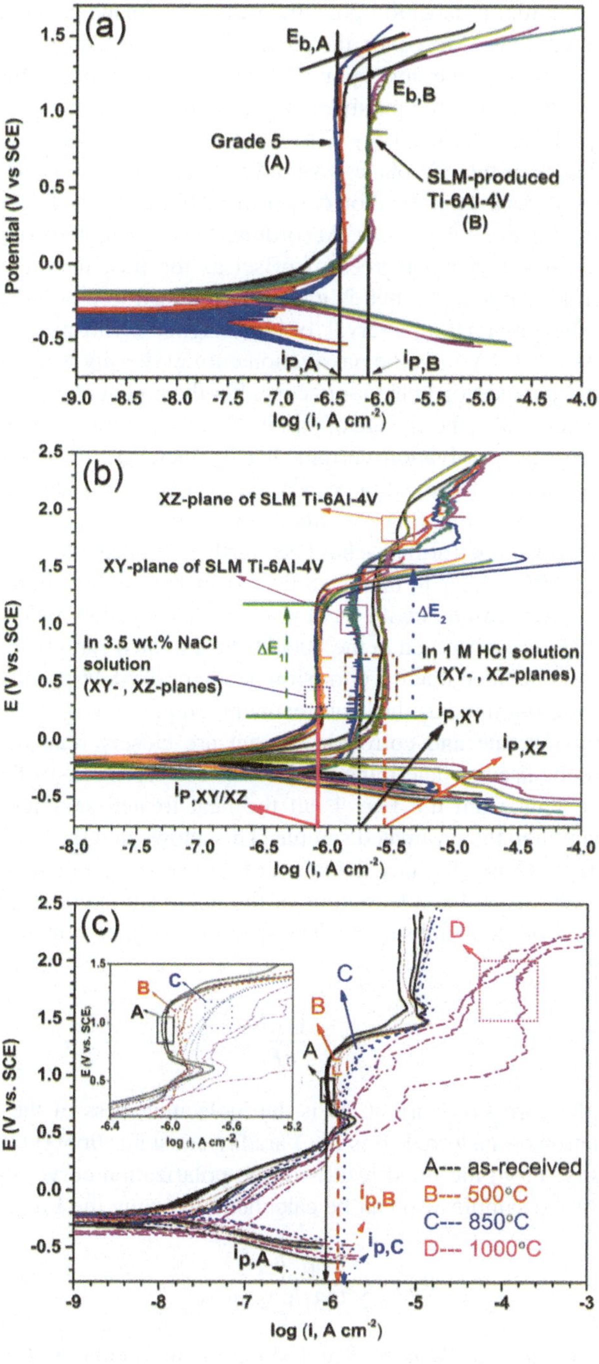


**Fig. 4** Potentiodynamic curves for the: (**a**) commercial Grade 5 and SLM-produced Ti-6Al-4V in 3.5wt.% NaCl, (**b**) SLM-produced Ti-6Al-4V in different planes (XY plane and XZ plane) in 3.5wt.% NaCl and 1M HCl, and (**c**) SLM-produced Ti-6Al-4V with heat treatment at 500°C, 850°C and 1000°C in 3.5wt.% NaCl [9–11]

potentials, as the initial passivation potential of the XY plane is significantly lower than that of the XZ plane. Moreover, the passivation current densities of the XY and XZ planes are similar to the corresponding ones in the sodium chloride solution (0.76 μA/cm$^2$) compared to the passivation current density in the hydrochloric acid solution ($i_{P,XY}$: 1.78 μA/cm$^2$, $i_{P,XZ}$ 2.81 μA/cm$^2$). It can be deduced that the corrosion resistance of the XY plane is slightly better than that of the XZ plane.

Figure 4c compares the polarization curves of SLM-produced Ti-6Al-4V alloys at different heat treatment temperatures. According to the figure, the polarization curve after 0.5 V also has varying degrees of offset as the heat treatment temperature increases, until the passivation area is gradually reduced to a non-detectable level. For example, the passivation interval of the original SLM-produced sample is between 0.65 V and 1.2 V, and the passivation current density is about 0.9 μA/cm$^2$. Also a secondary passivation behavior occurs after the potential exceeds 1.6 V. With the heat treatment to 500°, the change of passivation zone (between 0.6 and 1.25 V) is not apparent, but the passivation current density increases to about 1.3 μA/cm$^2$. Likewise, the secondary passivation occurred after 1.6 V. But the corrosion performance is reduced by comparing the fitted value of passivation current density. However, the corrosion performance has been further reduced after the heat treatment temperature of 850°, which is mainly reflected in the slightly higher passivation current density (~1.5 μA/cm$^2$) with a not noticable passivation area. Eventually, no passivation behavior is observed in the 1000°C heat treated sample. From the above results, it can be concluded that the corrosion behavior of SLM-produced Ti-6Al-4V will have a decreasing trend as the heat treatment temperature increases.

As the corrosion rate and corrosion current are closely related, the corrosion current can also be used to determine the relative corrosion rate of the sample [31–33]. As can be seen from the Fig. 4, all the heat treated samples have a larger corrosion current than the un-treated sample. This shows that the corrosion rate also has the same trend. Generally the calculation of the corrosion rate requires the fitting results of the cathode, and anode regions of the polarization curve; the density and equivalent weight of the metal sample. The specific calculation method as shown in Eq. 1 [34, 35]:

$$\mathrm{r} = \frac{\mathrm{j_{corr}M}}{nF\rho} \qquad \text{(Eq.1)}$$

where $j_{corr}$ is the corrosion current, M is the molecular mass of the alloy, n is the number of electrons transferred, F is the Faraday constant (96500 C/mol), and $\rho$ is the alloy density. From the fitted results of the polarization curve, the polarization resistance $R_p$ of the sample also can be calculated by using the Eq. 2 [30, 34]:

$$\mathrm{R_p} = \frac{\beta_a|\beta_c|}{2.303(\beta_a + |\beta_c|)}\frac{1}{\mathrm{j_{corr}}} \qquad \text{(Eq.2)}$$

where the $\beta_a$ and $\beta_c$ are the fitted Tafel slope of the anodic and cathodic curve, respectively. In addition, the polarization resistance has a similar meaning to the

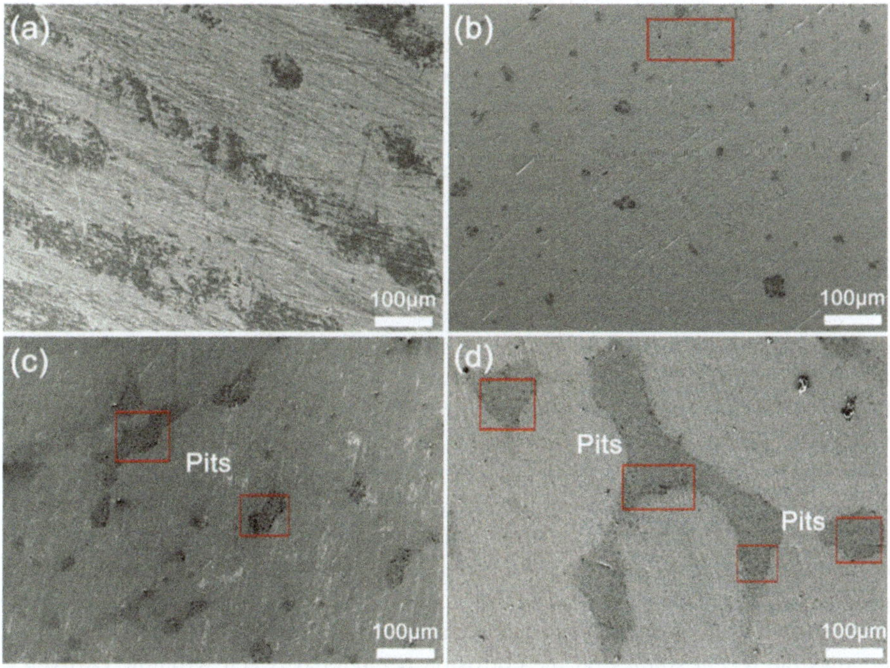


**Fig. 5** SEM images of SLM-produced Ti-6Al-4V and with heat treatment after the corrosion test in 3.5wt.% NaCl: (**a**) original state, (**b**) 500 °C heat treated, (**c**) 850 °C heat treated, and (**d**) 1000 °C heat treated [10]

polarization resistance appeared in EIS test. But the value could be different due to the different electrode surface conditions.

Overall, the XY plane of the SLM-produced Ti-6Al-4V in polarization tests shows a better corrosion performance than the XZ plane, and an inferior trend of the corrosion behavior as the heat treatment temperature increases. However, all the tested Ti-6Al-4V samples produced by SLM are not corrosion resistent than the commercial Grade 5 alloy. It can be ascribed to that the oxide film formed on the titanium alloys is very sensitive to the halogen ions (like $Cl^{-}$). So, the electrolyte will directly arrive the sample surface once the oxide film has broken by halogen ions [31, 32, 36]. In this case, a particular ion concentration difference will be formed between the oxide film and the generated pits on the oxide film will accelerate the corrosion rate in this area. Therefore, the corrosion morphology after the electrochemical test can be intuitive to help analyze the oxide film stability formed on the sample surface as well as the pitting. For example, the four pictures in Fig. 5 show the corrosion morphology of SLM-produced Ti-6Al-4V alloy and with 500 °C, 850 °C, and 1000 °C heat treatment, respectively. It can be seen from the figure that the oxide film on the surface of the four samples has different degrees of dissolution, in which the dark area is the oxide film dissolved region. The dissolution degree of the oxide film also increases with the higher heat treatment temperature.

The dissolution area increases to the final sheet-like dissolution morphology, the stability of the oxide film gradually reduced. Meanwhile, the number of pits increases continuously with the dissolved area is enlarged. It is proved that the heat treatment can reduce the corrosion resistance of SLM-produced Ti-6Al-4V alloy.

## 3 EBM-Produced Ti-6Al-4V

As one of the additive manufacturing technique, EBM has a more extensive power and faster scanning speed compared to SLM technique. Also, the energy of melting the powder in the EBM comes from the electron beam. This differs from the SLM technique by using the laser to direct melt the metal powder. The metal powder needs to be preheated in advance to avoid internal stress and cracks during a larger rapid cooling [37]. As such, EBM can manufacture a much broader range of microstructure of the finished alloy and offers different mechanical properties as well as corrosion properties from those by conventional processing or SLM. As the recent production of the EBM Ti-6Al-4V alloy is different in the scanning parameters, their microstructure and corrosion resistance may be also influenced. This section mainly reviews the recent studies on the corrosion behavior between wrought Ti-6Al-4V and EBM-produced Ti-6Al-4V with different fabrication planes.

Figure 6 shows the XRD patterns of wrought Ti-6Al-4V and Ti-6Al-4V produced by EBM. There is no difference in the phase constitution between the wrought Ti-6Al-4V and EBM-produced Ti-6Al-4V; both exhibite a $\alpha + \beta$ phase in phase constitution. The main difference is the peak intensity. Because the cooling rate of the molten metal powder in SLM is higher than that in the commercial manufacturing method, a significant amount of acicular α' phase is formed due to the thermal

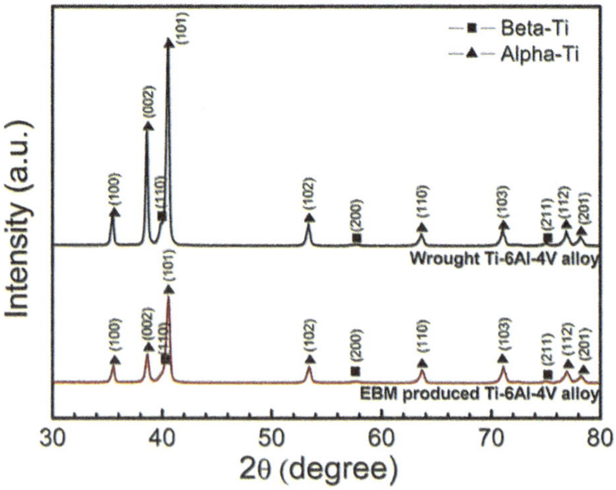


Fig. 6 The XRD patterns for wrought Ti-6Al-4V and EBM-produced Ti-6Al-4V in PBS solution [14]

**Table 2** The volume fraction ($V_f$) of EBM-produced Ti-6Al-4V alloy and wrought Ti-6Al-4V

| Sample | Cal. source | State | $V_{f,\alpha}$ | $V_{f,\beta}$ | Ref. |
|---|---|---|---|---|---|
| Wrought Ti-6Al-4V | XRD | | 95% | 5% | [14] |
| EBM-produced Ti-6Al-4V | XRD | Original | 89% | 11% | [14] |
| | EBSD | 0° | 98% | 2% | [37] |
| | EBSD | 45 °C | 95% | 6% | [37] |
| | EBSD | 55 °C | 98% | 2% | [37] |
| | EBSD | 90 °C | 96% | 4% | [37] |

diffusion [37], and the peak intensity is then reduced. However, as the metal powder was preheated at a constant of 800 K during EBM production, its cooling rate was much lower than that in SLM production [37], but it was still faster than wrought samples. Thus, the phase transition is related to the broadly cooling rate of the molten metal powder, and the phase volume fraction is also related to the cooling rate [38]. By fitting the XRD patterns, the volume fraction of each phases can also be calculated and summarized in Table 2. Since the AM technique is a fraction layer-wised production technology, such a influence in the microstructure is attributed to the temperature gradient of solidification process [7, 37]. Therefore, the distribution of its phase will not be consistent with the alloy by using the traditional production methods. The difference in volume fraction can be explained by using the micro-morphology and the calculated result of the XRD peak area. However, the β phase content is relatively low from the XRD pattern, so the corrosion behavior of the entire alloy is dominated by the α phase, which directly causes the 90°C sample with the most substantial peak (or the highest alpha phase content) to have the worst corrosion performance.

The peak intensity of the XRD can also reflect the relative phase width in the microstructure. As shown in Fig. 7, the EBM-produced samples have a much smaller α grain size due to the temperature difference formed in the molten pool. Similar to the SLM-produced samples, the SLM-produced samples with a higher cooling rate show grain refinement into an α' phase, as indistinguished by XRD. However, TEM observations confirm that the EBM samples have the similar microstructure to the wrought counterparts, and both are composed of α + β phases [14]. However, the width of α grain in the EBM sample measured in Fig. 7 is less than 1 μm, but between 2 and 6 μm in wrought samples.

The same phenomenon is also shown in SLM-produced Ti-6Al-4V sample, in which the size of α phase grains in SLM samples is significantly increased with increasing the heat treatment temperature [39], and their peak intensities are also enhanced. The EBM-produced Ti-6Al-4V samples at different building angles also exhibit a typical α phase after the thermal diffusion. The average length of the lamellar α phase is between 5 and 12 μm with an average width between 2 and 4 μm. Compared with the phase in Fig. 7, it is sufficiently demonstrated that EBM could produce a more varying microstructure through different manufacturing parameters. Furthermore, all EBM-produced Ti-6Al-4V samples are dominated by the α phase by EBSD analysis, while the β phase is sporadically distributed at the α

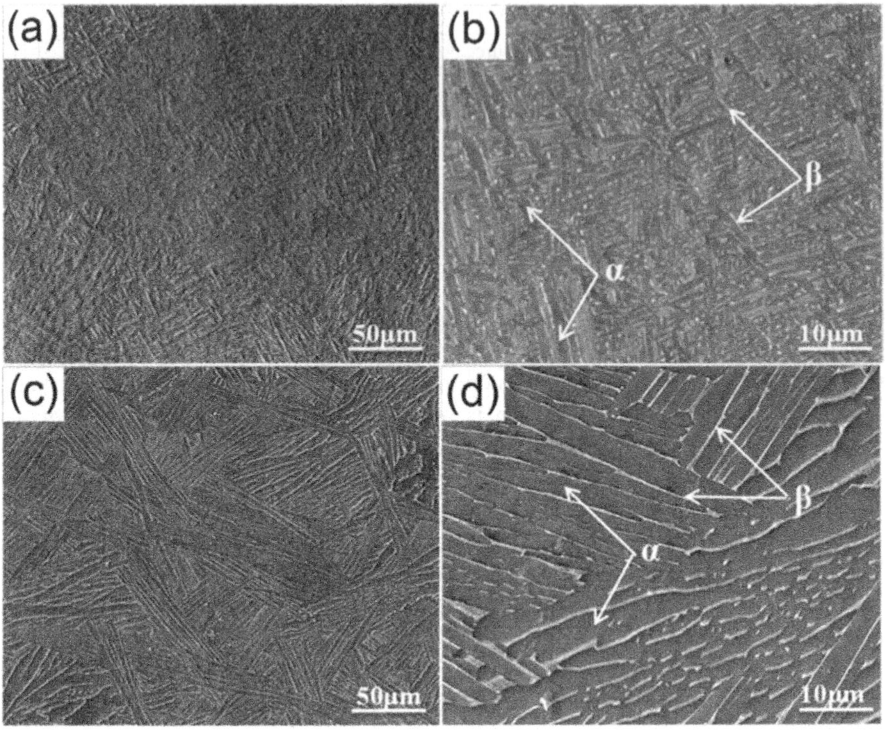


**Fig. 7** SEM images of: (**a**, **b**) EBM-produced Ti-6Al-4V, and (**c**, **d**) wrought Ti-6Al-4V [14]

phase boundary, the volume fractions are also calculated and summarized in Table 2 based on the EBSD map. It can be observed from Table 2 that the highest α phase volume fraction exists when the building angle is 0°, followed by 55°. The differences in residual β phase content in the various direction can be considered to be related to the anisotropy of its crystal orientation [40].

For the EIS test of the EBM-produced Ti-6Al-4V samples, a similar pattern is demonstrated in the different building planes. It is evident that the capacitance arc radius of 0° sample displayed in Fig. 8a is the largest. Followed by the 55° sample, and the 45° sample shows the smallest capacitance arc radius. Also the capacitance of all the samples display in the figure is close to a semi-circle. In the Bode phase diagram in Fig. 8b, the phase angles of all the samples in the high- and low-frequency regions are close to zero, indicating that the solution resistance is dominant. While the phase angle of the middle part has the same plateau port for all test samples, and the platform is near 80°. This reveals that there is a relatively dense oxide film formed at this region and close to the reaction of a capacitor [15, 37]. Also the 45° sample with the smallest impedance and the 0° sample with the largest impedance can be observed in the low-frequency range of Bode magnitude plot. Therefore, the impedance of the Ti-6Al-4V samples produced by EBM at different building angles can be analyzed and by order of 0° > 55° > 90° > 45°. Overall, the

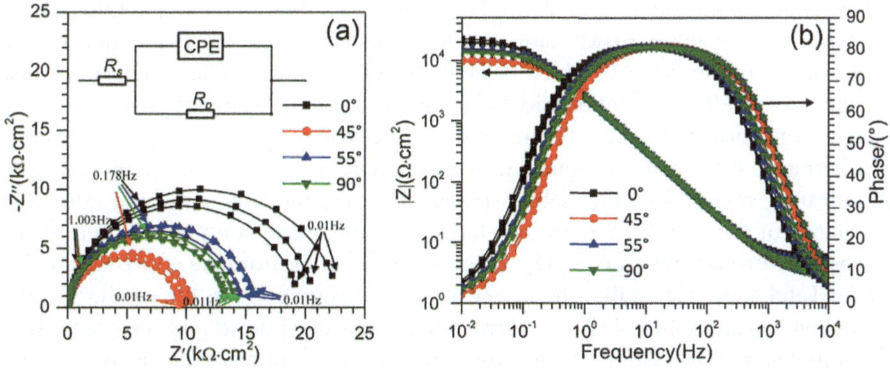


**Fig. 8** EIS test of EBM-produced Ti-6Al-4V in different building directions (1M HCl): (**a**) Nyquist plot with equivalent circuit diagram as illustration, and (**b**) Bode plots [37]

**Table 3** Summarization of corrosion current density and passive current density of wrought Ti-6Al-4V and EBM-produced Ti-6Al-4V with different building planes [14, 37]

| Sample | State | Test solution | Corrosion current density (μA/cm²) | Passive current density (μA/cm²) | Ref. |
|---|---|---|---|---|---|
| Wrought Ti-6Al-4V | – | PBS | ~0.012 | ~0.696 | [14] |
| EBM-produced Ti-6Al-4V | – | PBS | ~0.007 | ~0.733 | [14] |
| | 0° | 1 M HCl | ~0.14 | ~7.41 | [37] |
| | 45 °C | 1 M HCl | ~0.53 | ~26.79 | [37] |
| | 55 °C | 1 M HCl | ~0.26 | ~12.94 | [37] |
| | 90 °C | 1 M HCl | ~0.20 | ~15.67 | [37] |

equivalent circuit diagram with one time constant (illustration in Fig. 8a) can be applied to fit the test results. The charge transfer resistance of the fitted data shows the same result as the above analyses (0° > 55° > 90° > 45°).

The polarization curves of the EBM-produced Ti-6Al-4V and wrought Ti-6Al-4V in PBS solution are very similar. Both have a common passivation current density and breakdown potential. From the fitting data in Table 3, the corrosion current of EBM sample is slightly smaller than that of wrought ones, and the passivation current density is also higher than that of wrought ones but with a relatively small difference. However, the corrosion resistance of EBM sample is slightly better than that of wrought in terms of the EIS results discussed above. In contrast, the polarization curves of the EBM-produced samples at different building angles is significantly different in 1 M HCl solution. Since the previously formed oxide film has been completely removed in the cathode region (hydrogen evolution), the passivation of the anode region is basically controlled by the oxide film formed in the solution (oxidation behavior) [41]. The formation of the stabilized oxide form is on the anodic potential of ~0.3 V. So the passivation ability of the oxide film formed in 1 M HCl by these EBM samples can be judged from the magnitude of the passivation current density, which is summarized in Table 3. The conclusion is the smallest passivation current density of 0° samples, and 45° samples have the largest passivation current density; the results are consistent with the EIS test results. Furthermore, the corrosion potential of the sample at the junction of the cathodic polarization region and the anodic polarization region also show the corrosion resistance of the sample, and the substantially higher corrosion potential represents the better corrosion resistance [42]. There is also a fitting from the corrosion current in the Tafel zone. Generally, the larger of the corrosion current, the smallest of the corrosion resistance [41, 43]. The corrosion resistance extracted and compared from the fitting data of the Tafel region shows that the corrosion resistance is as the order of 0° > 55° > 90° > 45°.

The comparison of the corrosion morphology of the EBM-produced Ti-6Al-4V (Fig. 9b) and the morphology before the corrosion test (Fig. 9a) is shown in Fig. 9. It can be seen in Fig. 9a that all the samples basically have the same microstructure

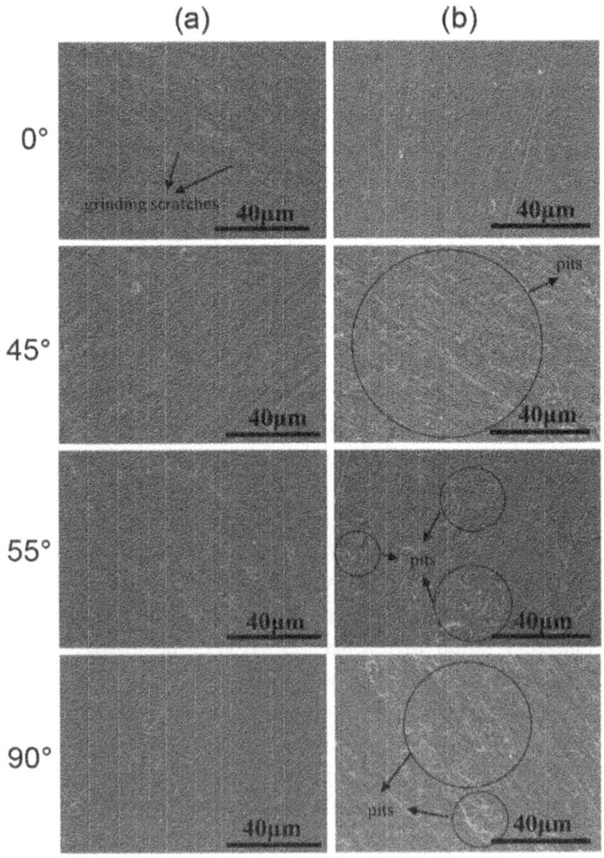


**Fig. 9** SEM image of EBM-produced Ti-6Al-4V with different building planes in 1M HCl solution. Row (**a**) is the sample before the corrosion test, row (**b**) is the sample after the polarization test [37]

state before the corrosion test. However, after the polarization test of 1.5 V in the anodic side, the number of corroded holes (pits) appeared on the sample surface and the pits vary with samples of different building orientations. For example, a significant amount of corrosion pits appeared in the 45° sample and occupied the full SEM image, which mixed with the vague grinding scratch. Such a substantial amount of corrosion illustrates that in contrast with other samples, the 45° degree sample is highly corrosive in 1 M HCl solution, and this conclusion also validates the results of the corrosion test. The best corrosion resistance of 0° sample in the corrosion test showed a few scattered pits on the micro-morphology, and the grinding scratch was also visible as the state before the corrosion test. Furthermore, the corrosion resistance of the 0° sample also proved to be the best of all other samples. The number of corrosion pits in the remaining 55° sample and 90° sample are consistent with the corrosion test results. The number of such corrosion pits is related to the grain boundary density. As a result of the lattice deformation inside the grains, the grain boundaries are more likely to be corroded. This non-uniform distribution will accelerate when mixed with the irregular lattice and atomic proportion separation at the grain boundaries [44, 45]. While the grain boundary densities distributions of these four samples are reported and identified by using EBSD in the literature [37], the magnitude is as the order of 45° > 90° > 55° > 0°. This arrangement is consistent with the corrosion test results that are the samples with large grain boundaries have poor corrosion resistance and more corrosion pits in their morphology.

In addition, as shown in Fig. 7, the EBM-produced Ti-6Al-4V exhibit a dual-phase structure alloy. Most of the alloy powders melted by electron beam, and only a small portion of β phase are deposited at the grain boundaries. Also the EBM manufacturing process has a relatively complex thermal history, such as the pre-heating of metal powder and the multiple melts and rapid cooling [4]. So the α phase is primarily formed by thermal diffusion from the β phase [37]. The content of the remaining small amount of fine β phase can be seen from Table 2, as the corrosion resistance decreases with the increase of the β phase content. Unlike the Ti-6Al-4V samples produced by SLM contains a significant amount of acicular α' phase and columnar prior β phase (Fig. 1). According to the analysis, the content of the columnar prior β phase affects the corrosion resistance of the SLM-produced Ti-6Al-4V alloy. The higher of the columnar prior β phase content (or the lower the acicular α' phase) means the better of the corrosion resistance [9–11, 28]. The β phase distribution of the EBM-produced Ti-6Al-4V alloy is a plurality of discontinuous point-like phase. These tiny β phases mixed with the large α phase will resulting to form a micro-galvanic couple. The β phase acts as a cathode in this system, and causes the α phase to be preferentially dissolved [15, 46]. While a vast potential difference is formed between the two-phase boundaries (the α-phase stable element Al and the β-phase stable element V), thereby increasing the chemical activity and reducing the corrosion resistance [15, 28]. In concluding of the EBM-produced Ti-6Al-4V, the EBM-produced 0° sample has the best corrosion resistance, followed by the 55° and 90° sample. Whereas the 45° sample has the worst corrosion resistance.

## 4 SLM-Produced Ti-TiB

Commercial pure titanium (CP-Ti) has been used extensively over the past few decades due to its favorable biocompatibility, proper mechanical properties [13, 47, 48], without causing allergies, and contamination in the body [49–51]. The application of CP-Ti has been widely used in human implants. Compared with the Co-Cr based alloys and stainless steel [52, 53], a spontaneously formed a $TiO_2$ oxide film on the surface of titanium will enhance the corrosion resistance due to its chemical stability and film compact properties [54–56]. However, there is also a lack of mechanical properties to CP-Ti like hardness, wear resistance and stiffness [57, 58], resulting in the limitation of biomedical applications. However, the properties can be improved by adding some reinforced material like TiC, SiC, CrB and $TiB_2$ [7, 59]. $TiB_2$ as one of the favorable reinforced material has a high hardness, elastic modulus and thermodynamic stability [60], and also biocompatible with element B, it makes up for the lack of mechanical properties of CP-Ti. The mechanical properties of the porous Ti-TiB composites fabricated by SLM have been summarized in the literature [7]. The measured yield strength and elastic modulus are similar to those of human bones. However, it is also an urgent concern for the long-term use of this material in the corrosion resistance compared to CP-Ti. Overall, this section reviews the corrosion behavior of SLM-produced CP-Ti and Ti-TiB composites in simulated human fluid (Hanks' solution) through electrochemical test and immersion test.

The result of Nyquist plot in electrochemical tests reveals that each sample has only one capacitor arc, and the arc radius of Ti-TiB are much larger than the CP-Ti, this phenomenon shows that the reaction impedance of Ti-TiB is higher than CP-Ti. In the Bode phase diagram, both samples show a similar passivation film densification, which can be determined from the overlapped plateau region where the mid-frequency region is greater than 70°. But the stability of passivation film formed on the surface of Ti-TiB is higher than that of CP-Ti, which can be obtained from the decreasing trend of phase angle in the low-frequency region. The tendency in phase angle diagram of CP-Ti is around 10° lower than that of Ti-TiB. So according to the previous EIS analysis, this EIS results can be using two-time constants equivalent circuit to fit. The fitting results of the oxide film resistance and the charge transfer resistance of CP-Ti and Ti-TiB are similar in numerical values, while the polarization resistance cannot be judged only by the charge transfer resistance. Therefore, the polarization resistance was then calculated to compare the system resistance. The fitted polarization resistance of Ti-TiB (~3.49 $M\Omega \cdot cm^2$) is 4 times larger than that of CP-Ti (~0.85 $M\Omega \cdot cm^2$). Overall, the reaction resistance of Ti-TiB is higher than that of CP-Ti from the EIS test results.

Figure 10 shows the polarization curves of CP-Ti and Ti-TiB fabricated by SLM. From the polarization curve can be seen that both samples have an evident passivation region in anodic polarization curve. In particular, the passivation interval of Ti-TiB (0.4–1.25 V) is nearly three times larger than CP-Ti (0.4–0.7 V). The passivation current density from the passivation region also reveals the same result.

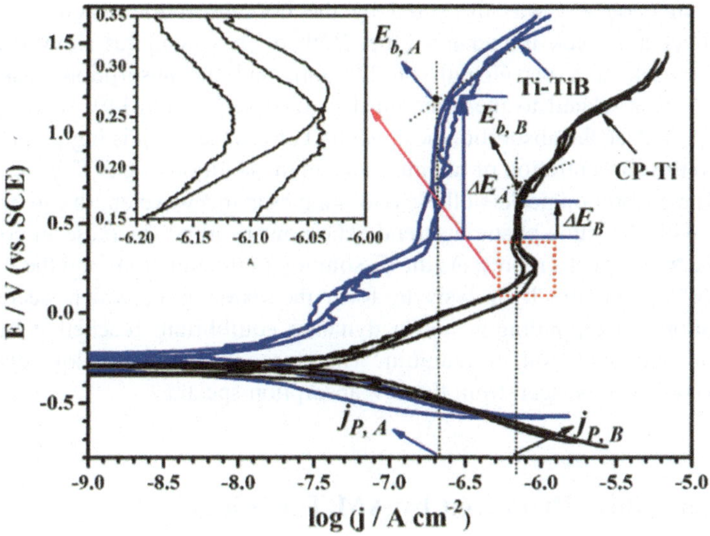


**Fig. 10** The potentiodynamic polarization curve of SLM-produced CP-Ti and Ti-TiB tested in Hanks' solution [15]

The passivation current density of CP-Ti is 0.75 μA/cm$^2$, which is much greater than Ti-TiB (0.22 μA/cm$^2$), this means the Ti-TiB samples have the ability to enter the passivation zone earlier. Furthermore, from the breakdown potential after the passivation region, the passivation film formed on the surface of Ti-TiB sample is more stable than CP-Ti.

In addition, as shown in the inset of Fig. 10, the active passivation behavior of anode region from 0.15 to 0.35 V is shown in SLM-produced CP-Ti, that is because of the passivation with unfinished open circuit potential [61]. However, due to the presence of $Cl^-$ in the solution, the CP-Ti sample will generate a large amount of $Ti_4^+$ [62], and spontaneously to form the $[TiCl_6]^{2-}$ [51]. The oxide film $TiO_2$ is formed when the hydrolysis of a certain amount $[TiCl_6]^{2-}$. These results demonstrate the $[TiCl_6]^{2-}$ is benefit to the passivation film formation of the titanium alloy [15, 63]. In contrast, the difference from SLM-produced CP-Ti is that the SLM-produced Ti-TiB goes directly into the passivation zone, which is more consistent with the passivation result of Ti-TiB explained earlier. While in the TiB system, since the unreacted $TiB_2$ and the converted TiB will form a corrosion microcell (micro galvanic couple) in the titanium matrix, the small $TiB_x$ regions with higher potential act as cathodes [64, 65], which greatly accelerate the Ti dissolution in the initial stage. This will promote the formation of $[TiCl_6]^{2-}$ to generate the stable $TiO_2$ oxide film to resist the further dissolution of the composite. Therefore, from the results of polarization curves that the corrosion resistance of Ti-TiB produced by SLM is better than that of CP-Ti in Hanks' solution.

The immersion test was measured by the UV absorption spectra of the two samples over a 7 week immersion time. Both of these samples showed a strong absorption peak at a wavelength of 275 nm, and the absorption peak at this wavelength was related to the titanium dissolved in solution [49] (corresponding to $[TiCl6]^{2-}$). Also the absorption peak of CP-Ti from the result is higher than that of Ti-TiB, so the concentration of titanium ions in the solution of CP-Ti is greater than Ti-TiB. The absorption peaks of these two samples at this wavelength can be directly observed with the same changing trend. The upward trend (increase in concentration) in the first 3 weeks represents the dissolution of titanium ions and the formation of $[TiCl6]^{2-}$; then the hydrolysis to form the oxide film, which decreases its concentration correspondingly, and a dynamic equilibrium reached between the dissolution and oxidation of titanium. Therefore, the relatively low dissolution ratio of Ti-TiB can be seen from the UV absorption spectra.

## 5 Other Alloys Produced by AM Technique

### *5.1 Co-Cr Based Alloys*

Ti-based and CoCr-based alloys are the principle application of better medical biomaterials. But the metal degradation caused by corrosion, friction and wear is inevitable in the application process, mainly due to the unfavorable effects of dissolved metal ions released in the human body. For example, a significant amount of Ni ions could cause allergy or even cancer. The released V ions from Ti-6Al-4V alloy could lead to osteoporosis and osteolysis [17, 66, 67]. While the CoCr-based alloys (such as CoCrW and CoCrMo) bid farewell to these elements that are harmful to the human body, because of their proper mechanical properties and corrosion properties can be used in the potential for bone implants [68–70]. With the maturity and popularity of AM technology, the mechanical properties and corrosion properties of CoCr alloys produced by AM technology have also been studied [17, 18, 71–73]. Therefore, this chapter mainly reviews the recent corrosion research of SLM-produced CoCr-based alloys.

The recent studies on the corrosion behavior of CoCr alloys produced by SLM mainly focus on CoCrW, CoCrWMo, CoCrMo and CoCeFeNiTi. The open circuit potential [18] was significantly affected when a small amount of F ions were added to the NaCl solution, and the potential value decreased from ~ −0.36 to ~ 0.42 V, which reveals that the SLM-produced CoCrW samples are greatly affected by the F ions in corrosion probability. From the charge transfer point of view, the all the test sample indicates only one capacitive arc under the three different test conditions. However, the radius of the capacitor arc slightly decreases with the addition of low concentration of F ions (0.05F). When the relatively large concentration of F ions (0.1F) is added to NaCl solution, the arc radius is further reduced, this phenomenon indicated that the charge transfer capacity decreases with the increase of F ion concentration. This is also confirmed in the Bode plot where the compactness of

**Table 4** Summarized EIS fitting results of CoCr based alloys [16, 18]

| Sample | Test solution | $R_s$ | $R_p$ | Ref. |
|---|---|---|---|---|
| | | ($\Omega \cdot cm^2$) | ($M\Omega \cdot cm^2$) | |
| SLM CoCrW | 0.9% NaCl | ~36 | ~1.18 | [18] |
| | NaCl+0.05F | ~35 | ~1.05 | [18] |
| | NaCl+0.1F | ~29 | ~0.88 | [18] |
| SLM CoCrWMo | Artificial saliva | – | ~6.21 | [16] |
| Cast CoCrWMo | Artificial saliva | – | ~3.74 | [16] |

**Table 5** Summarized Potentiodynamic polarization data of SLM-produced CoCrW. [17, 18]

| | | Test solution | Corrosion current density | Corrosion potential | Ref. |
|---|---|---|---|---|---|
| Sample | State | | ($nA/cm^2$) | (mV) | |
| CoCrW | – | PBS | ~20 | ~ −201 | [17] |
| | – | Hanks | ~12 | ~ −284 | [17] |
| | | 0.9% NaCl | ~42 | ~ −400 | [18] |
| | | NaCl+0.05F | ~44 | ~ −480 | [18] |
| | | NaCl+0.1F | ~148 | ~ −520 | [18] |

the oxide film formed on the surface of the CoCrW sample is the highest in the NaCl solution. With the increase of F ion concentration, the phase angle is gradually decreased, this results reflect the oxide film densification decreased. The oxide film is more likely to breakdown by F ions and Cl ions. After the EIS test data is fitted with an equivalent circuit diagram of one time constant, the resulting polarization impedance of the fitting component is also summarized in Table 4.

From the summarized data of polarization impedance (the impedance decreases with the increase of F ion concentration), a same result as the previous analysis can be obtained. The same conclusion is also validated in the polarization curve. The corrosion potential (data summarized in Table 5) is the highest in NaCl solution (−400 mV) without any F ions exists; with the addition of F ions, the corrosion potential is minimized (−520 mV) at a F ion concentration of 0.1F. In addition, the change trend of passivation current can also be observed in the figure, in which the passivation current density of NaCl is the smallest after the anode range of 0 V. Then gradually increases with the increase of F ion concentration. This shows that the addition of F ions inhibits the formation of the passive film, the degree of inhibition is related to F ion concentration. Furthermore, the corrosion resistance of heat treated SLM-produced CoCrW samples have not been improved but a substantial reduction [18]. In the test of Hanks' solution, the CoCrW samples fabricated by SLM have a better corrosion resistance than in the PBS solution [17], which can be obtained not only from the data of polarization curve but also from the ion concentration to judge.

The other CoCr-based alloys like CoCrWMo, its corrosion resistance compared to conventionally cast alloys can be determined from the polarization impedance in EIS fitting data [16], and the results are also shown in Table 4. The polarization impedance of CoCrWMo alloy produced by SLM is around 6.21 $M\Omega \cdot cm^2$, almost twice that of the cast sample (3.74 MΩ·cm$^2$). This result indicates that the CoCrWMo alloy manufactured by SLM is superior to the cast sample in corrosion resistance. In CoCrMo samples, similar results have also appeared. Combining the corrosion potential (−0.2 V), the corrosion current (−0.05 μA·cm$^{-2}$) and the polarization impedance (~1600 Ω·cm$^2$) of the polarization curve under artificial saliva solution with a pH of 5 [74]. It can be seen that the corrosion resistance of SLM-produced CoMrMo sample and the traditional as-cast sample are very close. Unfortunately, the SLM-produced CoCr alloy shows a pitting tendency after adding the elements of Fe, Ni and Ti [75]. But this pitting slightly improved after the heat treatment, which results in an increase in the pitting potential by ~0.4 V of SLM-produced CoCrFeNiTi.

## *5.2 SLM-Produced 316L Stainless Steel*

The corrosion resistance of stainless steel makes it have essential applications on industrial; the corrosion resistance of stainless steel is according to the contents of Cr (typically>11 wt.%) to forms a $Cr_2O_3$ oxide film on the surface [76, 77]. However, the stainless steel is not "always stainless", it may corrode in an aqueous environment, and this type of corrosion is driven by its microstructure [78]. Such as the impurity of alloys (MnS) can cause the second phase in the original single-phase structure and result in pitting [79–81]. However, the SLM-produced stainless steel samples due to different scanning parameters (especially scanning power and scanning speed), the porosity of the sample will be various. And optimize the size of the impurity (MnS), it is bound to cause their micro-topography and the difference will lead to reflect the corrosion behavior. In the literature [20], the corrosion behavior of stainless steel samples and wrought samples with different scan powers and scan rates has been investigated. In the literature [19], wrought samples, SLM samples, and annealed SLM samples were also studied. This section reviews the corrosion behavior (pitting) of 316L stainless steel based on a cyclic polarization curve and static polarization test method.

Figure 11 illustrates the cyclic polarization curve with highest porosity (Fig. 11a: ~0.4%) and lowest porosity (Fig. 11b: ~0.02%) produced by the SLM method [20]. It is clear from the Fig. 11 that the pitting potential of stainless steel manufactured by SLM is significantly higher than wrought samples. The pitting potential observed for wrought samples was ~0.4 V. However, the SLM samples were ~0.9 V, which was ~0.5 V higher than wrought samples. This result shows that SLM-produced stainless steel is more pitting-proof than wrought samples. A similar result was also obtained in literature [19], where the pitting potential of SLM-produced stainless steels was much higher than that of wrought stainless steels, but the pitting potential of the annealed sample was slightly lower than that of the original SLM samples. Besides, it can be seen from the anodic polarization curve before the pitting potential, that the sample 4 with the highest porosity represented in in Fig. 11a, is also the most unstable sample. The phenomenon illustrated that the

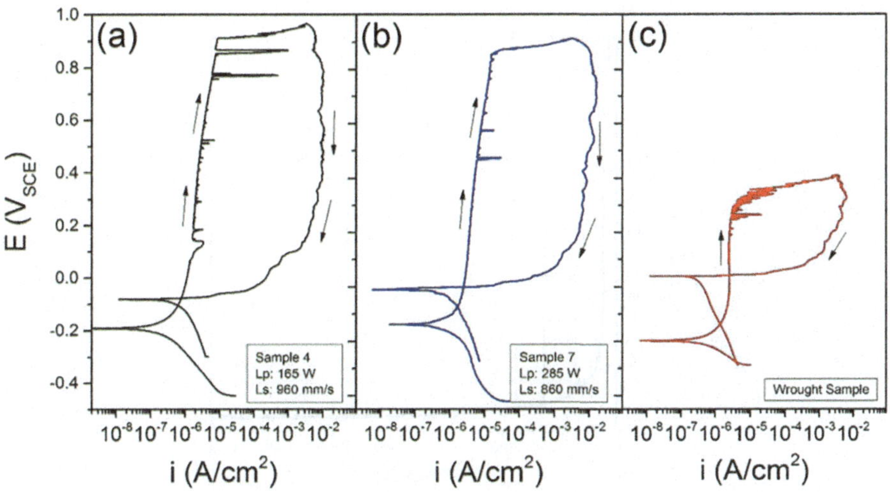


**Fig. 11** Cyclic potentiodynamic polarization of SLM-produced stainless steel with (**a**) largest porosity (sample 4) and (**b**) lowest porosity (sample 7); and (**c**) wrought stainless steel [20]

current density increases instantaneously and decreases rapidly to form the horizontally "spikes" shape in the curve while the potential remains essentially constant. This result in the stronger pitting corrosion on the oxide film when this peak is greater. In contrast, the number and intensity of such "spikes" in Fig. 11b are significantly reduced that indicating the porosity of SLM-produced stainless steel also affects the anodic polarization curve, where the smaller of the porosity leads to the stable curve. However, the samples prepared by SLM did not have much difference in the pitting potentials, and also were not directly related to the sample porosity; and the corrosion current and the corrosion potential were even not directly related to the porosity [20]. But the re-passivation potential did not show the same results (Fig. 11), where the wrought samples showed the highest reactivation potential (~0.04 V) compared to the SLM samples (below ~0 V). While the re-passivation potential of SLM-produced samples are gradually decreased from the highest ~0 V (Sample 7) to ~ −0.1 V (Sample 4) with the increase of porosity [20]. From this relationship, it can be seen that the growth of porosity is not conducive to the re-passivation properties of the SLM-produced samples. Once pitting occurs, the SLM samples have less re-passivation ability than the wrought samples, especially with the higher porous density of SLM samples is found more difficult to re-passive.

The metastable pitting characteristics can be predicted by the potentiostatic polarization test. The measured curve is shown in Fig. 12. Comparing the SLM sample with the wrought sample, the SLM sample fluctuated the most in the first 15 min, but the wrought sample fluctuated the most within the last 15 min. The amount of metastable pitting can be combined with the immersion time and sample surface area to calculate metastable pitting frequency. The calculation result is also shown in Fig. 12, indicated by λ. From the data comparison, the porosity present in

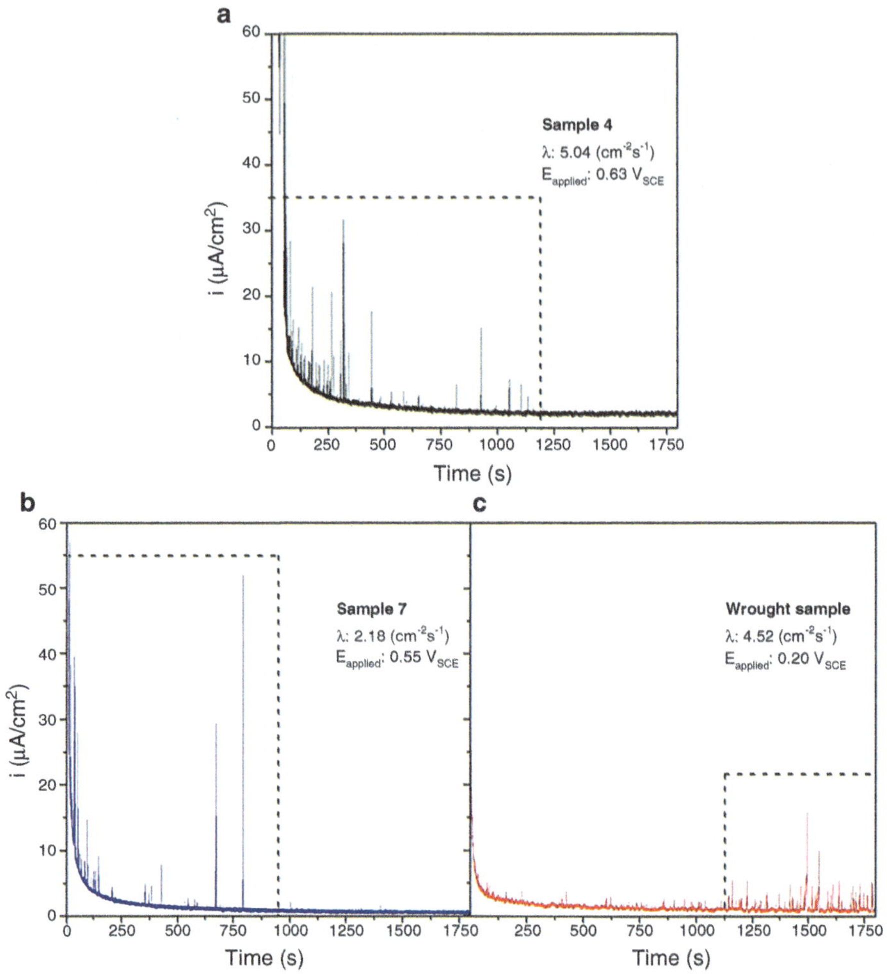


**Fig. 12** Potentiostaic testing of SLM-produced stainless steel with (**a**) largest porosity (sample 4) and (**b**) lowest porosity (sample 7); and (**c**) wrought stainless steel [20]

the SLM sample is related to the metastable pitting frequency. The sum of the peak areas that represent the cumulative passing charge of those metastable pits. For example, Sample 4 with the highest porosity produced by SLM contained the most considerable cumulative charge (3.67 mC/cm$^2$), and the maximum metastable pore frequency (5.04 cm$^{-2}$ s$^{-1}$) in the calculation. Sample 7 with the lowest porosity produced by SLM contained the smallest cumulative charge (0.1 mC/cm$^2$), and the lowest metastable pore frequency (2.18 cm$^{-2}$ s$^{-1}$). While for wrought samples, it has a higher metastable pore frequency (4.52 mC/cm$^2$) but a relatively more moderate metastable pore frequency (0.2 mC/cm$^2$). This shows that although wrought samples are most prone to pitting, whereas they suffer the least damage when metastable pores occur.

## 6 Concluding Remarks

This chapter reviews the recent corrosive properties of alloys made by AM technology (Ti alloys, CoCr alloys and stainless steels). Although the corrosion resistance of SLM-produced Ti-6Al-4V is not satisfactory, the corrosion resistance of EBM-produced Ti-6Al-4V can catch up with that of wrought samples. Furthermore, most the CoCr samples and stainless steel also show improved corrosion resistance (in particular the pitting corrosion of stainless steel is suppressed thus increase the pitting potential). This indicates that the corrosion resistance can be improved due to the unique additive manufacturing (AM) technology in the production. However, there is lack of the research on the corrosion behavior of the AM produced alloy in recent work, and as many researches are needed to consolidate the corrosion status of the alloy manufactured by AM technology.

**Acknowledgement** *This research was supported by the Australian Research Council's Discovery Projects (DP110101653)* and by the Project of Shanghai Science and Technology Commission (14DZ2261000). *The authors are grateful to N.W. Dai, Y. Chen, X.H. Gu,* P. Duan for their collaboration.

## References

1. Banerjee D, Williams JC (2013) Perspectives on Titanium Science and Technology. Acta Materialia 61 (3):844-879. doi:https://doi.org/10.1016/j.actamat.2012.10.043
2. Geetha M, Singh A, Asokamani R, Gogia A (2009) Ti based biomaterials, the ultimate choice for orthopaedic implants–a review. Progress in Materials Science 54 (3):397-425.
3. Zhang LC, Liu YJ, Li SJ, Hao YL (2018) Additive manufacturing of titanium alloys by electron beam melting: a review. Advanced Engineering Materials 20 (5):1700842.
4. Liu YJ, Li SJ, Wang HL, Hou WT, Hao YL, Yang R, Sercombe TB, Zhang LC (2016) Microstructure, defects and mechanical behavior of beta-type titanium porous structures manufactured by electron beam melting and selective laser melting. Acta Materialia 113:56-67. doi:https://doi.org/10.1016/j.actamat.2016.04.029
5. Ehtemam-Haghighi S, Cao G, Zhang LC (2017) Nanoindentation study of mechanical properties of Ti based alloys with Fe and Ta additions. Journal of Alloys and Compounds 692:892-897.
6. Liu YJ, Wang HL, Li SJ, Wang SG, Wang WJ, Hou WT, Hao YL, Yang R, Zhang LC (2017) Compressive and fatigue behavior of beta-type titanium porous structures fabricated by electron beam melting. Acta Materialia 126:58-66. doi:https://doi.org/10.1016/j.actamat.2016.12.052
7. Zhang LC, Attar H (2016) Selective laser melting of titanium alloys and titanium matrix composites for biomedical applications: a review. Advanced Engineering Materials 18 (4):463-475.
8. Zhang LC, Klemm D, Eckert J, Hao YL, Sercombe TB (2012) Manufacture by selective laser melting and mechanical behavior of a biomedical Ti-24Nb-4Zr-8Sn alloy. Scripta Materialia 65 (1):21-24.
9. Dai N, Zhang LC, Zhang J, Chen Q, Wu M (2016) Corrosion behavior of selective laser melted Ti-6Al-4V alloy in NaCl solution. Corrosion Science 102:484-489. doi:https://doi.org/10.1016/j.corsci.2015.10.041

10. Dai N, Zhang J, Chen Y, Zhang LC (2017) Heat treatment degrading the corrosion resistance of selective laser melted Ti-6Al-4V alloy. Journal of The Electrochemical Society 164 (7):C428-C434. doi:https://doi.org/10.1149/2.1481707jes
11. Dai N, Zhang LC, Zhang J, Zhang X, Ni Q, Chen Y, Wu M, Yang C (2016) Distinction in corrosion resistance of selective laser melted Ti-6Al-4V alloy on different planes. Corrosion Science 111 (2016):703-710. doi:https://doi.org/10.1016/j.corsci.2016.06.009
12. Attar H, Calin M, Zhang LC, Scudino S, Eckert J (2014) Manufacture by selective laser melting and mechanical behavior of commercially pure titanium. Materials Science and Engineering: A 593:170-177. doi:https://doi.org/10.1016/j.msea.2013.11.038
13. Liu YJ, Li XP, Zhang LC, Sercombe TB (2015) Processing and properties of topologically optimised biomedical Ti–24Nb–4Zr–8Sn scaffolds manufactured by selective laser melting. Materials Science and Engineering: A 642:268-278. doi:https://doi.org/10.1016/j.msea.2015.06.088
14. Bai Y, Gai X, Li SJ, Zhang LC, Liu YJ, Hao YL, Zhang X, Yang R, Gao YB (2017) Improved corrosion behaviour of electron beam melted Ti-6Al-4V alloy in phosphate buffered saline. Corrosion Science 123:289-296. doi:https://doi.org/10.1016/j.corsci.2017.05.003
15. Chen Y, Zhang J, Dai N, Qin P, Attar H, Zhang LC (2017) Corrosion Behaviour of Selective Laser Melted Ti-TiB Biocomposite in Simulated Body Fluid. Electrochimica Acta 232:89-97. doi:https://doi.org/10.1016/j.electacta.2017.02.112
16. Xin XZ, Chen J, Xiang N, Gong Y, Wei B (2014) Surface characteristics and corrosion properties of selective laser melted Co–Cr dental alloy after porcelain firing. Dental Materials 30 (3):263-270. doi:https://doi.org/10.1016/j.dental.2013.11.013
17. Lu Y, Wu S, Gan Y, Li J, Zhao C, Zhuo D, Lin J (2015) Investigation on the microstructure, mechanical property and corrosion behavior of the selective laser melted CoCrW alloy for dental application. Materials Science and Engineering: C 49:517-525. doi:https://doi.org/10.1016/j.msec.2015.01.023
18. Lu Y, Guo S, Yang Y, Liu Y, Zhou Y, Wu S, Zhao C, Lin J (2018) Effect of thermal treatment and fluoride ions on the electrochemical corrosion behavior of selective laser melted CoCrW alloy. Journal of Alloys and Compounds 730:552-562. doi:https://doi.org/10.1016/j.jallcom.2017.09.318
19. Chao Q, Cruz V, Thomas S, Birbilis N, Collins P, Taylor A, Hodgson PD, Fabijanic D (2017) On the enhanced corrosion resistance of a selective laser melted austenitic stainless steel. Scripta Materialia 141:94-98. doi:https://doi.org/10.1016/j.scriptamat.2017.07.037
20. Sander G, Thomas S, Cruz V, Jurg M, Birbilis N, Gao X, Brameld M, Hutchinson C (2017) On the corrosion and metastable pitting characteristics of 316L stainless steel produced by selective laser melting. Journal of The Electrochemical Society 164 (6):C250-C257.
21. Chen Y, Zhang JX, Gu XH, Dai NW, Qin P, Zhang LC (2018) Distinction of corrosion resistance of selective laser melted Al-12Si alloy on different planes. Journal of Alloys and Compounds 747:648-658.
22. Liu YJ, Liu Z, Jiang Y, Wang GW, Yang Y, Zhang LC (2018) Gradient in microstructure and mechanical property of selective laser melted AlSi10Mg. Journal of Alloys and Compounds 735:1414-1421.
23. Li XP, Kang CW, Huang H, Zhang LC, Sercombe TB (2014) Selective laser melting of an Al86Ni6Y4.5Co2La1.5 metallic glass: Processing, microstructure evolution and mechanical properties. Materials Science and Engineering: A 606:370-379. doi:https://doi.org/10.1016/j.msea.2014.03.097
24. Karimzadeh F, Heidarbeigy M, Saatchi A (2008) Effect of heat treatment on corrosion behavior of Ti–6Al–4V alloy weldments. Journal of Materials Processing Technology 206 (1):388-394.
25. Thijs L, Verhaeghe F, Craeghs T, Van Humbeeck J, Kruth J-P (2010) A study of the microstructural evolution during selective laser melting of Ti–6Al–4V. Acta Materialia 58 (9):3303-3312.

26. Vilaro T, Colin C, Bartout JD (2011) As-Fabricated and Heat-Treated Microstructures of the Ti-6Al-4V Alloy Processed by Selective Laser Melting. Metallurgical and Materials Transactions A 42 (10):3190-3199. doi:https://doi.org/10.1007/s11661-011-0731-y
27. Vrancken B, Thijs L, Kruth J-P, Van Humbeeck J (2012) Heat treatment of Ti6Al4V produced by Selective Laser Melting: Microstructure and mechanical properties. Journal of Alloys and Compounds 541:177-185. doi:https://doi.org/10.1016/j.jallcom.2012.07.022
28. Chen JR, Tsai WT (2011) In situ corrosion monitoring of Ti–6Al–4V alloy in H2SO4/HCl mixed solution using electrochemical afm. Electrochimica Acta 56 (4):1746-1751. doi:https://doi.org/10.1016/j.electacta.2010.10.024
29. Alves AC, Wenger F, Ponthiaux P, Celis JP, Pinto AM, Rocha LA, Fernandes JCS (2017) Corrosion mechanisms in titanium oxide-based films produced by anodic treatment. Electrochimica Acta 234:16-27. doi:https://doi.org/10.1016/j.electacta.2017.03.011
30. Bandeira RM, van Drunen J, Garcia AC, Tremiliosi-Filho G (2017) Influence of the thickness and roughness of polyaniline coatings on corrosion protection of AA7075 aluminum alloy. Electrochimica Acta 240:215-224. doi:https://doi.org/10.1016/j.electacta.2017.04.083
31. Atapour M, Pilchak A, Shamanian M, Fathi M (2011) Corrosion behavior of Ti–8Al–1Mo–1V alloy compared to Ti–6Al–4V. Materials & Design 32 (3):1692-1696
32. Qin P, Liu YJ, Sercombe TB, Li Y, Zhang C, Cao C, Sun H, Zhang LC (2018) Improved corrosion resistance on selective laser melting produced Ti-5Cu alloy after heat treatment. ACS Biomaterials Science & Engineering 20: in press. doi:https://doi.org/10.1021/acsbiomaterials.8b00319
33. Yang Y, Chen Y, Zhang J, Gu X, Qin P, Dai N, Li X, Kruth JP, Zhang LC (2018) Improved corrosion behavior of ultrafine-grained eutectic Al-12Si alloy produced by selective laser melting. Materials & Design 146:239-248.
34. Huttunen-Saarivirta E, Rajala P, Bomberg M, Carpén L (2017) EIS study on aerobic corrosion of copper in ground water: influence of micro-organisms. Electrochimica Acta 240:163-174. doi:https://doi.org/10.1016/j.electacta.2017.04.073
35. Feng R, Beck J, Ziomek-Moroz M, Lvov SN (2016) Electrochemical corrosion of ultra-high strength carbon steel in alkaline brines containing hydrogen sulfide. Electrochimica Acta 212:998-1009. doi:https://doi.org/10.1016/j.electacta.2016.07.070
36. Karimi S, Nickchi T, Alfantazi AM (2012) Long-term corrosion investigation of AISI 316L, Co–28Cr–6Mo, and Ti–6Al–4V alloys in simulated body solutions. Applied Surface Science 258 (16):6087-6096
37. Gong X, Cui Y, Wei D, Liu B, Liu R, Nie Y, Li Y (2017) Building direction dependence of corrosion resistance property of Ti–6Al–4V alloy fabricated by electron beam melting. Corrosion Science 127:101-109. doi:https://doi.org/10.1016/j.corsci.2017.08.008
38. Kherrouba N, Bouabdallah M, Badji R, Carron D, Amir M (2016) Beta to alpha transformation kinetics and microstructure of Ti-6Al-4V alloy during continuous cooling. Materials Chemistry and Physics 181:462-469. doi:https://doi.org/10.1016/j.matchemphys.2016.06.082
39. Appolaire B, Héricher L, Aeby-Gautier E (2005) Modelling of phase transformation kinetics in Ti alloys – Isothermal treatments. Acta Materialia 53 (10):3001-3011. doi:https://doi.org/10.1016/j.actamat.2005.03.014
40. De Formanoir C, Michotte S, Rigo O, Germain L, Godet S (2016) Electron beam melted Ti–6Al–4V: Microstructure, texture and mechanical behavior of the as-built and heat-treated material. Materials Science and Engineering: A 652:105-119. doi:https://doi.org/10.1016/j.msea.2015.11.052
41. Shi Y, Peng C, Feng Y, Wang R, Wang N (2017) Microstructure and electrochemical corrosion behavior of extruded Mg–Al–Pb–La alloy as anode for seawater-activated battery. Materials & Design 124:24-33. doi:https://doi.org/10.1016/j.matdes.2017.03.058
42. Wang N, Wang R, Feng Y, Xiong W, Zhang J, Deng M (2016) Discharge and corrosion behaviour of Mg-Li-Al-Ce-Y-Zn alloy as the anode for Mg-air battery. Corrosion Science 112:13-24. doi:https://doi.org/10.1016/j.corsci.2016.07.002

43. Shi Z, Liu M, Atrens A (2010) Measurement of the corrosion rate of magnesium alloys using Tafel extrapolation. Corrosion Science 52 (2):579-588. doi:https://doi.org/10.1016/j.corsci.2009.10.016
44. Osório WR, Freire CM, Garcia A (2005) The role of macrostructural morphology and grain size on the corrosion resistance of Zn and Al castings. Materials Science and Engineering: A 402 (1):22-32. doi:https://doi.org/10.1016/j.msea.2005.02.094
45. LuoJian Z (2012) Influence of grain size on corrosion resistant of commonly used metals. Corrosion&Protection 33 (4):349-352
46. Song G, Atrens A, Dargusch M (1998) Influence of microstructure on the corrosion of diecast AZ91D. Corrosion Science 41 (2):249-273. doi:https://doi.org/10.1016/S0010-938X(98)00121-8
47. Zaveri N, Mahapatra M, Deceuster A, Peng Y, Li L, Zhou A (2008) Corrosion resistance of pulsed laser-treated Ti–6Al–4V implant in simulated biofluids. Electrochimica Acta 53 (15):5022-5032. doi:https://doi.org/10.1016/j.electacta.2008.01.086
48. Yu F, Addison O, Davenport AJ (2015) A synergistic effect of albumin and H(2)O(2) accelerates corrosion of Ti6Al4V. Acta Biomaterialia 26:355-365. doi:https://doi.org/10.1016/j.actbio.2015.07.046
49. Alves VA, Reis RQ, Santos ICB, Souza DG, de F. Gonçalves T, Pereira-da-Silva MA, Rossi A, da Silva LA (2009) In situ impedance spectroscopy study of the electrochemical corrosion of Ti and Ti–6Al–4V in simulated body fluid at 25°C and 37°C. Corrosion Science 51 (10):2473-2482. doi:https://doi.org/10.1016/j.corsci.2009.06.035
50. Rahmati B, Sarhan AAD, Basirun WJ, Abas WABW (2016) Ceramic tantalum oxide thin film coating to enhance the corrosion and wear characteristics of Ti6Al4V alloy. Journal of Alloys and Compounds 676:369-376. doi:https://doi.org/10.1016/j.jallcom.2016.03.188
51. Saud SN, Hosseinian.S R, Bakhsheshi-Rad HR, Yaghoubidoust F, Iqbal N, Hamzah E, Ooi CHR (2016) Corrosion and bioactivity performance of graphene oxide coating on TiNb shape memory alloys in simulated body fluid. Materials Science and Engineering: C 68:687-694. doi:https://doi.org/10.1016/j.msec.2016.06.048
52. Sing SL, Yeong WY, Wiria FE (2016) Selective laser melting of titanium alloy with 50 wt% tantalum: Microstructure and mechanical properties. Journal of Alloys and Compounds 660:461-470. doi:https://doi.org/10.1016/j.jallcom.2015.11.141
53. Chen JR, Tsai, Wen Ta (2011) In situ corrosion monitoring of Ti–6Al–4V alloy in H2SO4/HCl mixed solution using electrochemical AFM. Electrochimica Acta 56 (4):1746-1751. doi:https://doi.org/10.1016/j.electacta.2010.10.024
54. Addison O, Davenport AJ, Newport RJ, Kalra S, Monir M, Mosselmans JF, Proops D, Martin RA (2012) Do 'passive' medical titanium surfaces deteriorate in service in the absence of wear? Journal of the Royal Society, Interface 9 (76):3161-3164. doi:https://doi.org/10.1098/rsif.2012.0438
55. Mabilleau G, Bourdon S, Joly-Guillou ML, Filmon R, Basle MF, Chappard D (2006) Influence of fluoride, hydrogen peroxide and lactic acid on the corrosion resistance of commercially pure titanium. Acta Biomaterialia 2 (1):121-129. doi:https://doi.org/10.1016/j.actbio.2005.09.004
56. Fei Yu OA, Alison J. Davenport, Stephen J Baker (2015) Lipopolysaccharide inhibits or accelerates biomedical titanium corrosion depending on environmental acidity. International Journal of Oral Science:1-8
57. Attar H, Bönisch M, Calin M, Zhang LC, Zhuravleva K, Funk A, Scudino S, Yang C, Eckert J (2014) Comparative study of microstructures and mechanical properties of in situ Ti–TiB composites produced by selective laser melting, powder metallurgy, and casting technologies. Journal of Materials Research 29 (17):1941-1950. doi:https://doi.org/10.1557/jmr.2014.122
58. Cheloui H, Zhang Z, Shen X, Wang F, Lee S (2011) Microstructure and mechanical properties of TiB–TiB2 ceramic matrix composites fabricated by spark plasma sintering. Materials Science and Engineering: A 528 (10-11):3849-3853. doi:https://doi.org/10.1016/j.msea.2011.01.096

59. Morsi K, Patel VV (2007) Processing and properties of titanium–titanium boride (TiBw) matrix composites—a review. Journal of Materials Science 42 (6):2037-2047. doi:https://doi.org/10.1007/s10853-006-0776-2
60. Attar H, Bönisch M, Calin M, Zhang LC, Scudino S, Eckert J (2014) Selective laser melting of in situ titanium–titanium boride composites: processing, microstructure and mechanical properties. Acta Materialia 76:13-22.
61. Atapour M, Pilchak A, Frankel GS, Williams JC (2010) Corrosion behaviour of investment cast and friction stir processed Ti–6Al–4V. Corrosion Science 52 (9):3062-3069. doi:https://doi.org/10.1016/j.corsci.2010.05.026
62. González JEG, Mirza-Rosca JC (1999) Study of the corrosion behavior of titanium and some of its alloys for biomedical and dental implant applications. Journal of Electroanalytical Chemistry 471 (2):109-115. doi:https://doi.org/10.1016/S0022-0728(99)00260-0
63. Wei D-X, Koizumi Y, Li Y, Yamanak K, Chiba A (2016) Submicron lamellar porous structure formed by selective dissolution of Ti-Al alloy. Materials & Design 98:1-11. doi:https://doi.org/10.1016/j.matdes.2016.02.096
64. Håkansson E, Hoffman J, Predecki P, Kumosa M (2017) The role of corrosion product deposition in galvanic corrosion of aluminum/carbon systems. Corrosion Science 114:10-16
65. Anes V, Pedro RS, Henriques E, Freitas M, Reis L (2016) Galvanic corrosion of aircraft bonded joints as a result of adhesive microcracks. Procedia Structural Integrity 1:218-225. doi:https://doi.org/10.1016/j.prostr.2016.02.030
66. Martín-Cameán A, Jos Á, Mellado-García P, Iglesias-Linares A, Solano E, Cameán AM (2015) In vitro and in vivo evidence of the cytotoxic and genotoxic effects of metal ions released by orthodontic appliances: A review. Environmental Toxicology and Pharmacology 40 (1):86-113. doi:https://doi.org/10.1016/j.etap.2015.05.007
67. Choubey A, Balasubramaniam R, Basu B (2004) Effect of replacement of V by Nb and Fe on the electrochemical and corrosion behavior of Ti–6Al–4V in simulated physiological environment. Journal of Alloys and Compounds 381 (1):288-294. doi:https://doi.org/10.1016/j.jallcom.2004.03.096
68. Ramirez-Ledesma AL, Lopez-Molina E, Lopez HF, Juarez-Islas JA (2016) Athermal ε-martensite transformation in a Co–20Cr alloy: Effect of rapid solidification on plate nucleation. Acta Materialia 111:138-147. doi:https://doi.org/10.1016/j.actamat.2016.03.047
69. Yamanaka K, Mori M, Chiba A (2015) Surface characterisation of Ni-free Co–Cr–W-based dental alloys exposed to high temperatures and the effects of adding silicon. Corrosion Science 94:411-419. doi:https://doi.org/10.1016/j.corsci.2015.02.030
70. Tkachenko S, Datskevich O, Kulak L, Jacobson S, Engqvist H, Persson C (2014) Wear and friction properties of experimental Ti–Si–Zr alloys for biomedical applications. Journal of the Mechanical Behavior of Biomedical Materials 39:61-72. doi:https://doi.org/10.1016/j.jmbbm.2014.07.011
71. Xiang N, Xin X-Z, Chen J, Wei B (2012) Metal–ceramic bond strength of Co–Cr alloy fabricated by selective laser melting. Journal of Dentistry 40 (6):453-457. doi:https://doi.org/10.1016/j.jdent.2012.02.006
72. Yamanaka K, Mori M, Chiba A (2014) Refinement of solidification microstructures by carbon addition in biomedical Co–28Cr–9W–1Si alloys. Materials Letters 116:82-85. doi:https://doi.org/10.1016/j.matlet.2013.10.109
73. Lu Y, Lu Y, Gan Y, Gan Y, Lin J, Lin J, Guo S, Guo S, Wu S, Wu S (2017) Effect of laser speeds on the mechanical property and corrosion resistance of CoCrW alloy fabricated by SLM. Rapid Prototyping Journal 23 (1):28-33
74. Zeng L, Xiang N, Wei B (2014) A comparison of corrosion resistance of cobalt-chromium-molybdenum metal ceramic alloy fabricated with selective laser melting and traditional processing. The Journal of Prosthetic Dentistry 112 (5):1217-1224. doi:https://doi.org/10.1016/j.prosdent.2014.03.018
75. Fujieda T, Shiratori H, Kuwabara K, Hirota M, Kato T, Yamanaka K, Koizumi Y, Chiba A, Watanabe S (2017) CoCrFeNiTi-based high-entropy alloy with superior tensile strength and

corrosion resistance achieved by a combination of additive manufacturing using selective electron beam melting and solution treatment. Materials Letters 189:148-151. doi:https://doi.org/10.1016/j.matlet.2016.11.026
76. Davis JR (1994) Stainless steels. ASM international,
77. Sedriks AJ (1996) Corrosion of stainless steel. 2nd edition edn. John Wiley and Sons, Inc., New York, NY (United States),
78. Ryan MP, Williams DE, Chater RJ, Hutton BM, McPhail DS (2002) Why stainless steel corrodes. Nature 415 (6873):770-774
79. Williams DE, Kilburn MR, Cliff J, Waterhouse GIN (2010) Composition changes around sulphide inclusions in stainless steels, and implications for the initiation of pitting corrosion. Corrosion Science 52 (11):3702-3716. doi:https://doi.org/10.1016/j.corsci.2010.07.021
80. Wijesinghe TLSL, Blackwood DJ (2007) Real time pit initiation studies on stainless steels: The effect of sulphide inclusions. Corrosion Science 49 (4):1755-1764. doi:https://doi.org/10.1016/j.corsci.2006.10.025
81. Jun J, Holguin K, Frankel G (2013) Pitting corrosion of very clean type 304 stainless steel. Corrosion 70 (2):146-155

# Effect of Process Parameters of Fused Deposition Modeling and Vapour Smoothing on Surface Properties of ABS Replicas for Biomedical Applications

**Jasgurpreet Singh Chohan, Rupinder Singh, and Kamaljit Singh Boparai**

## 1 Surface Finishing Techniques for FDM Parts

Recently, the rapid casting of biomedical implants has become matter of interest for many researchers as this technology allows development of fast, cost effective and tailor-made implants based on patient data to meet specific clinical and geometrical constraints [1–3]. However, the poor surface finish of FDM parts appears to be massive barrier against its functionality for rapid casting applications. The surface roughness, surface hardness and dimensional accuracy of FDM patterns have significant impact on surface quality of final castings [4]. Although the fact that poor surface finish of FDM parts could not be completely eliminated but different techniques have been developed to improve surface quality of FDM parts which are further divided into pre-processing and post-processing finishing techniques (Fig. 1).

### *1.1 Pre-processing Techniques of Surface Finishing*

The angle of orientation or deposition angle can be altered with respect to machine co-ordinate system in CAD model to achieve desired objectives [5]. Generally, it is focussed on minimum cost, time, support material usage and most importantly the surface finish required on specific plane. Kattethota and Henderson [6] initially

J. S. Chohan
Mechanical Engineering Department, Chandigarh University, Gharaun, Mohali, India

R. Singh (✉)
Production Engineering Department, Guru Nanak Dev Engineering College, Ludhiana, India

K. S. Boparai
Mechanical Engineering Department, MRSPTU, Bathinda, India



B. AlMangour (ed.), *Additive Manufacturing of Emerging Materials*,
https://doi.org/10.1007/978-3-319-91713-9_7

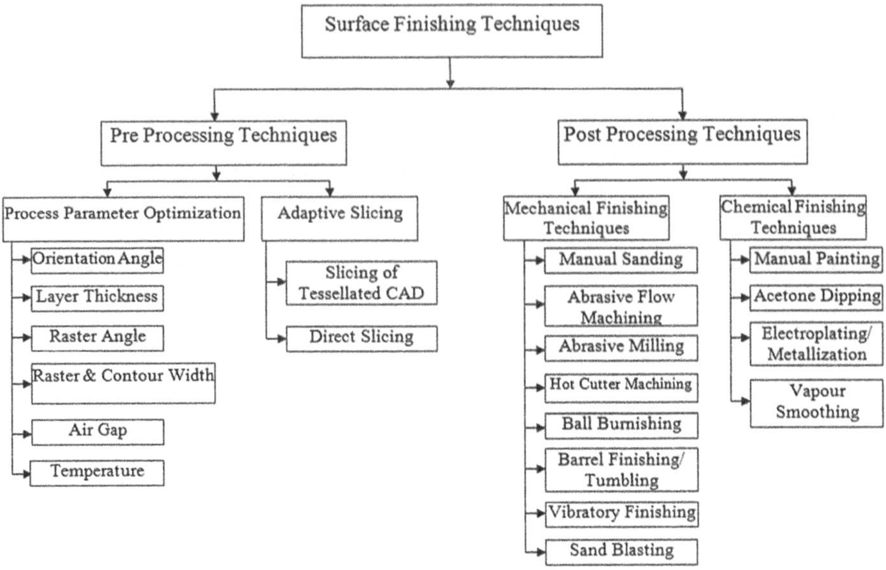


**Fig. 1** Classification of surface finishing techniques adopted for FDM parts (Source: Manufacturing Research Lab, GNDEC Ludhiana)

investigated the effects of build orientation on FDM parts and found that the angle of 0° yielded maximum surface finish. In most of the studies, the orientation angles of 0° and 90° were found most effective [7–11] for surface finish, build times and cost. Similarly, the layer thickness is second most significant parameter which influences the surface roughness of FDM parts. Generally, the smaller layer thickness results in smoother surface and higher dimensional accuracy but it requires longer duration for part fabrication [12–14]. On the other hand, larger layer thickness would cause sufficient surface irregularities, dimensional variability but needs lesser time to fabricate and thus trade-off must be done.

Although other pre-processing parameters such as air gap, contour width, raster width, raster angle and temperature have noticeable impact on surface roughness and dimensional accuracy of parts but the optimized settings of these parameters varies with part geometry, intricacy and manufacturing environment [15].

The adaptive slicing techniques have been developed which act as balancing agent between surface finish and build times. It has been observed that constant layer thickness leads to wastage of time in some situations as the layer (slice) thickness may not affect the surface roughness every time [16]. This is completely case specific depending upon dimensions and shape of parts. The adaptive slicing technique focuses on developing such algorithms which can automatically vary slice height and reduce production time and generate variable tool paths depending upon part geometry to achieve the minimum surface roughness [17]. Originally, the solid model is sliced by CAD software before transferring the STL data to FDM machine. The slicing of the CAD model is carried out either directly on a solid model or after tessellation. In slicing, the set of horizontal planes divide the CAD model in closed curves or polygons and the space between two successive horizontal planes is called as a slice [18].

## 1.2 Post-processing Techniques of Surface Finishing

Generally, the post-processing is required for every part regardless of additive manufacturing technique used for fabrication. The support structure flakes and extra plastic particles are needed to be removed, cleaned and washed before using parts for further applications. Also, the pre-processing techniques cannot improve the surface finish beyond certain limits and thus, there is utmost requirement of post-processing for rapid tooling and casting applications.

The mechanical finishing comprises traditional surface finishing techniques which require direct contact between tool and work piece such as abrasive flow machining, abrasive milling, barrel finishing, vibratory bowl finishing, ball burnishing and hot cutter machining [19–24]. The mechanical methods exhibit several challenges to finish the complex and intricate shapes. The abrasive action of media rounds the sharp edges and corners and distorts the part geometry and dimensional stability. Moreover, deeper lying surfaces within grooves, notches or other indentations are more difficult for the abrasives to reach.

The major precedence of chemical finishing over mechanical finishing techniques is that there is no contact of tools with work surface which ensure better dimensional and geometrical stabilities. Many authors used acetone both in liquid and vapour form to improve the surface finish of FDM parts [25–27]. The exposure of parts to the acetone environment enhanced compressive strength, flexural strength water tightness, humidity and wear resistance but slightly reduced the tensile strength [28, 29]. Moreover, parts undergone chemical treatment showed less dimensional changes as compared to mechanically brushed parts. Still, there is a risk of eroding and dissolving of small features of parts for longer durations using undiluted acetone while use of dilute solution would extend the immersion time. Also, the use of concentrated laser beam for surface finishing of FDM parts resulted in high temperature at small area which led to evaporation of plastic material [30].

An advanced finishing technique has been developed where FDM parts are exposed to chemical vapours in controlled environmental conditions to ensure uniform finishing, retain part intricacy and avoid overheating [31]. The Vapour Smoothing (VS) apparatus has been developed by Stratasys where ABS parts are alternatively cooled, heated and again cooled in dedicated chambers at constant temperature as shown in Fig. 2. Initially, the parts are allowed to cool for few minutes in cooling chamber (pre-cooling) at 0 °C and then placed in smoothing chamber (smoothing). The fixed amount of solvent (smoothing fluid) is heated in smoothing chamber for 10–30 s as suggested by manufacturer. The smoothing chamber is equipped with heater at bottom and maintained at 65 °C. Afterwards, the parts are again cooled in cooling chamber (post-cooling) and the whole cycle is

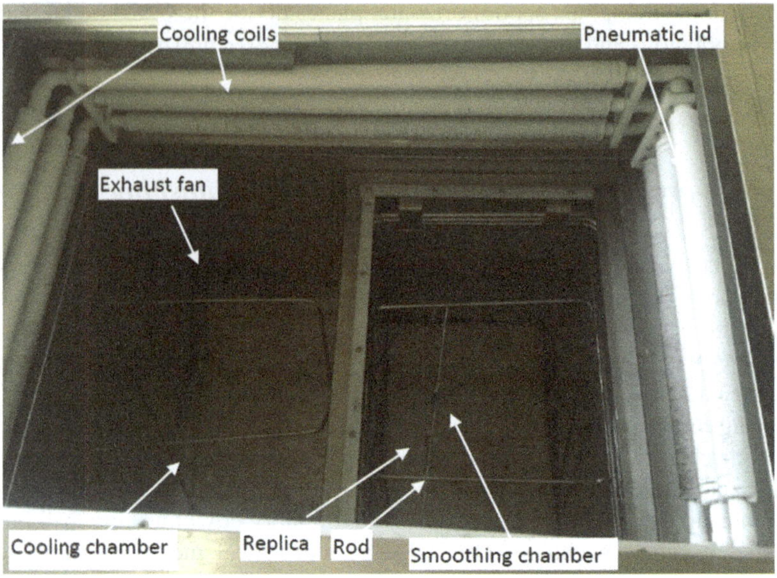


**Fig. 2** Cooling and smoothing chambers of vapour smoothing apparatus (Source: Manufacturing Research Lab, GNDEC Ludhiana)

repeated until desired finish is achieved. The preliminary studies on vapour smoothing technique showed good surface finish and dimensional stability of standard test parts. But, the surface characteristics of complex parts must be ascertained. Thus, next section would investigate the impact of various process parameters of coupled FDM and VS operations on surface finish, surface hardness and dimensional accuracy of replicas of biomedical implants. This analysis would act as base for the development of mathematical models so as to predict the surface properties of ABS replicas well before production phase.

## 2 Optimization Study of Process Parameters of FDM and Vs Processes

The ABS-P400 was used as model material along with P400SR as support material to manufacture hip joint replicas (Fig. 3) through commercial "u-Print SE" FDM apparatus. The previous literature and operation manuals of vapour smoothing apparatus were carefully analyzed to chalk out most prominent process parameters who would define the surface properties of ABS parts. Consequently, the impact of input parameters of combined FDM and VS processes on surface roughness, surface hardness and dimensional accuracy of hip implant replicas was ascertained. Since the orientation angle (α) has maximum influence on surface finish, it was considered as pre-processing input parameter. Although density (ρ) does not affects the surface

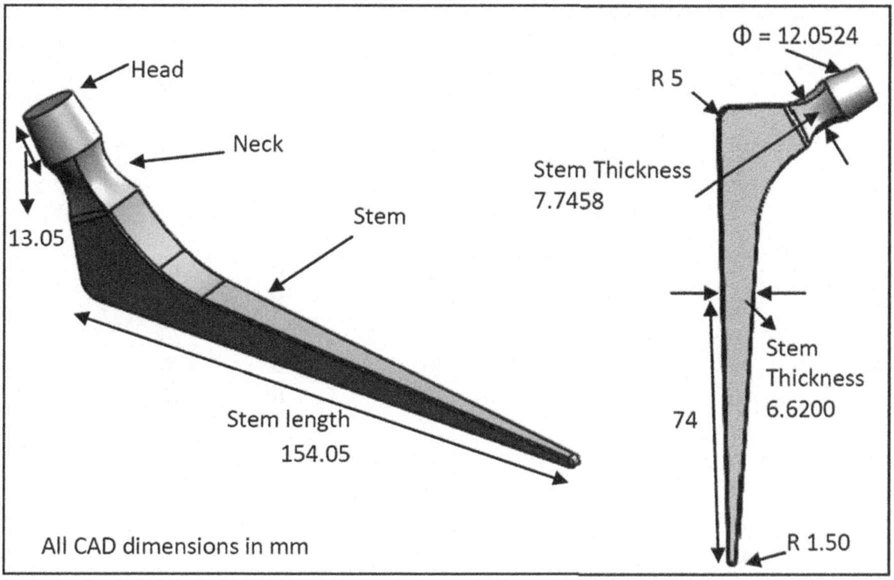


**Fig. 3** Benchmark component of hip joint (Source: Manufacturing Research Lab, GNDEC Ludhiana)

roughness of parts but it may influence the hardness of ABS replicas which play critical role during investment casting. The uPrint FDM apparatus has tendency to fabricate parts with different density (interior fill) settings which could significantly affect the mechanical properties, weight and manufacturing time.

Thus, two levels of orientation angle i.e. 0° and 90° along with three levels of density (i.e. low, high and solid) were selected as pre-processing parameters for investigation.

Similarly, four input parameters of VS apparatus i.e. pre-cooling time ($T_{PC}$), smoothing time ($T_S$) post-cooling time ($T_{PT}$) and number of cycles (N) were selected. The smoothing fluid used for experimentation comprises 30% decafluoropentane and 70% dichloroethylene. The volatile fluid vapourises at very low temperature which penetrates the part surface and is later exhausted by fan provided in the cooling chamber. The average surface roughness was measured before ($R_{ai}$) and after vapour smoothing ($R_{af}$) using Mitutoyo Surface Roughness Tester "SJ-210" as per ISO 4287 regulations [32]. Similarly, the surface hardness was measured with Durometer before ($H_{di}$) and after vapour smoothing ($H_{df}$) on shore D scale (unit less) using ASTM D 2240 standards [33]. For mathematical equations, the hardness was suitably converted into spring force as one shore D unit is equal to 0.4445 N. The thickness of stem section ($D_o$ = 6.6200 mm) at pre-specified location i.e. 74 mm above the tip was measured before ($D_b$) and after vapour smoothing ($D_a$) operations using Mitutoyo Crysta Apex C 163012 co-ordinate measuring machine adopting ISO 10360-2 standards [34]. Similarly, the Head diameter ($D_o$ = 12.0524 mm) located at 1.5 mm from top and Neck

**Table 1** Taguchi $L_{18}$ based control log of experimentation

| Exp. No. | α (°) | ρ × $10^{-3}$ (g/$cm^3$) | $T_{PC}$ (mins.) | $T_S$ (s) | $T_{PT}$ (mins.) | N |
|---|---|---|---|---|---|---|
| 1 | 0 | 614.77 | 10 | 10 | 10 | 1 |
| 2 | 0 | 614.77 | 15 | 15 | 15 | 2 |
| 3 | 0 | 614.77 | 20 | 20 | 20 | 3 |
| 4 | 0 | 820.52 | 10 | 10 | 15 | 2 |
| 5 | 0 | 820.52 | 15 | 15 | 20 | 3 |
| 6 | 0 | 820.52 | 20 | 20 | 10 | 1 |
| 7 | 0 | 945.78 | 10 | 15 | 10 | 3 |
| 8 | 0 | 945.78 | 15 | 20 | 15 | 1 |
| 9 | 0 | 945.78 | 20 | 10 | 20 | 2 |
| 10 | 90 | 614.77 | 10 | 20 | 20 | 2 |
| 11 | 90 | 614.77 | 15 | 10 | 10 | 3 |
| 12 | 90 | 614.77 | 20 | 15 | 15 | 1 |
| 13 | 90 | 820.52 | 10 | 15 | 20 | 1 |
| 14 | 90 | 820.52 | 15 | 20 | 10 | 2 |
| 15 | 90 | 820.52 | 20 | 10 | 15 | 3 |
| 16 | 90 | 945.78 | 10 | 20 | 15 | 3 |
| 17 | 90 | 945.78 | 15 | 10 | 20 | 1 |
| 18 | 90 | 945.78 | 20 | 15 | 10 | 2 |

Source: Manufacturing Research Lab, GNDEC Ludhiana

thickness ($D_o = 7.7458$ mm) at mid-section of neck have been specified as locations to evaluate dimensional accuracy of intricate part features. Finally, the percent error in dimensions before vapour smoothing ($\%\Delta D_b$) and after vapour smoothing ($\%\Delta D_a$) was compared with original CAD dimensions.

Taguchi $L_{18}$ design of experimentation (DOE) technique was used to execute the experiments (Table 1). The results were obtained in terms of percentage change in surface roughness and hardness and displayed in Table 2. The dimensional accuracy of head diameter, neck thickness and stem thickness has been measured before and after smoothing process. The measurements of part features are compared with original CAD dimensions to calculate $\%\Delta D_b$ and $\%\Delta D_a$ (Table 3). The initial data observation reveals significant reduction in average surface roughness of FDM replicas after vapour smoothing. Also, the hardness of ABS replicas has been enhanced as compared to unfinished parts. The response of vapour smoothing process on dimensional accuracy of each part feature is distinct from each other. The dimensional deviation of neck thickness and stem thickness has been significantly reduced while an increase in deviation has been noticed in head diameter.

The SN ratio plots have been acquired to evaluate the effect of individual parameter on response based on experimental data. The SN ratio plot in Fig. 4a indicates that smoothing time and number of cycles are most significant parameters which affect surface roughness.

Moreover, the increase in smoothing time and number of cycles has direct impact on percentage change in surface roughness. On the other hand, surface hardness is

**Table 2** Results obtained from control log of experimentation for surface roughness and hardness

| Exp. No. | $R_{ai}$ (μm) | $R_{af}$ (μm) | %$\Delta R_a$ | $H_{di}$(N) | $H_{df}$ (N) | %$\Delta H_d$ |
|---|---|---|---|---|---|---|
| 1 | 3.0125 | 0.8995 | 70.14 | 71.5 | 76 | 6.29 |
| 2 | 3.1070 | 0.5356 | 82.76 | 72 | 77.5 | 7.63 |
| 3 | 3.0685 | 0.2154 | 92.98 | 70.5 | 77 | 9.22 |
| 4 | 3.3932 | 0.8455 | 75.08 | 76.5 | 82.5 | 7.84 |
| 5 | 3.3895 | 0.4450 | 86.87 | 77 | 83.5 | 8.44 |
| 6 | 3.4956 | 0.4527 | 87.05 | 77.5 | 82.5 | 6.45 |
| 7 | 3.7897 | 0.4665 | 87.69 | 81.5 | 87.5 | 7.36 |
| 8 | 3.8975 | 0.5277 | 86.46 | 82 | 88 | 7.32 |
| 9 | 3.8195 | 0.9021 | 76.38 | 83 | 90 | 8.43 |
| 10 | 8.6807 | 0.9262 | 89.33 | 46.5 | 51 | 9.67 |
| 11 | 8.8012 | 1.5226 | 82.70 | 46.5 | 50 | 7.53 |
| 12 | 8.9575 | 2.2689 | 74.67 | 46.5 | 50.5 | 8.60 |
| 13 | 8.8059 | 2.130 | 75.81 | 52 | 57.5 | 10.57 |
| 14 | 8.4515 | 0.7640 | 90.96 | 52.5 | 57 | 8.57 |
| 15 | 8.4822 | 0.9381 | 88.94 | 52.5 | 57.5 | 9.52 |
| 16 | 8.5215 | 0.5411 | 93.65 | 58.5 | 63 | 7.69 |
| 17 | 8.6525 | 2.5023 | 71.08 | 58.5 | 64 | 10.25 |
| 18 | 8.5827 | 1.4136 | 83.53 | 60.5 | 66 | 9.09 |

Source: Manufacturing Research Lab, GNDEC Ludhiana

**Table 3** Results obtained from control log of experimentation for dimensional accuracy of different part features

| | Head diameter | | Neck thickness | | Stem thickness | |
|---|---|---|---|---|---|---|
| Exp. No. | %$\Delta D_b$ | %$\Delta D_a$ | %$\Delta D_b$ | %$\Delta D_a$ | %$\Delta D_b$ | %$\Delta D_a$ |
| 1 | 0.33 | 0.45 | 1.57 | 1.10 | 3.68 | 3.34 |
| 2 | 0.33 | 0.76 | 1.54 | 0.63 | 3.68 | 1.16 |
| 3 | 0.33 | 1.28 | 1.56 | 0.21 | 3.68 | 0.19 |
| 4 | 0.33 | 0.62 | 1.53 | 0.94 | 3.68 | 2.58 |
| 5 | 0.33 | 1.30 | 1.53 | 0.52 | 3.68 | 0.62 |
| 6 | 0.33 | 1.54 | 1.58 | 0.26 | 3.68 | 0.95 |
| 7 | 0.33 | 0.85 | 1.55 | 0.52 | 3.68 | 0.63 |
| 8 | 0.33 | 1.39 | 1.56 | 0.25 | 3.68 | 1.62 |
| 9 | 0.33 | 0.38 | 1.58 | 0.88 | 3.68 | 2.58 |
| 10 | 0.16 | 0.94 | 0.25 | 0.12 | 1.55 | 0.05 |
| 11 | 0.16 | 0.21 | 0.25 | 0.19 | 1.55 | 0.31 |
| 12 | 0.16 | 0.25 | 0.25 | 0.21 | 1.55 | 0.24 |
| 13 | 0.16 | 0.28 | 0.25 | 0.15 | 1.55 | 0.27 |
| 14 | 0.16 | 0.75 | 0.25 | 0.11 | 1.55 | 0.17 |
| 15 | 0.16 | 0.38 | 0.25 | 0.18 | 1.55 | 0.88 |
| 16 | 0.16 | 0.82 | 0.25 | 0.05 | 1.55 | 0.03 |
| 17 | 0.16 | 0.26 | 0.25 | 0.22 | 1.55 | 1.12 |
| 18 | 0.16 | 0.36 | 0.25 | 0.17 | 1.55 | 0.77 |

Source: Manufacturing Research Lab, GNDEC Ludhiana

a

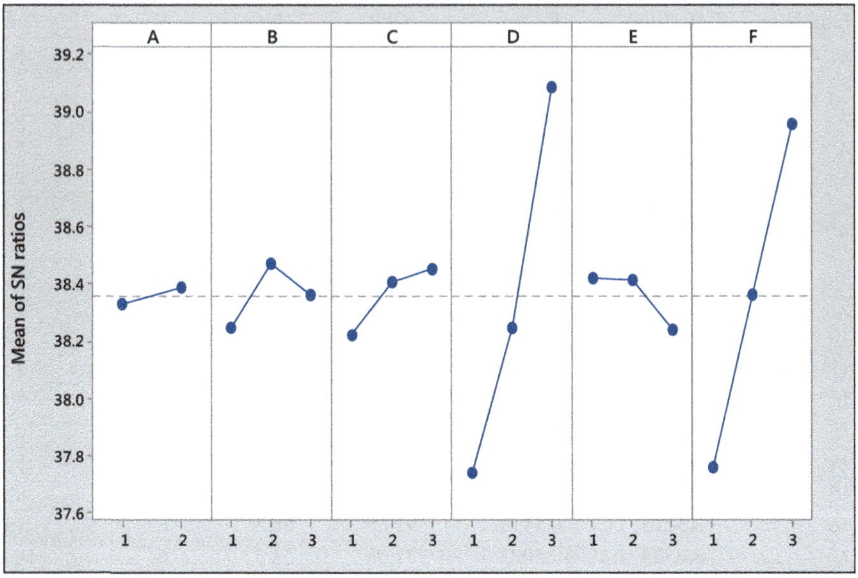


b

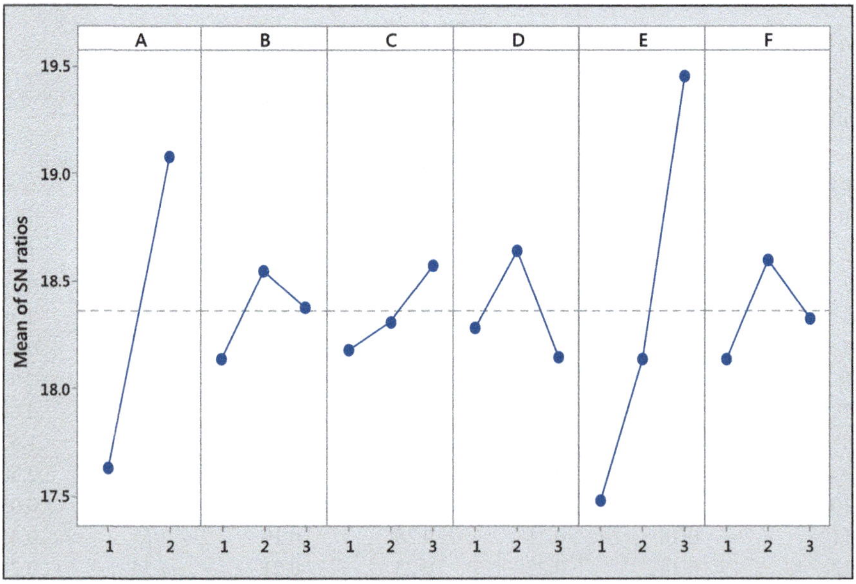


**Fig. 4** SN ratio plots for percentage change (**a**) surface roughness (**b**) hardness (Source: Manufacturing Research Lab, GNDEC Ludhiana)

highly enhanced by increase in orientation angle and post-cooling time (Fig. 4b). Similar inferences can be made regarding other response parameters i.e. dimensional accuracy of different part features (Fig. 5). The Analysis of variance (ANOVA)

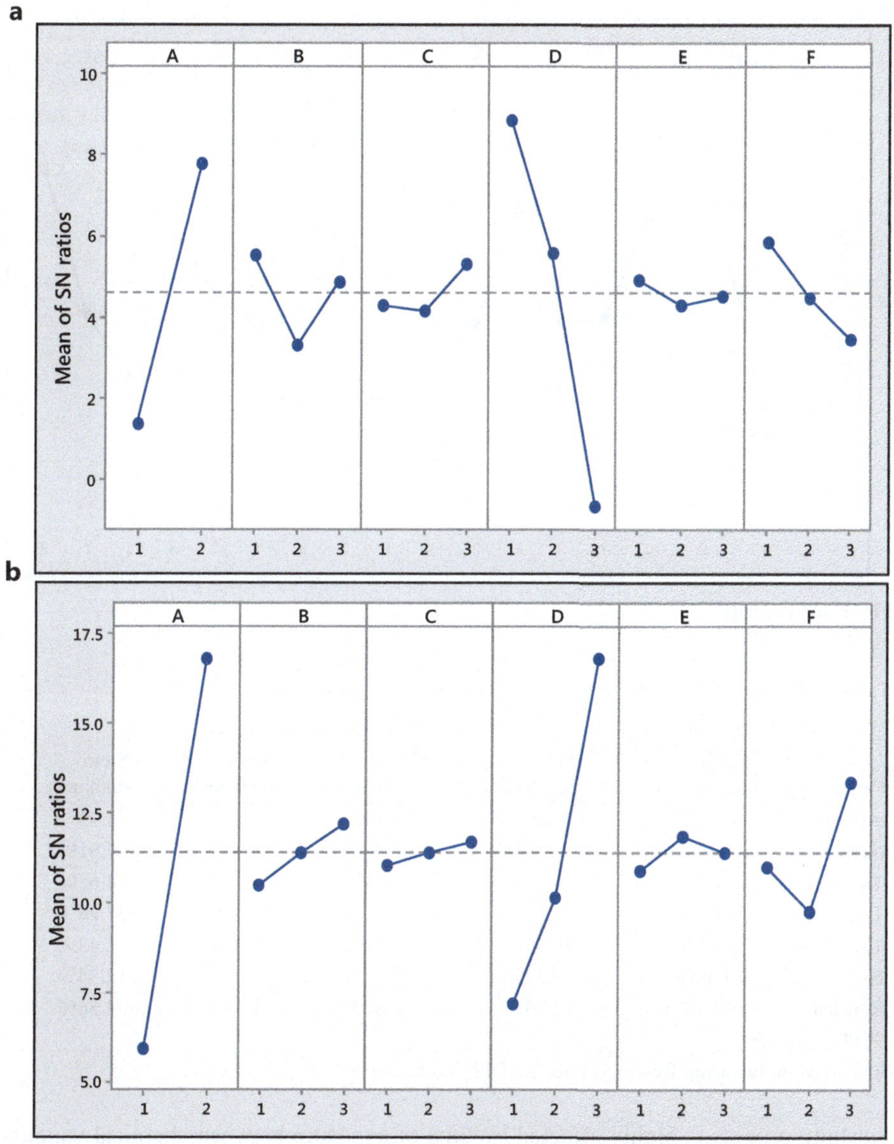


**Fig. 5** SN ratio plots for dimensional accuracy (**a**) Head diameter (**b**) Neck thickness (**c**) Stem Thickness (Source: Manufacturing Research Lab, GNDEC Ludhiana)

technique helped to estimate the percentage contribution of each parameter (Table 4) which further would lead to formulation of mathematical equations.

The dimensional accuracy is highly influenced by smoothing time and orientation angle while post-cooling time has maximum influence on surface hardness of hip replicas.

The response on surface finish is directly proportional to smoothing time and number of cycles which are most significant parameters. This indicates that the

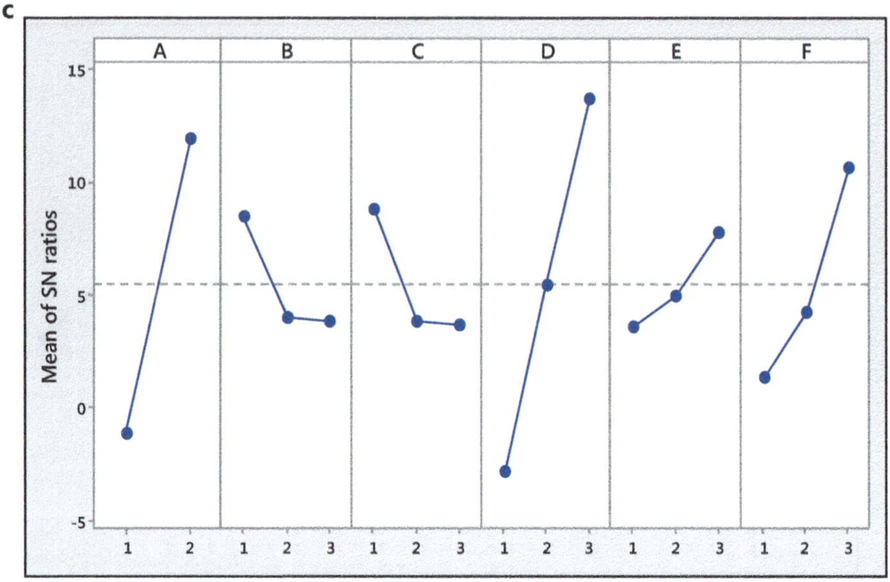


**Fig. 5** (continued)

**Table 4** Percentage contribution of each input parameter on response

| Parameters | Surface roughness | Surface hardness | Dimensional accuracy | | |
|---|---|---|---|---|---|
| | | | Head diameter | Neck thickness | Stem thickness |
| α (°) | 0.13% | 34.69% | 34.58% | 59.78% | 34.93% |
| ρ (g/cm$^3$) | 1.42% | 0.764% | 2.8% | 0.98% | 3.81% |
| $T_{PC}$ (mins.) | 1.61% | 1.71% | 0.85% | 0.14% | 4.65% |
| $T_S$ (s) | 51.07% | 2.90% | 51.74% | 32.57% | 37.28% |
| $T_{PT}$ (mins.) | 1.14% | 44.46% | 0.21% | 0.28% | 2.43% |
| N | 40.08% | 2.32% | 3.2% | 4.5% | 12.34% |
| Residual error | 4.55% | 13.15% | 6.62% | 1.75% | 4.56% |

Source: Manufacturing Research Lab, GNDEC Ludhiana

finishing process is highly affected by time of exposure between chemical vapours and plastic material. The larger exposure duration ensures complete contact between vapours with plastic surface. The vapours get sufficient time to penetrate the upper surface and cause localised melting of thin upper plastic layers. The layers of ABS thermoplastic undergo viscous mass transportation as semi-molten plastic flows from peaks into the valleys (Fig. 8). After cooling, the viscous material settles as smooth surface under the effect of surface tension. The surface tension forces tend to attain minimum surface area which is only possible with smooth surface i.e. absence of peaks and valleys [35]. The upper and transverse view of surface has been viewed

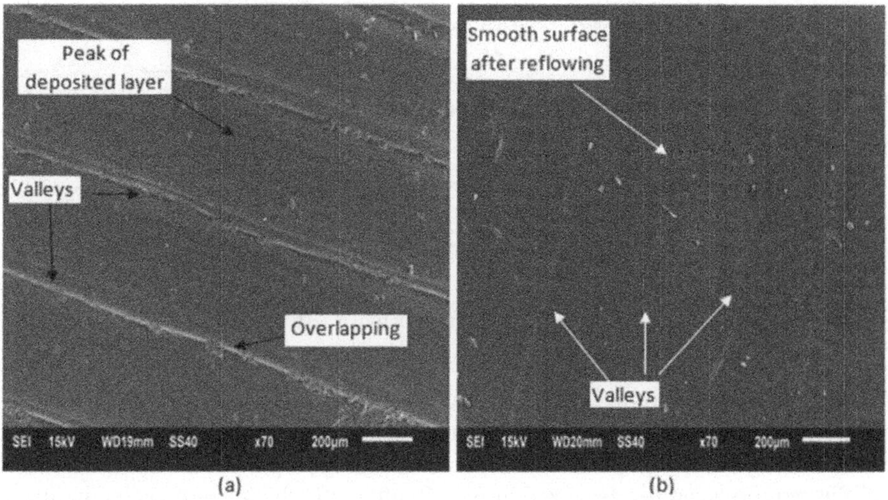


**Fig. 6** SEM micrograph top view (**a**) before VS (**b**) after VS (Source: Manufacturing Research Lab, GNDEC Ludhiana)

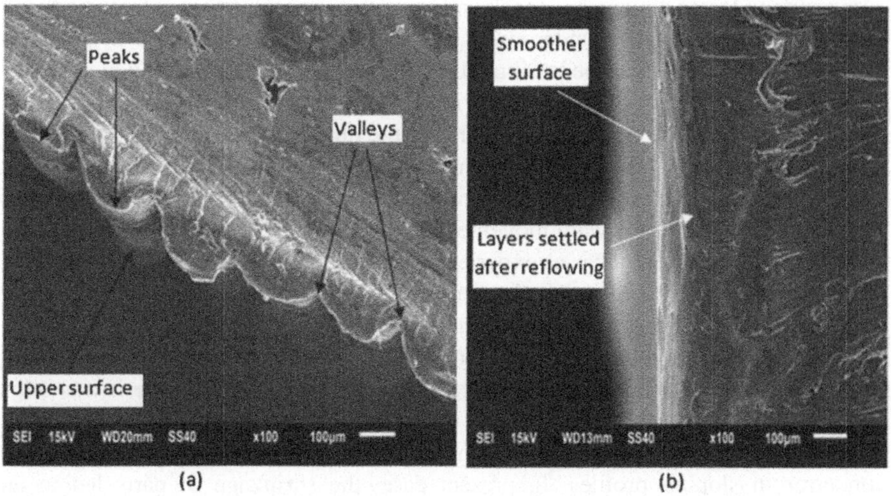


**Fig. 7** SEM micrograph transverse view (**a**) before VS (**b**) after VS (Source: Manufacturing Research Lab, GNDEC Ludhiana)

under Scanning Electron Microscope (SEM) are shown in Figs. 6 and 7 respectively. The peaks and valleys are clearly visible and distinct lines can be noticed where two roads join to make valleys. The overlapping of plastic layers can be seen before VS as parts are manufactured with zero air gap by FDM. The disappearance of peaks after vapour smoothing process also resulted in decrease in peak height and thus shrink age is noticed during dimensional measurements.

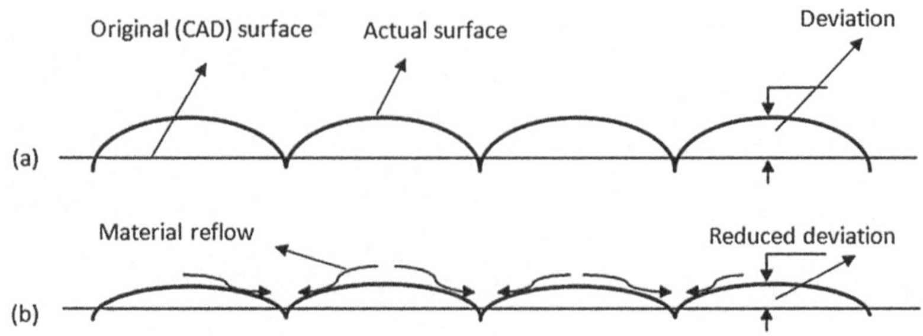


**Fig. 8** Dimensional accuracy and surface finish (**a**) before VS (**b**) after VS

After vapour smoothing, the smooth surfaces is achieved with glossy finish due to uniform re-settlement of plastic layers after cooling. The over-lapping is also absent in SEM images photographed after vapour smoothing process. Also, the small traces of plastic reflow can be visualized in Fig. 7b after VS which is not visible in image before finishing.

The percentage deviation before and after smoothing in head diameter, neck thickness and stem thickness has been shown in Table 3. The FDM apparatus caused over-sizing in linear dimensions (neck and stem sections) while radial features (head diameter) are found to be under-sized which has also been reported in previous studies [36, 37]. This can be attributed to intricacy of part features which led to approximation errors. Moreover, the deviation is higher in case of replicas fabricated at 0° and thus, it is recommended to fabricate the hip replicas at vertical positioning (orientation angle 90°) to attain better dimensional stability. After vapour smoothing, partial melt-down of layers induced shrinkage in parts as layers get re-arranged as smooth surface. The downward movement of upper plastic layers tend to reduce the overall dimensions of replicas as experienced by previous researchers [35, 36]. This shrinkage induces dimensional changes in different part features after vapour smoothing. Also, the dimensional accuracy after vapour smoothing ($\%\Delta D_a$) is highly influenced by accuracy before smoothing ($\%\Delta D_b$). The head diameter measurements after vapour smoothing indicated further shrinkage because it was originally produced under-sized.

On the other hand, over-sizing was observed in stem section due to approximation error in sloping profiles. In present case, the shrinkage of parts led to an improvement in dimensional accuracy of stem thickness and stem thickness but deteriorated the accuracy of head diameter. The comparison between original CAD dimension (Do) and actual rough surface before vapour smoothing (Db) has been conceptualized in Fig. 8a. After fabrication under FDM (before vapour smoothing), the significant deviation between CAD and actual surface is visible.

There is shrinkage in part dimensions as surface becomes smoother after vapour smoothing process (see Fig. 8b). As the smoothing time is increased, the material reflowing rate increases which increase the shrinkage and resulted in smoother surface. The SEM micrographs of stair-step have been acquired to validate the

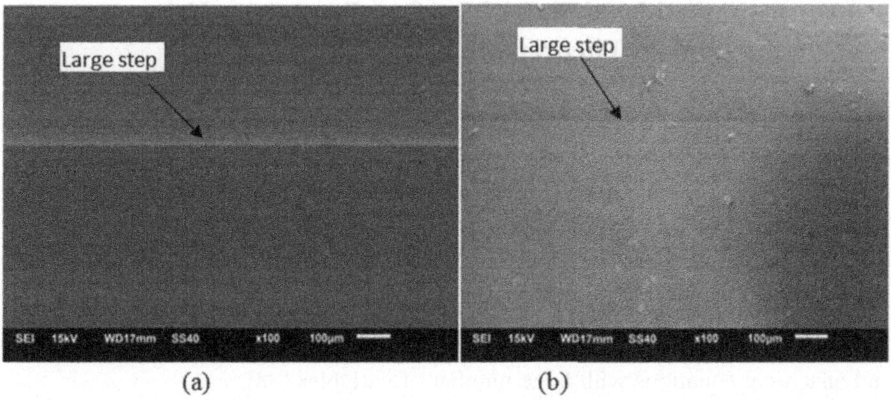


**Fig. 9** SEM micrograph of stair-step (**a**) before VS (**b**) after VS (Source: Manufacturing Research Lab, GNDEC Ludhiana)

**Table 5** Optimum parameter settings for different response terms

| Parameters | Surface roughness | Surface hardness | Dimensional accuracy: Head diameter | Dimensional accuracy: Neck thickness | Dimensional accuracy: Stem thickness |
|---|---|---|---|---|---|
| α (°) | 90° | 90° | 90° | 90° | 90° |
| ρ (g/cm$^3$) | High | High | Low | Solid | Low |
| $T_{PC}$ (mins) | 20 min | 20 min | 20 min | 20 min | 10 min |
| $T_S$ (s) | 20 s | 15 s | 10 s | 20 s | 20 s |
| $T_{PT}$ (mins) | 10 min | 20 min | 10 min | 15 min | 20 min |
| N | 3 | 2 | 1 | 3 | 3 |

Source: Manufacturing Research Lab, GNDEC Ludhiana

smoothing phenomenon and dimensional changes occurring during finishing process (Fig. 9). The difference between magnitude of stair-step before and after smoothing can be clearly differentiated.

The importance of post-cooling has been highlighted after analysis of surface hardness of ABS replicas. The surface hardness is enhanced by increase in post-cooling time which most significant parameter. Second most influential parameter is orientation angle which affects distribution of internal layer deposition based on part geometry.

After exposure to hot vapours, the immediate cooling is required for resettlement and hardening of surface. As the FDM parts are cooled for longer duration, the surface becomes harder. Also, the overall impact of vapour smoothing process is to increase the surface hardness of hip replicas. The indenter enters more easily when surface layers are uneven and can be easily pierced. After VS process, the smooth roughness profiles were achieved which efficiently restricted the piercing of indenter.

As concluding remarks to this section, the optimum level of parameters for each response have been tabulated which can be used to achieve desired effect during vapour smoothing (Table 5).

# 3 Mathematical Modeling of Surface Properties Using Buckingham Pi Theorem

This sections aims to develop the mathematical equations which could precisely predict the response based on given conditions. These equations can be recommended to biomedical manufacturing industry to achieve desired results without performing further experiments. The inferences made through Taguchi and ANOVA results were used to develop a mathematical relationship between input parameters and response. The Buckingham $\pi$ theorem of dimensional analysis was applied in present context for mathematical modelling. The Dimensional analysis has proved to be an effective tool for generating the dimensionally homogenous and analytical equations with large number of variables [38].

Although, the most significant parameters govern the result but all the input parameters with fundamental units (MLT) were included during calculations. The input parameters selected based on optimization experiments along with fundamental dimensions are written as:

(i) Initial Roughness ($R_{ai}$) in μm $M^0 L^1 T^0$
(ii) Final Roughness ($R_{af}$) in μm $M^0 L^1 T^0$
(iii) Surface Hardness $H_d$ in N $M^1 L^1 T^{-2}$
(iv) Initial deviation $\Delta D_b$ in mm $M^0 L^1 T^0$
(v) Final deviation $\Delta D_a$ in mm $M^0 L^1 T^0$
(vi) Density ($\rho$) in g/mm$^3$ $M^1 L^{-3} T^0$
(vii) Pre-cooling Time ($T_{PC}$) in minutes $M^0 L^0 T^1$
(viii) Smoothing Time ($T_S$) in seconds $M^0 L^0 T^1$
(ix) Post-cooling Time ($T_{PT}$) in minutes $M^0 L^0 T^1$

The generalised mathematical models for prediction of response can be generated using process parameters of FDM and VS processes through Buckingham Pi theorem. After generating generalised equation, further experiments were conducted with by varying significant parameters in case of each response to complete the individual equation.

## 3.1 Mathematical Model of Surface Roughness

The general equation for percentage change in surface roughness has been derived as:

$$\%\Delta Ra = \frac{100.KN.Tpc}{TPT^2} \tag{1}$$

Where $K_N$ is constant of proportionality which depends upon smoothing time ($T_S$) for given number of cycles i.e. N = 1, 2, 3 since these are most significant parameters. The values of constant of proportionality have been calculated experimentally by varying smoothing time as 0, 5, 10, 15, 20, 25 and 30 s. The individual value of $K_N$ has been calculated from best fitted curve equations of trend lines for each cycle and given as:

$$K_N = -0.071\,(T_S^2) + 4.175\,T_S + 30.38 \quad \text{if, } N = 1$$
$$K_N = -0.075\,(T_S^2) + 4.164\,T_S + 37.63 \quad \text{if, } N = 2$$
$$K_N = -0.079\,(T_S^2) + 4.101\,T_S + 46.27 \quad \text{if, } N = 3$$

The mathematical equations were formulated and finally, the validity of equation was confirmed assuming following conditions:

$\alpha = 90°$, $T_{PC} = 1$ min, $T_{PT} = 10$ min, $T_S = 15$ s, $N = 1$.

Since for $N = 1$, value of $K_N$ is given as $-0.071\,(T_S^2) + 4.175\,T_S + 30.38$. Thus Eq. 1 can be re-written as:

$$\%\Delta R_a = \frac{100.\left[-0.071\,(T_S)^2 + 4.175\,(T_S) + 30.38\right].T_{PC}}{T_{PT}^2}$$

Inserting values:

$$\%\Delta R_a = \frac{100 \times \left[-0.071\,(15)^2 + 4.175\,(15) + 30.38\right] \times 1}{10^2}$$

Solving the equation the predicted value of $\%\Delta R_a$ was 77.03%, whereas at same conditions, the experimental results yielded 76.48% improvement. A little deviation was noticed between predicted and experimental results which confirmed the efficacy and reliability of mathematical model.

## *3.2 Mathematical Model of Surface Hardness*

The equation for surface hardness can be written as:

$$H_d = \frac{K_N \cdot T_{PC} \cdot T_{PT}.\rho.(R_{ai})^2}{T_s^4} \tag{2}$$

Here, the value of $K_N$ is given as:

$K_N = -0.004(T_{PC})^2 + 0.505(T_{PC}) + 21.68$

After inserting the values of $K_N$ in Eq. 2, the final equation can be written as:

$$H_d = \frac{\left[-0.004(T_{PT})^2 + 0.505(T_{PT}) + 21.68\right] \;.\; T_{PC}.\rho.(R_{ai})^2}{T_S^4} \tag{3}$$

The performance of mathematical Eq. (3) was evaluated by comparing the experimental values with predicted data. The example has been shown below to demonstrate the calculations at post-cooling for 10 min. The other conditions are:

$\rho = 0.82092$ g/cm$^3$, $T_{PC} = 20$ min., $T_{PT} = 10$ min., $R_{ai} = 7.2115$ µm, $T_S = 15$ s

Inserting vales in Eq. 3, $H_d = \frac{\left[-0.004(10)^2 + 0.505\,(10) + 21.68\right] \times 20 \times 60 \times 0.82092 \times (7.2115)^2}{(15)^4}$

After solving the hardness comes out to be 26.62 N which is predicted by model while experimental data shows 27.34 N in terms of spring force. Thus, the negligible error (2.63%) was found between modelled and experimental data.

### 3.3 Mathematical Model of Dimensional Accuracy

The generalised equation to predict the dimensional accuracy (percentage change) of FDM parts after vapour smoothing is given as:

$$\%\Delta D_a = \left(\frac{\%\Delta D_b \cdot K_N . T_S \cdot T_{PT}}{T_{PC}^2}\right) \tag{4}$$

The value of constant of proportionality $K_N$ are given as:

$K_N = 0.054\ (T_s)^2 – 0.243\ (Ts) + 5.364$ for head diameter
$K_N = -0.003\ (T_s)^2 – 0.081\ (Ts) + 6.238$ for neck thickness
$K_N = 0.003\ (T_s)^2 – 0.269\ (Ts) + 5.341$ for stem thickness

The mathematical equation for prediction of percentage deviation in measurement of stem thickness after vapour smoothing has been given as:

$$\%\Delta D_a = \left[\frac{\%\Delta D_b . \left(0.003\ Ts^2 - 0.269\ Ts + 5.341\right) T_S \cdot T_{PT}}{T_{PC}^2}\right] \tag{5}$$

The above equation was used to confirm the performance of mathematical model to calculate the percentage deviation in stem thickness at different conditions. The calculations were made assuming following data:

$\%\Delta D_b = 1.5$, $T_{PC} = 15$ min., $T_S = 10$ s, $T_{PT} = 5$ min.

Inserting value in Eq. 5, we get

$$\%\Delta D_a = \left[\frac{\%\Delta D_b . \left(0.003\ Ts^2 - 0.269\ Ts + 5.341\right) T_S \cdot T_{PT}}{T_{PC}^2}\right]$$

$$\%\Delta D_a = \left[\frac{1.5 . \left(0.003\ 10^2 - 0.269 . 10 + 5.341\right) 10 \cdot 5}{15^2}\right]$$

$\%\Delta D_a = 1.5 \times 2.951 \times 10 \times 5/225 = 0.983$

The percentage change after vapour smoothing has been calculated as 0.983% while experiments showed 0.95% error in dimensions. The very minute variation in dimensional accuracy results proved efficient applicability of mathematical model for manufacturing industry.

# 4 Multi-response Optimization

The multi-response optimization is performed to achieve single set of optimum parameters to achieve maximum surface finish, hardness and dimensional accuracy. The data of SN ratios during 18 experiments performed in previous section has been used.

The SN ratios of all the response terms along with experimental orthogonal array (OA) have been shown in Table 6 which were given as input data for multi-response optimization.

The increase in SN ratios would lead to robust design and eliminate effect of noise factors in addition to improvement in surface finish, hardness and dimensional accuracy. The details of goals of individual response along with constraints and range have been shown in Table 6. The equal importance and weight has been given to achieve equal and uniform enhancement in surface finish, hardness and dimensional accuracy.

The optimum set of parameters suggested by multi-response optimization tool has been displayed in Table 7. Along with levels of input parameters, the SN ratios at suggested input parameters have been predicted for each response.

The 90° orientation angle has been recommended with low density settings which would save manufacturing time and cost. The low density settings would consume

**Table 6** Constraints and goals set for multi-response optimization

| Parameters | Goal | Lower limit | Upper limit | Weight | Importance |
|---|---|---|---|---|---|
| $\alpha$ (°) | Constraint to region | 0° | 90° | 1 | 1 |
| $\rho$ (g/cm$^3$) | Constraint to region | Low | Solid | 1 | 1 |
| $T_{PC}$ (mins) | Constraint to region | 10 min | 20 min | 1 | 1 |
| $T_S$ (s) | Constraint to region | 10 s | 20 s | 1 | 1 |
| $T_{PT}$ (mins) | Constraint to region | 10 min | 20 min | 1 | 1 |
| N | Constraint to region | 1 | 3 | 1 | 1 |
| SN Ratio (%$\Delta R_a$) | Maximize | 36.9193 | 39.4302 | 1 | 1 |
| SN Ratio (%$\Delta H_d$) | Maximize | 15.9730 | 20.4815 | 1 | 1 |
| SN Ratio (%$\Delta D_a$ Head diameter) | Maximize | −3.7504 | 13.5556 | 1 | 1 |
| SN Ratio (%$\Delta D_a$ Neck diameter) | Maximize | −0.8279 | 26.0206 | 1 | 1 |
| SN Ratio (%$\Delta D_a$ Stem thickness) | Maximize | −10.4749 | 30.4576 | 1 | 1 |

Table 7 Optimum settings of parameters and predicted SN ratios

| Parameter | Value |
|---|---|
| Orientation angle (A) | 90° |
| Density (B) | Low |
| Pre-cooling time (C) | 20 min. |
| Smoothing time (D) | 20 s |
| Post-cooling time (E) | 20 min. |
| Number of cycles (F) | 3 |
| SN Ratio ($\%\Delta R_a$) | 39.5784 |
| SN Ratio ($\%\Delta H_d$) | 19.9472 |
| SN Ratio ($\%\Delta D_a$ Head diameter) | 3.04608 |
| SN Ratio ($\%\Delta D_a$ Neck thickness) | 23.7073 |
| SN Ratio ($\%\Delta D_a$ Stem thickness) | 29.1218 |
| Desirability | 0.7891 |

minimum material and minimum fabrication time as experienced from previous experiments. On the other hand, replicas with 90° orientation angle would yield maximum enhancement in surface finish, hardness and accuracy. The overall desirability of optimization calculations comes out to be 0.7891. The relation between input parameters and each response has been plotted as a consequence of multi-response optimization analysis (Fig. 10). The high, medium and low levels of each parameter are shown whereas levels with red colour line indicate optimum (current) level. The vertical red lines indicate optimum level of each parameter and black dots represent response. The point of co-incidence of blue line and red line can be used to calculate SN ratio for respective response. Thus, it can be concluded that multi-response optimization module has successfully optimized the analytical parameters of combined FDM-VS operations. The predicted optimum set of parameters for combined FDM-VS process can be further used to finish the patterns for investment casting.

## 5 Differential Scanning Calorimetry

The Differential Scanning Calorimetry (DSC) tests were performed on samples extracted from hip replicas to explore the thermal modifications appeared after VS. The DSC is most commonly used technique for thermal analysis of plastic components.

It is used to study the material characteristics and response of heat on polymers as a function of temperature and time. In DSC analysis, the samples undergo physical transformation or phase transitions when heated at controlled conditions. The tests were performed using Mettler Toledo-DSC1 apparatus and comparison was made between samples before vapour smoothing and after smoothing. Figure 11a illustrates the temperature scans and it is evident that the green shaded area is smaller for

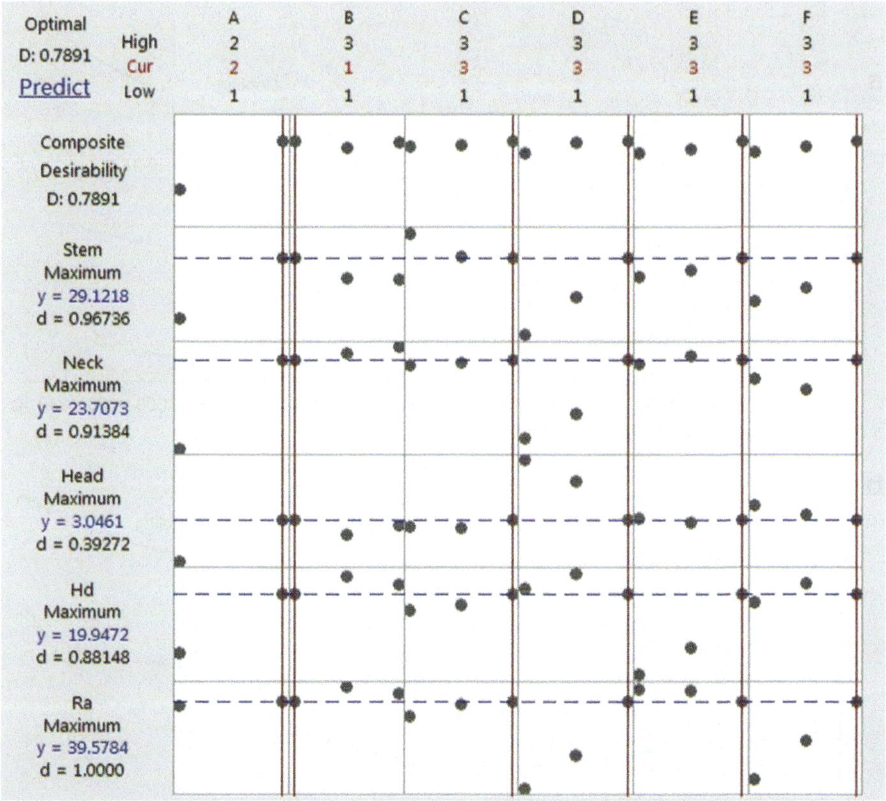


**Fig. 10** Multi-response optimization plot for combined FDM-VS (Source: Manufacturing Research Lab, GNDEC Ludhiana)

sample extracted before VS. The green shaded depicts the melting enthalpy of ABS material which is larger for samples after VS (Fig. 11b) as compared to before smoothing. The initial minute depression in the curve signifies glass transition temperature where mechanical properties of material change significantly due to heating. This is followed by the significant depression in curve along with a distinct peak which calculates the melting enthalpy. The vapour smoothing increases the glass transition temperature and melting enthalpy which indicates the there is an increase in heat bearing capacity of samples. Thus, the parts exposed to vapours are able to withstand higher temperature as compared to unexposed.

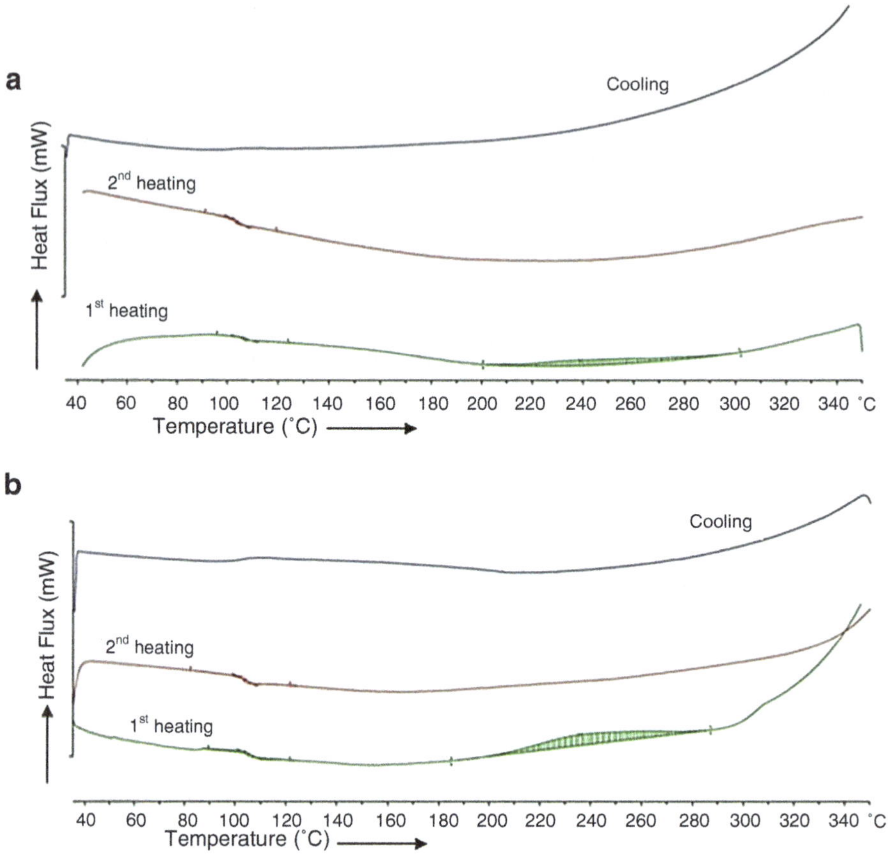


**Fig. 11** DSC thermal scans (**a**) before VS (**b**) after VS (Source: Manufacturing Research Lab, GNDEC Ludhiana)

## 6 Summary

The VS technique leads to an improvement in surface finish, surface hardness of FDM parts as layers settle as smooth surface after re-melting. This phenomenon induced shrinkage which caused dimensional instability in parts. Thus, parts must be manufactured slightly over-sized to compensate the shrinkage due to vapour smoothing. Based on significant process parameters, the mathematical models for each response i.e. surface roughness, surface hardness and dimensional accuracy were formulated using Buckingham Pi theorem of dimensional modelling. These outcomes could prove beneficial for the production of patient specific implants via FDM-VS-IC route which could significantly reduce the production time and cost.

**Acknowledgement** The authors are thankful to DST (GOI) for financial support and Manufacturing Research Lab (Dept. of Production Engg., Guru Nanak Dev Engg. College, Ludhiana (India) for technical support.

# References

1. Horáček M, Charvát O, Pavelka T, Sedlák J, Madaj M, Nejedlý J. Medical implants by using RP and investment casting technologies. In 69th WFC 2011 Feb:107-111
2. Gibson I, editor. Advanced manufacturing technology for medical applications: reverse engineering, software conversion and rapid prototyping. John Wiley & Sons; 2006 Jun 14
3. Rao AS, Dharap MA, Venkatesh JV. Development of Rapid Tooling for Investment Casting using Fused Deposition Modeling Process. Advanced Materials Research. 2014 Jul 9;970
4. Hanus A, Špirutová N, Beňo J. Surface quality of foundry pattern manufactured by FDM method-rapid prototyping. Archives of Foundry Engineering. 2011;11(1):15-20.
5. Sreedhar P, MathikumarManikandan C, Jothi G. Experimental investigation of surface roughness for fused deposition modeled part with different angular orientation. International Journal of Advanced Design and Manufacturing Technology. 2012 Jan 1;5(3):21-8
6. Kattethota G, Henderson M. A visual tool to improve layered manufacturing part quality. InProceedings of Solid Freeform Fabrication Symposium 1998 Aug 10: 327-334
7. Vasudevarao B, Natarajan DP, Henderson M, Razdan A. Sensitivity of RP surface finish to process parameter variation. InSolid Freeform Fabrication Proceedings 2000 Aug 8 (pp. 251-258). Austin: The University of Texas
8. Byun HS, Lee KH. Determination of the optimal build direction for different rapid prototyping processes using multi-criterion decision making. Robotics and Computer-Integrated Manufacturing. 2006 Feb 28;22(1):69-80
9. Noriega A, Blanco D, Alvarez BJ, Garcia A. Dimensional accuracy improvement of FDM square cross-section parts using artificial neural networks and an optimization algorithm. The International Journal of Advanced Manufacturing Technology. 2013 Dec 1;69(9-12):2301-13.
10. Gurrala PK, Regalla SP. Multi-objective optimisation of strength and volumetric shrinkage of FDM parts: a multi-objective optimization scheme is used to optimize the strength and volumetric shrinkage of FDM parts considering different process parameters. Virtual and Physical Prototyping. 2014 Apr 3;9(2):127-38
11. Rattanawong W, Masood SH, Iovenitti P. A volumetric approach to part-build orientations in rapid prototyping. Journal of Materials Processing Technology. 2001 Dec 20;119(1):348-53
12. Nancharaiah T, d Ranga Raju VR, Raju R. An experimental investigation on surface quality and dimensional accuracy of FDM components. 2010:1 (2), 106-111
13. Ahn SH, Montero M, Odell D, Roundy S, Wright PK. Anisotropic material properties of fused deposition modeling ABS. Rapid prototyping journal. 2002 Oct 1;8(4):248-57
14. Arumaikkannu G, Maheshwaraa NU, Gowri S. A genetic algorithm with design of experiments approach to predict the optimal process parameters for FDM. InProceedings of Solid Freeform Fabrication Symposium 2005 (pp. 150-161).
15. Chohan JS, Chohan JS, Singh R, Singh R. Pre and post processing techniques to improve surface characteristics of FDM parts: a state of art review and future applications. Rapid Prototyping Journal. 2017 Apr 18;23(3):495-513
16. Yang Y, Fuh JY, Loh HT, Wong YS. A volumetric difference-based adaptive slicing and deposition method for layered manufacturing. Transactions-American Society of Mechanical Engineers Journal of Manufacturing Science and Engineering. 2003 Aug 1;125(3):586-94
17. Pandey PM, Reddy NV, Dhande SG. Slicing procedures in layered manufacturing: a review. Rapid prototyping journal. 2003 Dec 1;9(5):274-88

18. Pandey PM, Reddy NV, Dhande SG. Real time adaptive slicing for fused deposition modelling. International Journal of Machine Tools and Manufacture. 2003 Jan 31;43(1):61-71
19. Pandey PM, Reddy NV, Dhande SG. Improvement of surface finish by staircase machining in fused deposition modeling. Journal of materials processing technology. 2003 Jan 10;132 (1):323-31
20. Leong KF, Chua CK, Chua GS, Tan CH. Abrasive jet deburring of jewellery models built by stereolithography apparatus (SLA). Journal of Materials Processing Technology. 1998 Nov 1;83(1):36-47
21. Galantucci LM, Dassisti M, Lavecchia F, Percoco G. Improvement of fused deposition modelled surfaces through milling and physical vapor deposition. Available at http://www.poliba.it/Didattica/docs/scorepoliba2014 _submission_187.pdf (Accessed on December 22, 2014)
22. Boschetto A, Bottini L, Veniali F. Microremoval modeling of surface roughness in barrel finishing. The International Journal of Advanced Manufacturing Technology. 2013 Dec 1;69 (9-12):2343-54.
23. Fischer M, Schöppner V. Some investigations regarding the surface treatment of Ultem* 9085 parts manufactured with fused deposition modeling. In Proceedings of 24 th Annual International Solid Freeform Fabrication Symposium, Austin, Aug 2013 (pp. 12-14)
24. Vinitha, M., Rao, A.N. and Mallik, M.K. (2012) "Optimization of Speed Parameters in Burnishing of Samples Fabricated by Fused Deposition Modeling", International Journal of Mechanical and Industrial Engineering, Vol. 2, No. 2, pp. 10-12
25. Galantucci LM, Lavecchia F, Percoco G. Experimental study aiming to enhance the surface finish of fused deposition modeled parts. CIRP Annals-Manufacturing Technology. 2009 Dec 31;58(1):189-92.
26. Kuo CC, Mao RC. Development of a precision surface polishing system for parts fabricated by fused deposition modeling. Materials and Manufacturing Processes. 2016 Jun 10;31(8):1113-8
27. Garg A, Bhattacharya A, Batish A. On surface finish and dimensional accuracy of FDM parts after cold vapor treatment. Materials and Manufacturing Processes. 2016 Mar 11;31(4):522-9
28. Percoco G, Lavecchia F, Galantucci LM. Compressive properties of FDM rapid prototypes treated with a low cost chemical finishing. Research Journal of Applied Sciences, Engineering and Technology. 2012 Oct 1;4(19):3838-42
29. Galantucci LM, Lavecchia F, Percoco G. Quantitative analysis of a chemical treatment to reduce roughness of parts fabricated using fused deposition modeling. CIRP Annals-manufacturing technology. 2010 Dec 31;59(1):247-50
30. Taufik M, Jain PK. Laser assisted finishing process for improved surface finish of fused deposition modelled parts. Journal of Manufacturing Processes. 2017 Dec 31;30:161-77
31. Priedeman WR, Smith DT. Smoothing method for layer manufacturing process. United States Patent No. US8123999B2, USA. 2011
32. ISO 4287 (1997), Geometrical Product Specification (GPS) – Surface Texture: Profile Method – Terms, Definition and Surface Texture Parameters, International Organization for Standardization (ISO), Geneva
33. ASTM D2240 (2010). Standard Test Method for Rubber Property - Durometer Hardness, American Society for Testing and Materials, 2010
34. ISO 10360-2 (2009). Geometrical product specifications (GPS) - Acceptance and reverification tests for coordinate measuring machines (CMM) - Part 2: CMMs used for measuring linear dimensions, 2009
35. Chohan JS, Singh R, Boparai KS. Parametric optimization of fused deposition modeling and vapour smoothing processes for surface finishing of biomedical implant replicas. Measurement. 2016 Dec 31;94:602-13
36. Chohan JS, Singh R, Boparai KS, Penna R, Fraternali F. Dimensional accuracy analysis of coupled fused deposition modeling and vapour smoothing operations for biomedical applications. Composites Part B: Engineering. 2017 May 15;117:138-49

37. Singh J, Singh R, Singh H. Repeatability of linear and radial dimension of ABS replicas fabricated by fused deposition modelling and chemical vapor smoothing process: a case study. Measurement. 2016 Dec 31;94:5-11
38. Dym C. Principles of mathematical modeling. Elsevier Academic press, Burlington; 2004 Aug 10.

# Development of Rapid Tooling Using Fused Deposition Modeling

**Kamaljit Singh Boparai and Rupinder Singh**

## 1 Introduction

Fused deposition modeling (FDM) process has been emerged as a revolution in the field of additive manufacturing (AM)) where complex 3D structures can fabricate in less time without the assistance of conventional tooling. FDM can be commonly used for modeling, prototyping, and batch production applications. Among applications, it include concept modelling, functional prototypes, manufacturing tooling and end use parts. From the identification of component, prototypes of acrylonitrile butadiene styrene (ABS) plastic material were prepared for commercial end user applications in vacuum molding, vacuum casting, investment casting etc. [1] Generally, the fabricated parts are used for design verification, functional testing, medical applications and patterns for casting process [2].

## 2 Development of Low Cost Composite Material Feedstock Filament

The detailed experimental study has been conducted at Manufacturing Research Lab (MRL), Guru Nanak Dev Engineering College (GNDEC), Ludhiana, Punjab to develop feed stock filament of composite material for FDM process in order to increase the application range of FDM parts. The various steps for the development of feed stock filaments are shown in Fig. 1.

K. S. Boparai
MRS Punjab Technical University Bathinda, Bathinda, India

R. Singh (✉)
Guru Nanak Dev Engineering College Ludhiana, Ludhiana, India



B. AlMangour (ed.), *Additive Manufacturing of Emerging Materials*,
https://doi.org/10.1007/978-3-319-91713-9_8

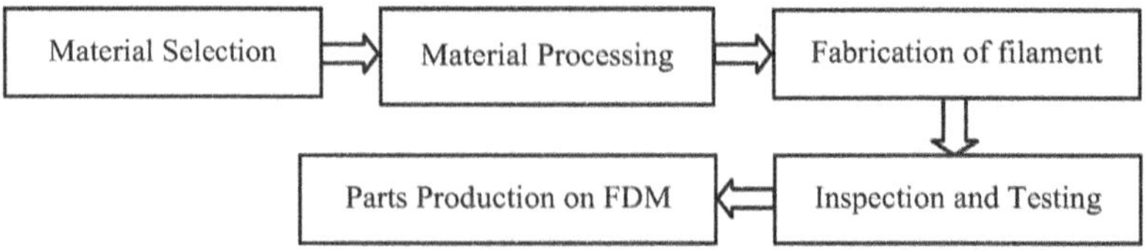


**Fig. 1** Flow chart of feedstock filament development [3]

**Table 1** Properties of filler materials

| S No. | Property | Al | $Al_2O_3$ |
|---|---|---|---|
| 1 | Melting point | 660.32 °C | 2075 |
| 2 | Density | 2.7 gm/cm$^3$ | 3.95–4.1 gm/cm$^3$ |
| 3 | Electrical resistivity | At 20 °C 28.2nΩ-m | ------ |
| 4 | Thermal conductivity | 237 W-m$^{-1}$ k$^{-1}$ | 30 W-m$^{-1}$ k$^{-1}$ |
| 5 | Thermal expansion | At 25 °C 23.1 μm-m$^{-1}$ k$^{-1}$ | ------- |
| 6 | Young's modulus | 70 GPa | ------- |

**Table 2** MFI of compositions

| | Composition (weight %) | | | | |
|---|---|---|---|---|---|
| S No | ABS | Nylon 6 | Al | $Al_2O_3$ | MFI (gm/10 min) |
| 1 | 100 | – | – | – | 2.415 |
| 2 | – | 100 | – | – | 10.61 |
| 3 | – | 70 | 30 | – | 6.825 |
| 4 | – | 60 | 40 | – | 3.975 |
| 5 | – | 50 | 50 | | 2.375[a] |
| 6 | – | 80 | – | 20 | 6.995 |
| 7 | – | 70 | – | 30 | 6.53 |
| 8 | – | 60 | – | 40 | 2.44[a] |
| 9 | – | 50 | – | 50 | 1.855 |

[a]Indicates MFI value of composite prepared near to standard ABS

In this study Nylon 6 E-35 grade was selected as a binder material. This is due to its outstanding properties as compared to other grades of nylon and having much suitability for extrusion applications.

The material was supplied by Gujrat state fertilizer limited in granular form having average particle size 4–5 mm and crushed to powder size (500–800 μm) with cryogenic grinding process. Pure Aluminum metal (Al) and aluminum oxide ($Al_2O_3$) were selected as a filler materials, having average particle size 4–5 mm and crushed to powder size (500–800 μm). The various important properties of Al and $Al_2O_3$ are listed below in Table 1.

The material flow properties are greatly influenced with the reinforcement of fillers and additives and it is necessary to investigate the flow of the composite material in liquefier head of FDM [4]. Ten observations were taken for each composition and the average value is shown in Table 2. This is generally a comparative study of flow rate under similar processing conditions and limits the practical levels at which they can be incorporated into the thermoplastic polymers.

Depending upon the application, the various properties can be tailor made by varying the loading of the filler material and binder material. From MFI test the index value of pure ABS is 2.415 gm/10 min and pure nylon6 is 10.61 gm/10 min. Nylon 6 can be loaded up to such limit that it's MFI should commensurate with pure ABS material. The MFI value of 40% loading of aluminum powder is 3.975 gm/10 min and 50% loading is 2.375 gm/10 min. So, approximately 45% loading is possible in nylon6 in order to obtain same flow rate and flexibility as the ABS material has during the processing of in FDM process. Similarly for $Al_2O_3$, less than 40% loading is possible.

For this experiment, up to 40% reinforcement of filler materials in nylon 6 was selected.

The various ingredients of composite material was heated in vacuum oven up to 50 °C for 10 h at absolute zero pressure in order to remove moisture and oil traces from the material.

The ingredients were mixed in tumbler mixture by rotating at the speed of 200 rpm for 2 h. No surfactants or plasticizers were added in the mixture, due to the self-lubricating property of Al and good binding properties of Nylon 6 material.

## *2.1 Fabrication of Filament on Single Screw Extruder*

### 2.1.1 Rheological Behavior

The successful processing of new filament material in FDM process depends upon fabrication of a strong and spool-able feed stock filament, having required consistent diameter, by careful selection of proportions of binder materials and fillers materials [5]. It is necessary to investigate the rheological behavior of composite material with respect to temperature, velocity and pressure drop during processing in liquefier head of FDM [4]. The reinforcement of filler in Nylon6 modified its structure and properties and it is necessary to investigate the effect, particularly for the analysis and design of processing operations [6]. Shenoy et al. [7] conducted melt flow test of polymer composites to generate curves showing the relationship between viscosity, temperature and shear rate. Ramanath et al. [8]; Bellini and Bertoldi [9]; Zhang and Chou [10]; Nikzad et al. [4] studied the thermal and flow behavior of biopolymers by modelling with finite element analysis. MFI is an alternative measure of viscosity for comparative purpose and defined as the amount of polymer in 'gm' extruded in '10' min through a capillary of standard dimensions (Diameter $2.0955 \pm 0.0051$ mm and length $8.0 \pm 0.025$ mm), as per ASTM D1238–73 standard. The apparatus consists of heated cylinder, piston, weight and capillary die as shown in Fig. 2a. MFI is an assessment of average molecular mass and is an inverse measure of the melt viscosity. In other words, higher the MFI, the more polymers flows under the processing conditions. The test was conducted by maintaining cylinder temperature

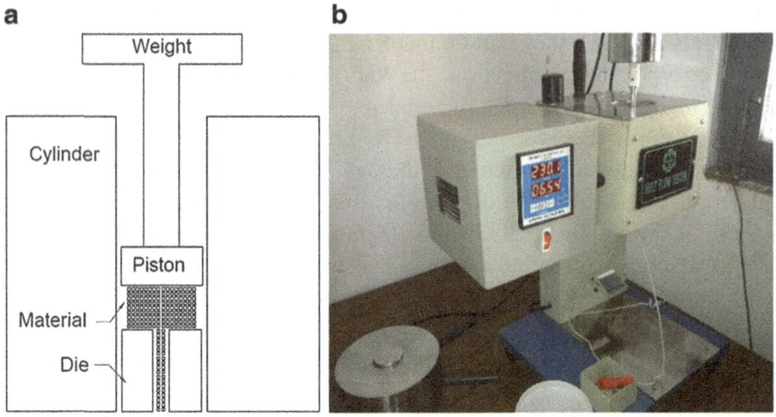


**Fig. 2** Melt flow test (**a**) Schematic of melt flow test; (**b**) Melt flow Tester

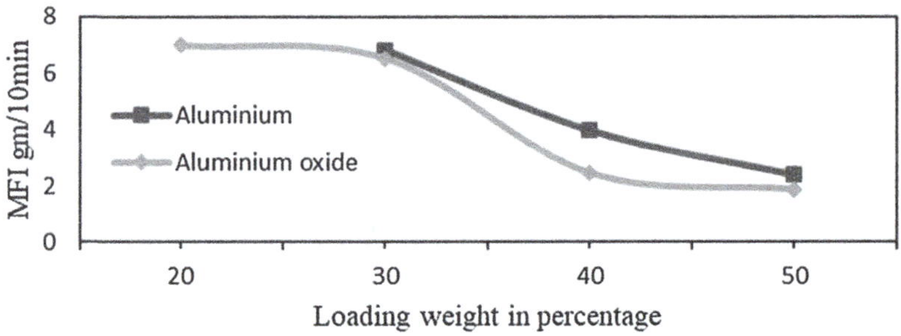


**Fig. 3** MFI of Al and $Al_2O_3$

of melt flow tester (Fig. 2b) at 230 °C and weight 3.8 Kg as specified in ASTM D 1238–73 standard. The average value of '10' observations was recorded for each composition. Figure 3 shows loading weight percentage of Al and $Al_2O_3$ in Nylon6 material. This is a comparative study of flow rate under similar processing conditions and limits the practical level, up to which filler materials may be incorporated into the thermoplastic polymers.

Depending upon the application, the tailor made properties may be inculcated by varying the loading of the filler and binder materials. From MFI test, the indices of pure ABS, pure Nylon6 are 2.415 gm/10 min and 10.61 gm/10 min respectively. Nylon6 may be loaded up to a limit that its MFI should commensurate with pure ABS material under similar processing conditions. The MFI value of 40% loading of Al powder is 3.975 gm/10 min and 50% loading is 2.375 gm/10 min. So approximately 45% loading of Al is possible in Nylon 6 in order to obtain same flow rate and flexibility as the ABS material have during the processing of in FDM process. Similarly for $Al_2O_3$, less than 40% loading is possible.

For this experiment, up to 40% loading of filler materials in Nylon 6 was selected. Table 3 summarized the weight proportions of compositions and their respective

**Table 3** Weight proportion of compositions

| Composition | Nylon 6 | Al | $Al_2O_3$ |
|---|---|---|---|
| A | 60 | 26 | 14 |
| B | 60 | 28 | 12 |
| C | 60 | 30 | 10 |

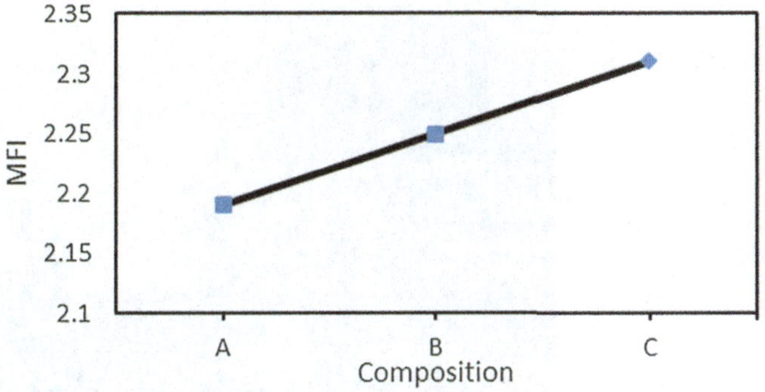


**Fig. 4** MFI values of compositions

MFI values are shown in Fig. 4. With the increase of $Al_2O_3$ content and decrease of Al in composition, MFI value decreases. This is due to increase in the proportion of large particle size in the matrix.

## 2.2 *Fabrication of Filament on Single Screw Extruder*

The 3D view, various specifications of single screw extruder are shown in Fig. 5 and Table 4 respectively. The variable parameters of extrusion process are barrel temperature, screw speed, die temperature, take up unit speed, water tank temperature and die nozzle diameter. The die nozzle diameter is 4 mm and the diameter of extruded filament was controlled by adjusting take up unit speed and water tank temperature. The barrel was hard tempered in order to reduce the friction and wear during the processing of composite materials. The screw and die nozzle was chromed plated so that material flows continuously without sticking with screw and nozzle surface.

Initially, number of trails were conducted and one set of parameters is selected which gives the favorable responses such as uniform distribution of filler materials, desired filament diameter with minimum deviation and strength. Table 5 shows the value of selected combination of single screw extruder parameters.

**Fig. 5** Screw extrusion process

**Table 4** Specifications of single screw extruder

| Screw diameter (mm) | L/D ratio | No of heaters | | Screw speed (rpm) | Take up unit speed (rpm) |
|---|---|---|---|---|---|
| | | Barrel | Die | | |
| 25 | 26 | 3 | 1 | 0–40 | 0–40 |

**Table 5** Process parameters

| Water tank temperature | Barrel temp. (°C) | Die temp. (°C) | Screw speed (rpm) | Take up speed (rpm) | Material composition |
|---|---|---|---|---|---|
| Room temperature | 170 | 205 | 35 | 27 | A/B/C |

### 2.2.1 Inspection and Testing

The suitability of composite material feedstock for FDM process is dependent upon various factors such as processing conditions of material, during fabrication of filament, homogenous particle dispersion of filler materials and binder material, rheological and adhesion behavior of constituents. It is necessary to mention here that the fabricated wire (Fig. 6a) was kept in dry place, due to hygroscopic nature of

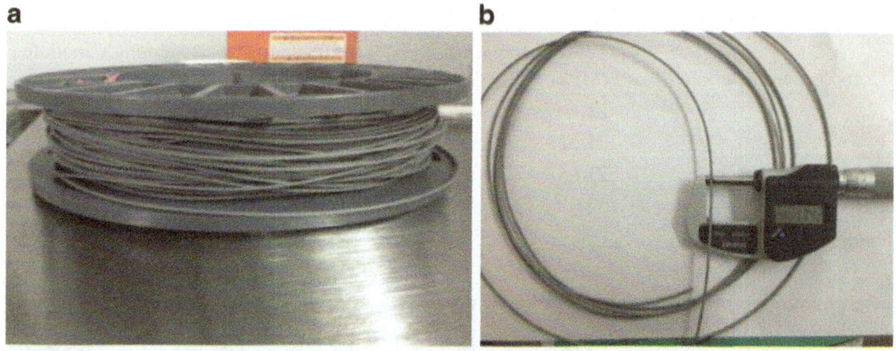


Fig. 6 Inspection of Filament (**a**) Filament spool; (**b**) Inspection with micrometer

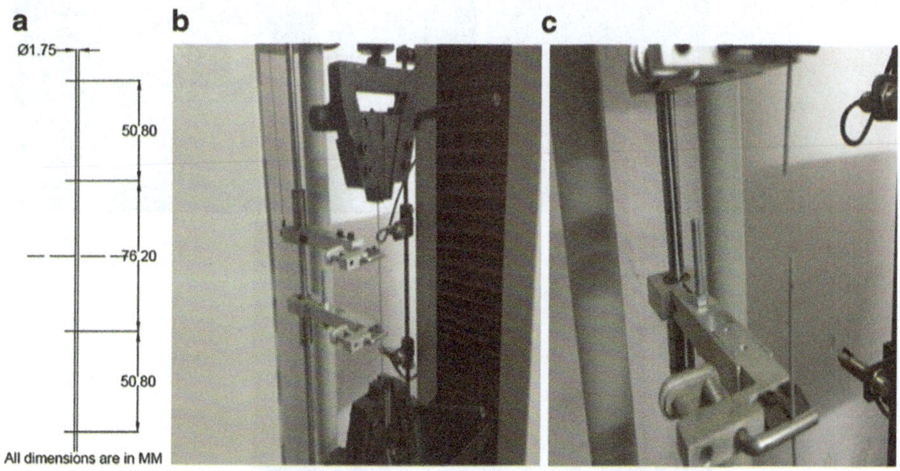


Fig. 7 Tensile testing of filament (**a**) specimen sample; (**b** and **c**) testing on universal tensile testing machine

material. The inspection of filament diameter was done with digital micrometer having range 0–25 mm and the least count 0.001 mm as shown in Fig. 6b. During the fabrication of filament, it was found that the diameter remained within the range of 1.75 ± 0.03 mm.

### 2.2.2 Mechanical Testing

The tensile properties of composite material feedstock filaments were tested according to ASTM-638 standard. Specimen sample for tensile testing is shown in Fig. 7a and were tested on universal testing machine. The specimen, before and after the testing is shown in Fig. 7b, c respectively. Five samples of each specimen were taken and average of five measurements is summarized in Table 6.

**Table 6** Tensile properties of various compositions

| Composition/material | Tensile strength (MPa) | % Elongation @ break | Young's modulus (MPa) |
|---|---|---|---|
| A | 21.40 | 18.62 | 582 |
| B | 21.53 | 12.74 | 760 |
| C | 21.65 | 8.56 | 1165 |
| ABS[a] | 22 | 6.0 | 1627 |

**Note:** [a]supplier's data

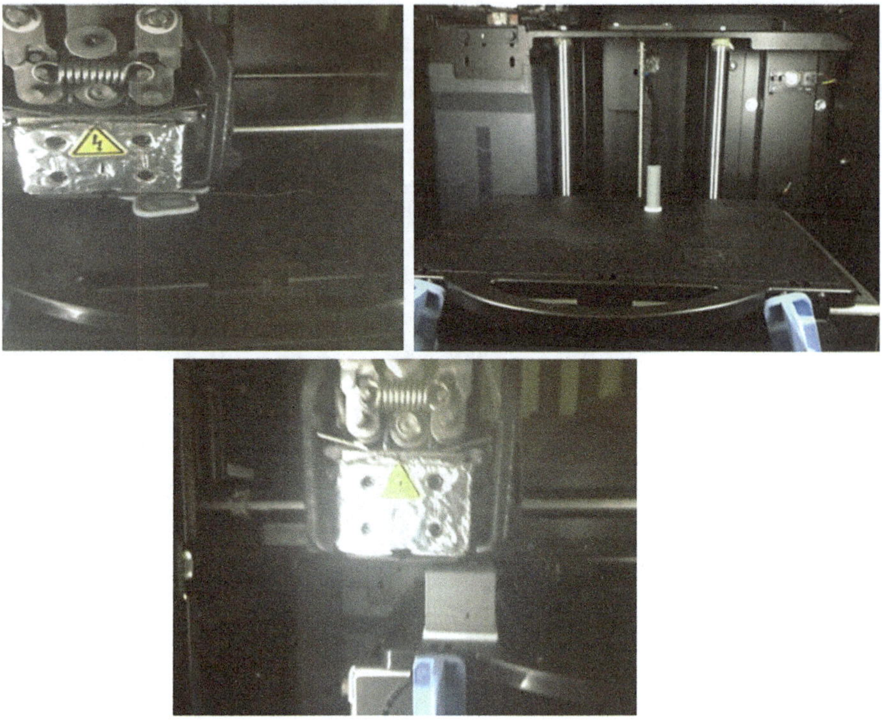

**Fig. 8** Parts fabrication on commercial FDM setup

### 2.2.3 Fabrication of Parts On FDM

After inspection, the composite material feedstock filament was spooled and then loaded directly into the commercial FDM setup (u-PrintSE, Stratasys, USA). The filaments of above selected three compositions were run without any interruption. No buckling of material took place at the entrance of liquefier head, which shows that the material has enough required strength and stiffness to force the material through nozzle. This is because of its consistency, accurate diameter and uniform distribution of filler materials. Sometimes it was observed that nozzle chocking has taken place during the fabrication of parts as shown in Fig. 8. It should be noted that

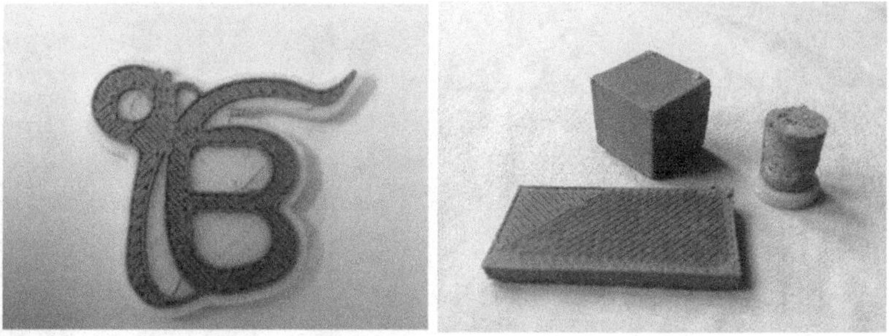

**Fig. 9** Fabricated parts on FDM

with the increase in proportion of abrasive material ($Al_2O_3$) content and particle size, there will be effect on the life of liquefier head nozzle, the same has been controlled by gradually selecting the smaller size abrasive particles so that material easily flow through the passage of nozzle (Ø0.425 mm approx.) without chocking and eroding its walls. During pilot run, it has been noticed that FDM nozzle chocked with the increase of content and particle size of $Al_2O_3$. For the present case, the particle size of Al and $Al_2O_3$ were 325 mesh (AFS No.) and 125 mesh (AFS No.) respectively and the content of $Al_2O_3$ were limited up to 14% by weight. The Al metal has self-lubricating property and produced no abrasion effect on nozzle. The nozzle designed for FDM system was used to fabricate parts and there is no requirement of specially designed nozzle for this work. The available FDM system can run Ø1.75 mm filament only. For Ø 3 mm filament, the content of $Al_2O_3$ will increase and may affect the life of nozzle. Further there are more chances of nozzle chocking with the decrease of its diameter and only Nano scale filler materials may be accomplished for un-interrupted flow of material through nozzle. Figure 9 shows fabricated parts.

### 2.2.4 Dynamic Mechanical Analysis

The DMA provides the information of polymeric material in dynamic state and generally used to determine the stiffness and damping behavior of the material. The tanδ, which is the ratio of storage modulus and loss modulus can be used as a comparative estimate of the damping property of the composite material [11]. The specimens for DMA analysis were fabricated with developed alternative material feedstock filaments on FDM system without any modification in its hardware and software. The rectangular shaped samples having specification 20 mm × 9.5 mm × 1.75 mm were for DMA analysis were prepared with solid density mode having raster angle 45°/45° as shown in Fig. 10.

The measurements for viscoelastic properties of developed composite materials and ABS material test specimens have been carried out in three point bending mode on Netzesh DMA (model 242e, Germany) under air atmosphere with a temperature range of 25–150 °C. The tests were performed at a fixed frequency of 1 Hz and

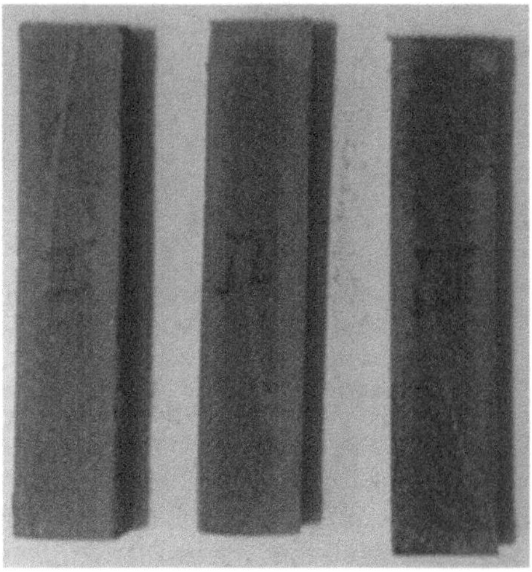

**Fig. 10** FDM printed samples for DMA

heating rate was 2 °C/min with a maximum dynamic force of 10 N. The specimens were initially dried in a vacuum oven at a temperature of 50 °C for 8 h. The characterization includes storage modulus, loss modulus and tanδ for qualitatively and quantitatively investigating the effect of fillers (Al and $Al_2O_3$) in Nylon6 matrix. Additionally, study highlights the shift of glass transition temperature and stiffness of composite materials.

# 3 Results and Discussion

## *3.1 Rheological Properties*

The rheological study of different compositions of Nylon6-Al-$Al_2O_3$ were carried out according to ASTM D1238–73 standard in order to investigate the flow rate at processing temperature in liquefier head of FDM system. It has been observed that the incorporation of filler material in the matrix of Nylon 6 increases its viscosity as compare to neat Nylon 6. Moreover it also increases with increase the particle size of filler material. The MFI value of Al and $Al_2O_3$ was plotted on weight percentage loading scale as shown in Fig. 3. The curve depicts that, for Al metal up to 45% loading and for $Al_2O_3$ up to 40% loading in Nylon 6 matrix is possible. This is a comparative result in order to predict the flow behavior, as ABS material has during the processing in FDM system. It was observed from Fig. 4 that composition 'C' have high MFI value (2.31) and composition A has least (2.19). The flow behavior of composition 'C' is almost same as the ABS material has, under same temperature and pressure conditions according to ASTM D1238-73 standard. When flow rate increase then desired value it causes improper building of part model and decreases surface finish.

## 3.2 *Mechanical Properties*

It has been established that as the proportion and size of filler material in polymeric material increases, mechanical strength decreases [11–13]. The effect of reinforcement of filler material on Young's modulus is shown in Table 6. The Young's modulus and tensile strength of composite material is less than ABS material but percentage elongation is more. This is due to the weak interaction of metallic material with polymeric materials and formation of voids around metal particles inside the structure. The Young's modulus strongly and tensile strength slightly decrease from the ABS material to those of composite materials and it increases with increase in Al content. Moreover it should be noted that $Al_2O_3$ particle size is larger than Al metal. The elongation at break shows significant change and increase with filler content and particle size. It was realized from the test that tensile strength of alternative material (composition 'C') filament has slightly lesser value than ABS filament but no buckling of filament took place at the entrance of liquefier head of FDM system during the fabrication of parts. The tensile strength decreases with the increase of weight proportions of $Al_2O_3$. The reason is the poor compatibility of $Al_2O_3$ material with binder material due to its abrasive nature. The filaments of composition 'A' and 'B' frequently buckled due to poor mechanical properties and causes nozzle chocking. So tensile testing of filament successfully determined the compatibility of alternative material filament with FDM system.

## 3.3 *Dynamic Mechanical Analysis*

### 3.3.1 Viscoelastic Behavior of Composite Materials

The viscoelastic behavior of three different compositions of Nylon6-Al-$Al_2O_3$ were studied in order to investigate the change in glass transition temperature and stiffness of material at processing temperature in liquefier head of FDM system. The parameters such as storage modulus characterize the stiffness, loss modulus highlights the energy dissipated in one loading cycle and loss factor (tan δ) represents the internal resistance or mechanical damping of the polymeric based materials.

#### Storage Modulus (E′)

The reinforcement of filler material in Nylon 6 matrix increases its viscosity as compare to neat Nylon 6. Moreover it also affected with the increase of filler material particle size. The variation of storage modulus with temperature for composition A, B and C are shown in Fig. 11. The higher value of storage modulus for composition C is due to the increase in intermolecular bonding resulting from the presence of large weight % proportion of Al than $Al_2O_3$ in Nylon 6 matrix. As already

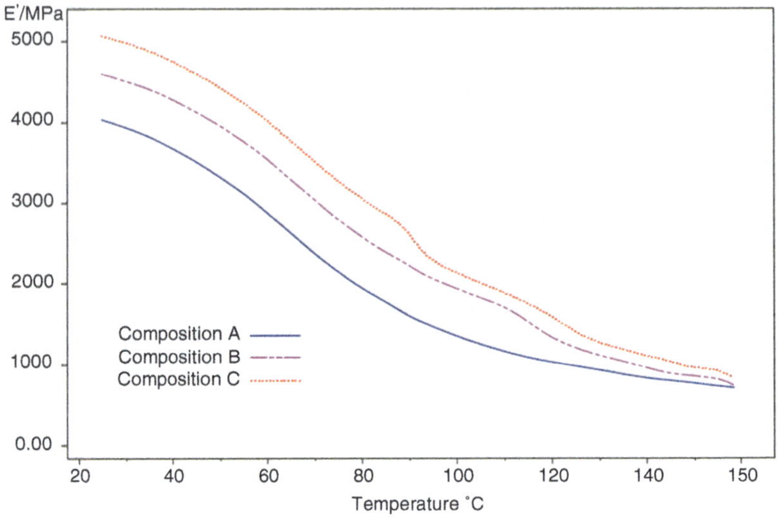


**Fig. 11** Schematic diagram of typical storage modulus curves vs. temperature

mentioned that the composition C contains 30% Al and 10% $Al_2O_3$ particles. Moreover average particle size of Al and $Al_2O_3$ are 50 μm and 150 μm respectively. Due to large particle size of $Al_2O_3$, the increase of its weight % proportions in Nylon 6 matrix causes poor interfacial bonding as in the case of composition A and B. In addition to above Al metal have self-lubricating property and high thermal conductivity than $Al_2O_3$. The higher reinforcement of Al cause better heat and stress transfer and retains better matrix adhesion. The higher value of storage modulus exhibits high stiffness which shows that composition C has high stiffness as compared to composition A and B. It is noteworthy that the increase in storage modulus due to filler reinforcement is more noticeable in the rubbery region than glassy region as compare to ABS material.

As the storage modulus indicates the capacity of material to store the input mechanical energy, it decrease with the increase of temperature [14]. The storage modulus is generally used for determining elastic properties of a polymeric based materials. The curves shows that, initially at low temperature, the storage modulus has high value and with the increase of temperature its value decreases gradually. In the intermediate region of the curves, the transition from the glassy state or energy elastic state to rubbery state or entropy elastic state occurs. This region indicates the onset of chain mobility [14]. This trend remains the same for all the compositions and is due to reinforcement of filler materials in the matrix. At elevated temperature, the storage modulus drops which exhibits loss of stiffness. Among various compositions, composition C have maximum value of storage modulus in rubbery region than other composites.

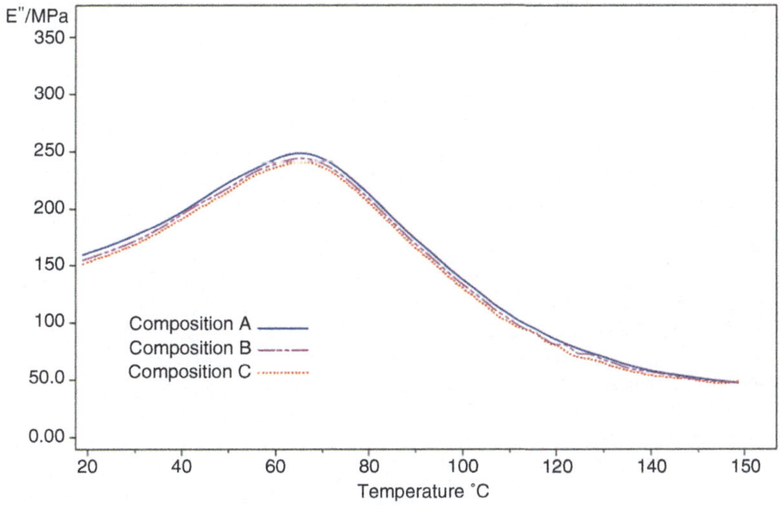


**Fig. 12** Schematic diagram of typical loss modulus curves vs. temperature

## Loss Modulus (E″)

The loss modulus highlights the viscous properties of a polymeric based materials and represents energy lost as heat or dissipated during one cyclic load. Typical loss modulus curves of above said three compositions as a function of temperature are illustrated in Fig. 12. It should be noted that loss modulus was approximately steady at low temperature and rises to peak value at the onset temperature, which indicates the maximum heat dissipated per unit deformation [15]. The loss modulus of composition A is higher followed by composition B and C. There are two explanations for the change in peaks of loss modulus with the variation in weight proportions of fillers. First the spherical shape of filler materials can enhance the energy dissipation under the influence of dynamic load [16]. Secondly the polymer metal slippage or filler particle- particle slippage are resisted at low temperature due to the difference of coefficient of thermal expansion of polymer and fillers. The particle-particle slippage dissipated more heat. As the amount $Al_2O_3$ in Nylon6 matrix increases, the peak values of loss modulus increases. The high peak value of loss modulus indicates that the intermolecular bonding is destroyed [15]. As Figure shows that the effect of type of filler and amount of filler are found to influence more on the value of loss modulus above the temperature of high peak value [17]. The curve become broaden after peak value which indicates that difference in physical state of matrix surrounding the filler materials (Al and $Al_2O_3$). The trend remains same for all the compositions.

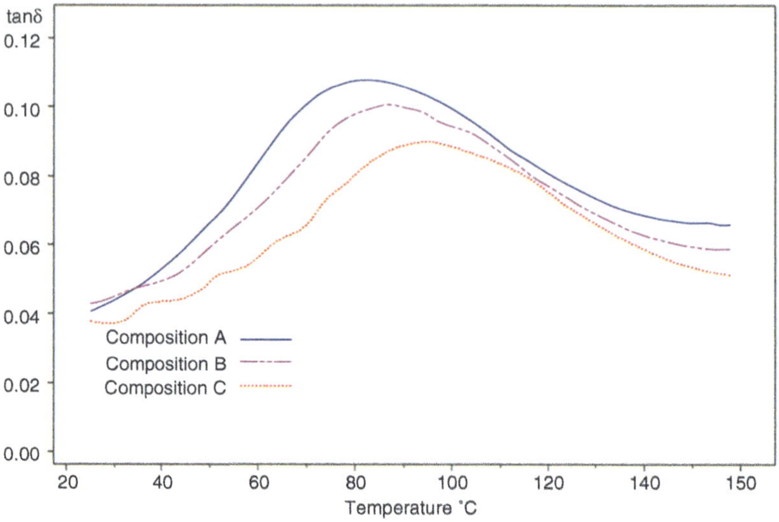


**Fig. 13** Schematic diagram of typical tan δ curves vs. temperature

Loss Factor or tan δ (E″/E′)

The tan δ which is a ratio of loss modulus and storage modulus (ISO 6721-1) is plotted as a function of temperature for the specimens prepared with three different compositions (Fig. 13). A high value of loss factor of a polymeric material is significative of non-elastic strain component, while low value reveals one that is more elastic. The DMA determines not only damping or modulus but also for suitable for glass transition temperature ($T_g$) measurements. As the change in modulus under the influence of temperature is much more noticeable in DMA than other techniques such as differential scanning calorimeter (DSC), so it is more suitable for $T_g$ measurements. Although $T_g$ can be evaluated from the peaks of loss factor and loss modulus but it is easy to measure from the peak of loss factor (ASTM D 4065-2001). The maximum peak points of damping factor (tan δ) highlights the glass transition temperature ($T_g$), which was found to be 82.5 °C, 87.6 °C and 93.5 °C of composition A, B and C respectively. Although the tan δ results highlight that the composition C material has high stiffness followed by composition B and A, but all these compositions have adequate stiffness to force the material through liquefier head of FDM system. It was observed during the fabrication of samples for DMA analysis that feedstock filaments prepared by composition C and B, runs successfully without buckling and chocking in liquefier head of FDM system. However in case of composition A sometime material buckled during the processing on FDM system. The reason is due to large particle size and amount of $Al_2O_3$ filler particles in matrix. The reinforcement of filler material reduces the tan δ curve peak by restricting the movement of polymer molecules [5]. With the filler loading in Nylon 6 matrix, tan δ value decreases and $T_g$ shows a shifting toward higher value. The shifting of $T_g$, toward higher temperatures indicates the decreased mobility of the chains with the reinforcement of fillers. The lowering of tan δ value indicates the improvement in interfacial bonding within the alternative material matrix [18]. Additionally, Tg, is considered as a measure of interfacial interaction among the molecules. The composites with poor interface bonding tend to dissipate more energy than those with good interfacial bonding [19]. Thus poor interface bonding results in high tan δ value. So, it can be concluded that the alternative material having composition C have more stiffness and glass transition temperature than the rest compositions which indicates that good interfacial adhesion between filler and matrix and is more suitable for FDM system.

### 3.3.2 Viscoelastic Behavior of ABS Material

The variation of the storage modulus, loss modulus and loss factor as a function of temperature for ABS material is shown in Fig. 14. As explained above, in low temperature region, the material is in glassy state. However, in this region ABS material have storage modulus value less than alternative composite materials and this difference remains throughout the experimentation. In the transition region, curve shows a rapidly decrease in storage modulus to a constant value with the increase of temperature but composite materials shows gradually decrease in its value. This result indicates that ABS material have less stiffness than composite materials. The loss modulus and loss factor curves of ABS material exhibited higher peak value compared to those of composite materials. At temperature below glass transition, ABS material have low loss modulus than those of composite materials and the intensity of corresponding peak is also lower. For ABS, the glass transition

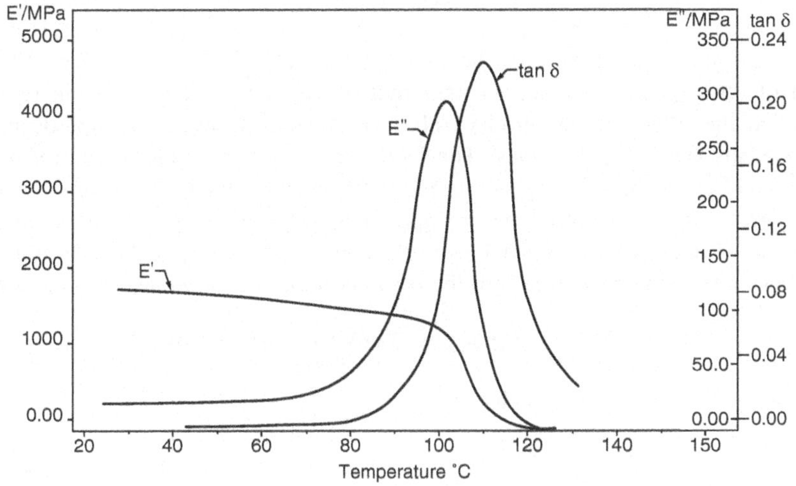


**Fig. 14** Schematic diagram of typical DMA curves of ABS material

**Table 7** Process parameters

| Level | Material composition ($P_1$) | Barrel temp. ($P_2$) | Die temp. ($P_3$) |
|---|---|---|---|
| 1 | A | 160 | 185 |
| 2 | B | 170 | 195 |
| 3 | C | 180 | 205 |

temperature corresponding to tan δ peak is 108 °C and its peak value is 0.228 which is higher than composite materials. This is another reason that highlights low stiffness due to higher molecular mobility in ABS material.

It had been established that the amount and size of filler materials in polymeric material affects the dynamic mechanical properties. From the above discussion it is realized that the composite materials possess good dynamic mechanical properties than standard ABS material.

## 4 Design of Experiments

During pilot experimentation, a number of trails were conducted by some hit and trail method and output of all these trails, gives a working range with in which critical parameters such as composition, barrel temperature and die temperature can be varied for the processing of alternative material on single screw extruder. Table 7 shows the levels of selected single screw extruder parameters. During the trial experiment, it was worked out that, variation of parameters within these limits produces a strong spool able filament suitable for FDM system and outside these limits causes interrupted flow of material through screw extruder which causes defects such as inconsistent diameter, ovality and poor strength of filament. However in actual working conditions, the behavior of particular parameter is affected by the level of other parameters due to their mutual interaction. Therefore, in order to visualize the effect of variation of all factors, it is necessary to opt design of experiments (DOEs) techniques. The DOEs techniques provide the simultaneous variation of all the influencing factors and systematically evaluate the significance of each factor. In full factorial design approach, large number of experiments are required, which results in wastage of time, material and efforts [20]. The minimum number of experiment required for the parametric optimization are calculated as:

$$[(\mathrm{L}-1)\times \mathrm{P}]+1=7\approx 9\,\text{Experiment runs}$$

Where L = levels, P = parameters

**Table 8** L9 OA experiment design

| Experiment runs | Composition/MFI | Barrel temperature °C | Die temperature °C |
|---|---|---|---|
| 1 | A/2.19 | 160 | 185 |
| 2 | A/2.19 | 170 | 195 |
| 3 | A/2.19 | 180 | 205 |
| 4 | B/2.25 | 160 | 195 |
| 5 | B/2.25 | 170 | 205 |
| 6 | B/2.25 | 180 | 185 |
| 7 | C/2.31 | 160 | 205 |
| 8 | C/2.31 | 170 | 185 |
| 9 | C/2.31 | 180 | 195 |

## 4.1 Taguchi's Approach

In this work, Taguchi's approach was applied for design and analysis of experiments. It is a most important statistical tool for designing effective systems at low cost by optimizing the process parameters [21]. Taguchi's L9 orthogonal array (L9 OA) experiment design was considered for optimizing the screw extruder parameters. This method provides a set of well-balanced minimum number of experiments and optimum results are predicted by analyzing Signal to Noise ratios (S/N). Three parameters such as material composition, barrel temperature and die temperature were selected. The water tank temperature was kept at room temperature (35 °C). Screw speed and take up unit speed was adjusted according to the diameter of filament so that diameter should be remains within the specified limits (1.75 ± 0.03 mm).

As mentioned earlier, three factors each have three levels were used in the investigation. The orthogonal array corresponding to three factors and three levels is L9 as shown in Table 8. Each column represents a test parameters and test condition is represented by each row, which is formed by the combination of different levels of the selected parameters.

The inspection of filament diameter was done with digital micrometer having range 0–25 mm and least count 0.001 mm as shown in Fig. 7b. During fabrication of filament, it was found that diameter remains within the range of 1.75 ± 0.03 mm.

## 4.2 Data Analysis Using ANOVA

ANOVA generally considered for the statistical validation of results [21]. It accounts the variation of all sources as well as error sources. The experimental errors are due to the replication of experiments. The ANOVA analysis was done on MINITAB 17 software. The signal to noise measures the sensitivity of quality characteristics from desired value. Signals highlight the effect on average responses and noises measures the deviations from output [22]. Based upon criterion, smaller is best for

**Table 9** Experiment outputs

| Experiment no. | Responses diameter deviation (mm) R1 | R2 | R3 | Mean | S/N ratio |
|---|---|---|---|---|---|
| 1 | 0.048 | 0.049 | 0.049 | 0.0486667 | 26.2550 |
| 2 | 0.049 | 0.048 | 0.049 | 0.0486667 | 26.2550 |
| 3 | 0.050 | 0.049 | 0.049 | 0.0493333 | 26.1368 |
| 4 | 0.050 | 0.050 | 0.049 | 0.0496667 | 26.0783 |
| 5 | 0.050 | 0.050 | 0.050 | 0.0500000 | 26.0206 |
| 6 | 0.050 | 0.049 | 0.050 | 0.0496667 | 26.0783 |
| 7 | 0.050 | 0.050 | 0.050 | 0.0500000 | 26.0206 |
| 8 | 0.049 | 0.050 | 0.050 | 0.0496667 | 26.0783 |
| 9 | 0.050 | 0.050 | 0.050 | 0.0500000 | 26.0206 |

Note: R1, R2 and R3 are the three repetitions of experiments in order to reduce the experimental error

**Table 10** ANNOVA for SN ratios

| Source | DF | Seg SS | Adj SS | Adj MS | F | P | % Contribution |
|---|---|---|---|---|---|---|---|
| **$P_1$** | 2 | 0.055745 | 0.055745 | 0.027873 | 71.74 | 0.014 | 80.2 |
| **$P_2$** | 2 | 0.003103 | 0.003103 | 0.001551 | 3.99 | 0.200 | 4.48 |
| **$P_3$** | 2 | 0.009869 | 0.009869 | 0.004935 | 12.70 | 0.073 | 14.2 |
| **Residual error** | 2 | 0.000777 | 0.000777 | 0.0003888 | | | 1.12 |
| **Total** | 8 | 0.069494 | | | | | |

performance characteristics were selected in this work. Table 9 shows the observed values of filament diameter deviation after conducting three trails of nine experimental runs.

$$\text{Smaller is best (minimize) } S/N = -10\log\left[1/n\sum\left(y^2\right)\right]$$

Where ‘y’ is the output responses (diameter deviation), ‘n’ is the observations.

## 4.3 *Analysis of Variance for SN ratios*

The ANOVA with all the effects are shown in Table 10. The experimental error or residual error have contribution 1.12% and as per thumb rule it should be <15%. This shows that no important parameter have been excluded from the experiment. Material composition ($P_1$) have highest contribution followed by die temperature ($P_3$) and barrel temperature ($P_2$).

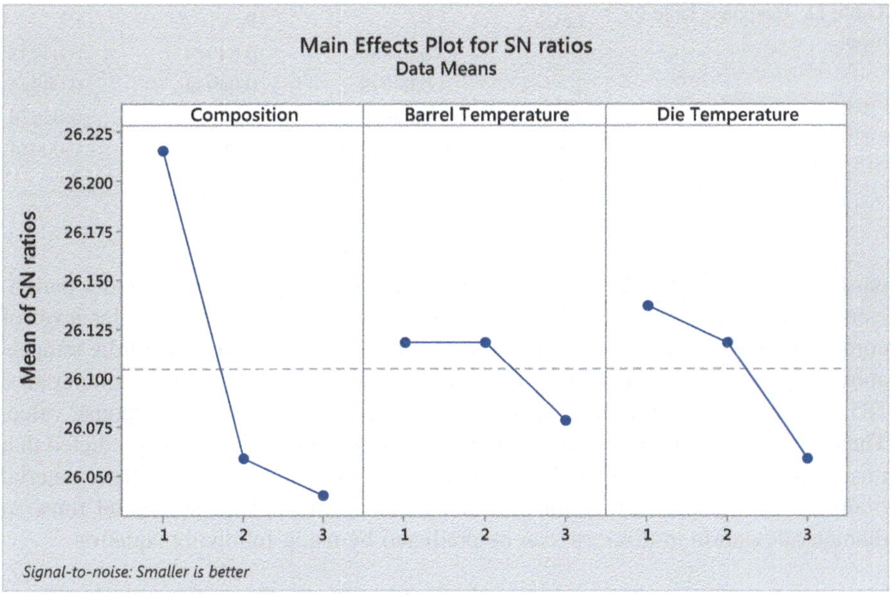


**Fig. 15** Main effects plot for SN ratios

## *4.4 Significance of Process Parameters*

As already mentioned earlier, the considered screw extruder parameters have three levels. The main effects plot for SN ratios shows that two parameters such as barrel temperature and die temperature shows relatively less changes as compare to material composition with the variation from level 1 to level 3 as shown in Fig. 15. In other words, it can be said that material composition is a significant parameter in terms of its influence on response, whereas, barrel temperature and die temperature are insignificant. The same trend is endorsed by ANNOVA for SN value results (Table 10). The analysis was carried out with 95% confidence level. The $F_{0.05\ (2,\ 2)}$ value from the standard table (critical values of F-distribution at 5%) is ‘19’. Material composition has F value more than table value and remaining parameters have less than table value. Therefore parameter having F value more than table value is significant and remaining are insignificant. Additionally, P value of material composition is $<0.05$, while barrel temperature and die temperature have $>0.05$.

## *4.5 Optimum Parameters*

As the objective is to minimize the response, the optimum conditions were selected based on lowest means of SN ratio value of each factor. The main effects plot for SN ratios was used to determine the optimum combination of single screw extruder

**Table 11** Response table for mean

| Level | $P_1$ | $P_2$ | $P_3$ |
|---|---|---|---|
| 1 | 0.04889 | 0.04944 | 0.04933 |
| 2 | 0.04978 | 0.04944 | 0.04944 |
| 3 | 0.04989 | 0.04967 | 0.04978 |
| Delta | 0.00100 | 0.00022 | 0.00044 |
| Rank | 1 | 3 | 2 |

parameters. The plot (Fig. 15) shows the influence of single screw extruder parameters on the tensile strength of filament. The optimum parameters are 1st level of composition ($P_{11}$), 2nd level of barrel temperature ($P_{22}$) and 1st level of die temperature ($P_{31}$) i.e. composition 'A' with barrel temperature 170 °C and die temperature 185 °C. The optimal levels of factors are selected to minimize the response value. The optimum process parameters combination is $P_{11}$ $P_{22}$ $P_{31}$ and it is predicted that this combination have to give lowest diameter deviation of alternative material filament. Since this combination was not included in the 9 experimental runs, so diameter deviation in this case can be predicted by using following equation.

$$\text{Predicted deviation} = \bar{T} + \left(\bar{P}_{11} - \bar{T}\right) + \left(\bar{P}_{22} - \bar{T}\right) + \left(\bar{P}_{31} - \bar{T}\right)$$

Where $\bar{T}$ is the mean deviation, $\bar{P}_{11}$, $\bar{P}_{22}$ and $\bar{P}_{31}$ represent mean deviation value as shown in Table 11. Substituting the values in above equation, then Predicted deviation is

$$\begin{aligned} &= 0.04951778 + (0.04889 - 0.04951778) + (0.04944 - 0.04951778) \\ &\quad + (0.04933 - 0.04951778) \\ &= 0.04862\,\text{mm} \end{aligned}$$

The response table for mean summarized the mean deviation value at different parameters levels. The delta shows the difference between highest and lowest mean value of parameters at different levels. The rank of parameters signifies relative importance of parameters in terms of their influence on response i.e. higher the delta value, higher is the influence of parameter on response. So parameter $P_1$ is the highest influencing parameter followed by $P_3$ and $P_2$ respectively.

## 4.6 *Empirical Relationship*

The above ANNOVA analysis realized that only material composition is significant parameter and remaining two (barrel temperature and die temperature) are insignificant. Thus the filament diameter deviation in the current investigation is a function of parameter $P_{1,}$ $P_2$ and $P_3$ and it can be expressed as

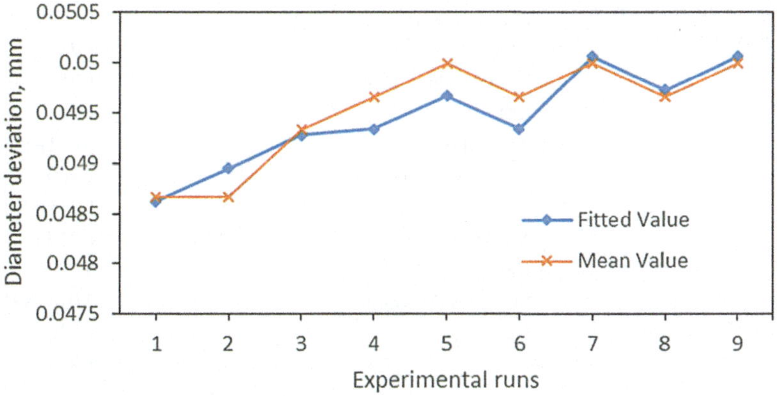


**Fig. 16** Experimental mean value and equation fitted value

$$\text{Tensile strength} = f(P_1, P_2, P_3)$$

By conducting linear regression analysis on MINITAB 17 software at 95% confidence level, the regression equation is expressed as

$$\begin{aligned}\text{Diameter deviation} &= C_1 + C_2P_1 + C_3P_2 + C_4P_3 \\ &= 0.02455 + 0.00833\,P_1 + 0.000011\,P_2 + 0.000022\,P_3\end{aligned}$$

Where $C_1$, $C_2$, $C_3$ and $C_4$ are constants and their value depends upon the interaction effect of main parameters.

The comparison of diameter deviation of alternative material filament between experimental value and equation fitted value obtained from linear regression model is shown in Fig. 16. For all experimental runs it can be realized that equation value is close to the experimental value. The deviation of <1% was observed for nine different combination of parameters runs. The developed equation accurately predict the value of diameter deviation with the magnitude of variation <1%. Thus the proposed equation is statistical suitable and accurately predict the influence of considered major parameters and their levels on diameter deviation of Nylon6-Al-$Al_2O_3$ alternative material filament.

## 5 Process Capability Study

Process capability study was conducted to compares the output responses of an in-control process with the specified limits. This is generally used to realize the capability of the process by comparing the actual process spread with the allowable process spread and measured by process σ (standard deviation) levels. Higher is the σ levels, better is the process performance. In this work, process capability was

**Table 12** Statistical analysis of filament diameter

| | |
|---|---|
| No of observations | 10 |
| USL | 1.78 |
| Target value, mean | 1.75 |
| LSL | 1.72 |
| Normality test | AD Test Passed |
| Maximum value | 1.769 |
| Average | 1.7516 |
| Minimum value | 1.737 |
| Range | 0.032 |
| Standard deviation | 0.010458 |
| $C_p$ | 1.430 |
| $C_{pk}$ | 1.354 |

visualized through the statistical measurements of $C_p$ and $C_{pk}$, with the help of Process Capability Wizard and QI Micros software.

$$C_p = (USL - LSL)/6\sigma$$
$$C_{pk} = \min[(USL - \mu)/3\sigma, (\mu - LSL)/3\sigma]$$

Where

USL = Upper specification limit
LSL = Lower specification limit
$\mu$ = Mean deviation
$\sigma$ = Standard deviation
T = Target value

From the above Taguchi based analysis, it was concluded that, the alternative material filament fabricated on optimum combination of single screw extruder parameters ($P_{11}$ $P_{22}$ and $P_{31}$), shows minimum deviations in filament diameter. For the successful loading of this filament in FDM system, it is necessary that, the diameter should be consistent, with minimum ovality and within specified range. The filament fabricated on the established optimum combination of parameters shows consistency in diameter. Ten observations of filament diameter were taken and analyzed with the help of 'process capability wizard' software as shown in Table 12.

The $C_p$ and $C_{pk}$ values of the process are 1.43 and 1.354 respectively, which is >1.33. The $C_p$ and $C_{pk}$ value >1.33 are generally considered for industry benchmark [10]. Figure 17 shows process capability histogram to highlight that the process has minimum variations and it consistently meets the target value. The X chart and R chart (see Figs. 18 and 19) respectively shows that the process varies within the range of 0.032 mm, which is less than the predicted value (0.0486 mm). Figure 20 shows normal probability plot with coefficient of determinant ($R^2$) = 0.9771 for acceptability of process (for batch and mass production).

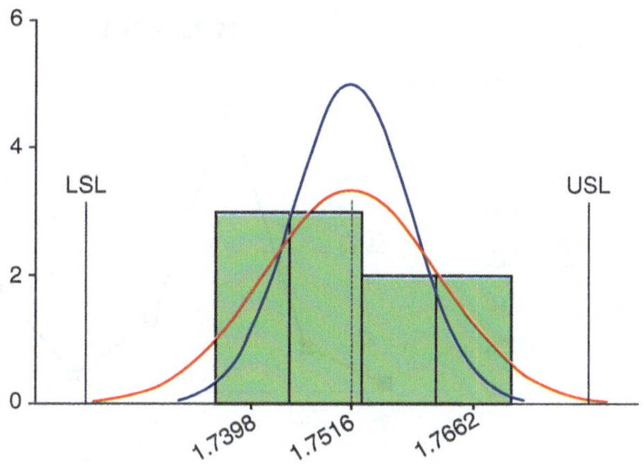


Fig. 17 Process capability Histogram for filament diameter

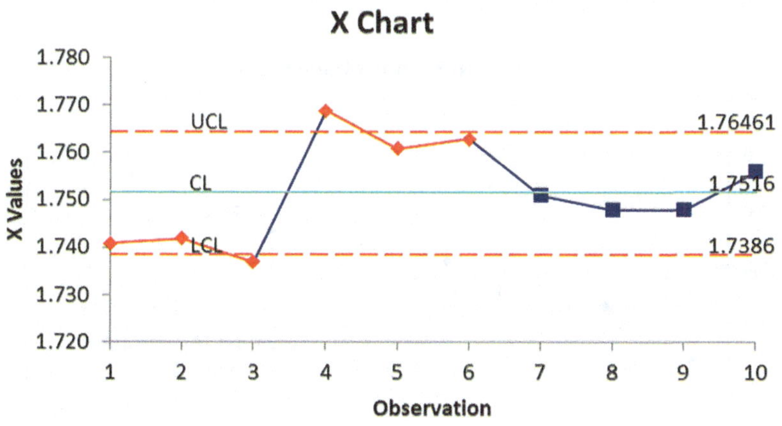


Fig. 18 X- Chart for filament diameter

## 6 Conclusions

The significance of this study is to present a frame work for development of filament for FDM system in terms of both academic research and industrial applications. The study demonstrates the high potential of alternative material for FDM process with its compatibility with FDM system and tailor made feature for RT and RM. This novel methodology helps to produce parts having complex geometry with customized properties through FDM process. The experimentation outputs are summarized as follows:

(i). A new alternate material containing Nylon6-Al-Al2O3 composite was successfully developed for FDM process.

(ii). A comparative study of flow rate under similar processing conditions was done by MFI and limits the reinforcement of filler content in polymer.

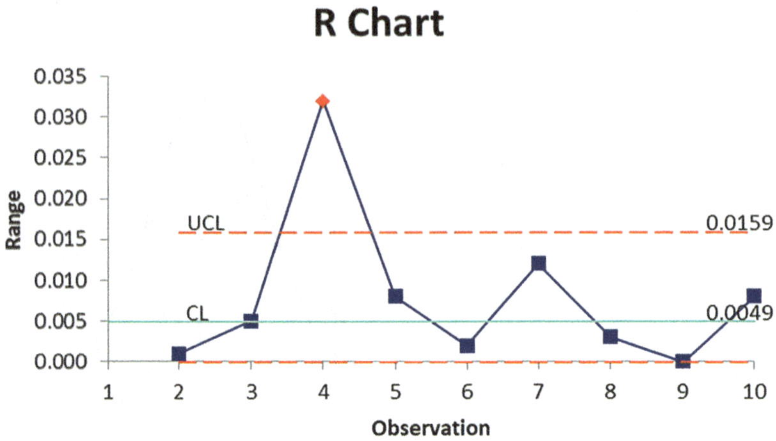


**Fig. 19** R- Chart for filament diameter

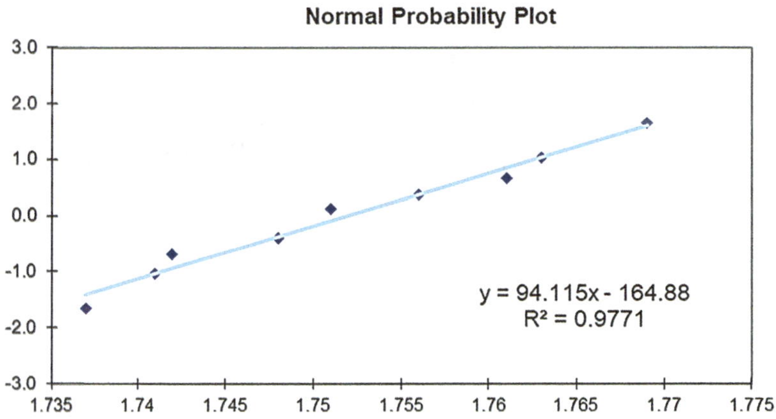


**Fig. 20** Normal Probability plot for filament diameter

Loading up to 40% by weight of Al + Al2O3 in Nylon 6 matrix is possible for uninterrupted processing in FDM system.

(iii). The tensile strength, tensile modulus of composition A, B and C are 64%, 53%, 28% and 2.7%, 2.1%, 1.5% respectively lower than ABS material. The elongation at break increases with filler content and particle size.

(iv). Dynamic mechanical analysis realized the structural investigations. The basic evaluated viscoelastic properties were storage modulus, loss modulus and loss factor.

(v). The storage modulus increases with the increase of filler amount in the Nylon6 matrix and also causes shift of glass transition temperature toward higher side. The increase of storage modulus highlights the good interfacial adhesion between polymer filler particles. Among various compositions, the composition C have rigid matrix.

(vi). The loss modulus curves which are indicative of the dissipated energy (heat or stress) under cyclic load were found to be shifted to higher position with the increase in Al2O3 content in matrix. Due to the abrasive nature of Al2O3 particles, the internal friction among particles increase with increase of particle contents. The reinforcement of filler content also causes broadening of the loss modulus curve after peak which causes the suppression of the relaxation process within the composite material. Higher value of loss modulus signifies increase in internal friction and enhancement of energy dissipation. The composition A have high peak value followed by composition B and C.

(vii). The glass transition temperature shifted toward higher side with reinforcement of filler material in matrix. This is because of constrains imposed on the mobility of molecules within the matrix. In case of composition C, tan δ curve shows large shift toward right side in the graph, which indicates the evidence of good interfacial bonding. Tan δ found to be lowest peak for composition C compared to composition B and A.

(viii). Optimized process parameters were selected by using Taguchi approach and ANNOVA method with the help of MINITAB 17 software. The analysis realized the influence of major process parameters and their levels on the response (diameter deviation) with the confidence level of 95%. Composition of alternative filament material found to be most influential parameter followed by die temperature and barrel temperature. The combination of optimum parameters is $P_{11}$ $P_{22}$ $P_{31}$.

(ix). Based upon the above criteria of experimental design, a linear regression equation was developed by the software and found to be valid for predicting the diameter deviation of alternative feed stock FDM filament. The equation predict the value of diameter deviation with the deviation of <1% as compared to experimental value.

(x). The diameter of filament fabricated on optimum parameters of single screw extruder have $C_p$ and Cpk values 1.43 and 1.354 respectively, which is >1.33. The process lies within the spread of $\pm 4\sigma$ and rejects <64 ppm (parts per million). The process uses <75% specifications. Moreover, the study highlights that, process has minimum variations and suitable for the fabrication of alternative material filament.

(xi). The parts fabricated with these composite materials may be used for RT applications (such as for development of grinding wheel with tailor made contour in dentistry applications) or RM applications (such as development of functionally graded material (Ref. Patent filed 2847/DEL/2013) prepared by $Al_2O_3$ reinforced FDM based pattern in investment casting applications). The Al parts casted by this process, contains $Al_2O_3$ particles on their outer surface and were highly wear resistant than pure Al parts.

## References

1. Singh R. Some investigations for small-sized product fabrication with FDM for plastic components. Rapid Prototyping Journal. 2013 Jan 11;19(1):58–63.
2. Nikzad M, Masood SH, Sbarski I. Thermo-mechanical properties of a highly filled polymeric composites for fused deposition modeling. Materials & Design. 2011 Jun 30;32(6):3448–56.
3. Singh Boparai K, Singh Boparai K, Singh R, Singh R, Singh H, Singh H. Experimental investigations for development of Nylon6-Al-Al2O3 alternative FDM filament. Rapid Prototyping Journal. 2016 Mar 21;22(2):217–24.
4. Mostafa N, Syed HM, Igor S, Andrew G. A study of melt flow analysis of an ABS-Iron composite in fused deposition modelling process. Tsinghua Science & Technology. 2009 Jun 30;14:29–37.
5. Masood SH, Song WQ. Development of new metal/polymer materials for rapid tooling using fused deposition modelling. Materials & design. 2004 Oct 31;25(7):587–94.
6. Wang M, Wang W, Liu T, Zhang WD. Melt rheological properties of nylon 6/multi-walled carbon nanotube composites. Composites Science and Technology. 2008 Sep 30;68 (12):2498–502.
7. Shenoy AV, Saini DR, Nadkarni VM. Rheology of nylon 6 containing metal halides. Journal of Materials Science. 1983 Jul 1;18(7):2149–55.
8. Ramanath, H.S., Chua, C.K., Leong, K.F. Melt flow behavior of poly-ε caprolactone in fused deposition modeling J Mater Sci: Mater Med .2008 19: 2541.
9. Bellini A, Shor L, Guceri SI. New developments in fused deposition modeling of ceramics. Rapid Prototyping Journal. 2005 Sep 1;11(4):214–20.
10. Zhang Y, Chou YK. Three-dimensional finite element analysis simulations of the fused deposition modelling process. Proceedings of the Institution of Mechanical Engineers, Part B: Journal of Engineering Manufacture. 2006 Oct 1;220(10):1663–71.
11. Bose S, Mahanwar PA. Effect of flyash on the mechanical, thermal, dielectric, rheological and morphological properties of filled nylon 6. Journal of Minerals & Materials Characterization & Engineering. 2004 Jan 1;3(2):65–89.
12. Spierings AB, Herres N, Levy G. Influence of the particle size distribution on surface quality and mechanical properties in AM steel parts. Rapid Prototyping Journal. 2011 Apr 26;17 (3):195–202.
13. Luyt AS, Molefi JA, Krump H. Thermal, mechanical and electrical properties of copper powder filled low-density and linear low-density polyethylene composites. Polymer Degradation and Stability. 2006 Jul 31;91(7):1629–36.
14. Benmesli S, Riahi F. Dynamic mechanical and thermal properties of a chemically modified polypropylene/natural rubber thermoplastic elastomer blend. Polymer Testing. 2014 Jun 30;36:54–61.
15. Sawpan MA, Holdsworth PG, Renshaw P. Glass transitions of hygrothermal aged pultruded glass fibre reinforced polymer rebar by dynamic mechanical thermal analysis. Materials & Design. 2012 Dec 31;42:272–8.
16. Qiao J, Amirkhizi AV, Schaaf K, Nemat-Nasser S. Dynamic mechanical analysis of fly ash filled polyurea elastomer. Journal of Engineering Materials and Technology. 2011 Jan 1;133 (1):011016.
17. Shanmugam D, Thiruchitrambalam M. Influence of alkali treatment and layering pattern on the tensile and flexural properties of Palmyra palm leaf stalk fiber (PPLSF)/jute fiber polyester hybrid composites. Composite Interfaces. 2014 Jan 2;21(1):3–12.
18. Rajesh C (2007) Development and characterization of Nylon Fiber Reinforced NBR composites. Thesis, Department of Chemistry, University of Calicut.
19. Murayama T (1999) Dynamic Mechanical Analysis of Polymeric Materials. Elsevier, New York.

20. Arora HS, Singh H, Dhindaw BK. Some observations on microstructural changes in a Mg-based AE42 alloy subjected to friction stir processing. Metallurgical and Materials Transactions B. 2012 Feb 1;43(1):92–108.
21. Jayaraman M, Sivasubramanian R, Balasubramanian V, Lakshminarayanan AK. Optimization of process parameters for friction stir welding of cast aluminium alloy A319 by Taguchi method.
22. Tamizharasan T, Kumar S. Optimization of cutting insert geometry using DEFORM-3 D: numerical simulation and experimental validation. International Journal of Simulation Modelling. 2012 Jun 1;11(2):65–76.

# Development of ABS-Graphene Blended Feedstock Filament for FDM Process

**Gurleen Singh Sandhu and Rupinder Singh**

## 1 Evolution of Conducting Polymers

Polymers have been utilized as coating materials for different applications, for example metal products are coated with plastic for protection against corrosion. Polymers have been considered as an insulator at most of the times [1]. Since the discovery of polyacetylene in 1977 by Shirakawa et al. different vital Conductive polymers (CPs) have been researched continuously, including polypyrrole (PPy), polyaniline (PANI), polythiophene (PT), and poly(p-phenylenevinylene) (PPV) [2]. Moreover, polymeric conductive composites have been explored because of their innumerable advantages over metals in various applicationsbecause they are lightweight, have better processability, and are shock resistant. Power gadgets, electric engines, generators, and thermal exchangers are some exemplary applications of conductive composites.

## 2 Acrylo Nitrile Butadiene Styrene and Graphene

Thermoplastics, or polymers, are among the least expensive materials that can be utilized in addictive manufacturing (AM) processes.Some of the commonly used thermoplastics in AM are: acrylonitile butadiene styrene (ABS), polylactic acid (PLA), polyvinyl alcohol (PVA) and polycarbonate. The ABSis most widely used

G. S. Sandhu
Mechanical Engineering Department, Guru Nanak Dev Engineering College Ludhiana, Ludhiana, India

R. Singh (✉)
Production Engineering Department, Guru Nanak Dev Engineering College Ludhiana, Ludhiana, India



B. AlMangour (ed.), *Additive Manufacturing of Emerging Materials*,
https://doi.org/10.1007/978-3-319-91713-9_9

material for 3D printing, since it is lightweight, has great impact strength, easily affordable, hard, rigid, has great chemical resistance and highly dimensionally stable [3]. Further, carbon fillers, which have a high thermal and electrical conductivity, can be reinforced in ABS polymer to make CP's [4]. In this chapter, the ABS plastic has been reinforced with two-dimensional (2D) carbon filler, commonly known asGr. The Gr is an allotrope of carbon having single layer $sp^2$ hybridization arranged in a 2D hexagonal lattice. The material has been of intense interest because of its mechanical, thermal, and electrical properties [5], which have manyfascinating electronic and optoelectronic properties [6]. Therefore, reinforcement of Gr in polymer matrices acts as an advantage for enhancing its mechanical, physical, electrical and its optical properties.

## 3 Extraction of Gr

The Gr has attracted attention as a promising material to create new polymer nano-composite due to its excellent properties and readily availability of its precursor, Graphite [7]. Different techniques have been adopted commercially to achieve a fine dispersion of Gr and polymer [8, 9].

In order to extract Gr from graphite powder a pilot experimentation has been conducted. Firstly, chemical exfoliation of graphite powder has been performed with chemical mixing of graphite with naphthalene powder in N-methyl pyrrolidone (NMP) arrangement taken after by holding in ultrasonic bath for 90 min at 500 °C, followed by centrifugation for 30 min at 3000 rpm at room temperature. The naphthalene serves as a molecular wedge to intercalate into the edge of graphite, which plays a key role during sonication (see Fig. 1a) and significantly improves the production yield of Gr (see Fig. 1b, c). Continuous maturing for 24 h was contributed for detachment of graphene layer from graphite. Pipetting was performed to gather the disintegrated graphene layer. The graphene so obtained was then dried in hot air

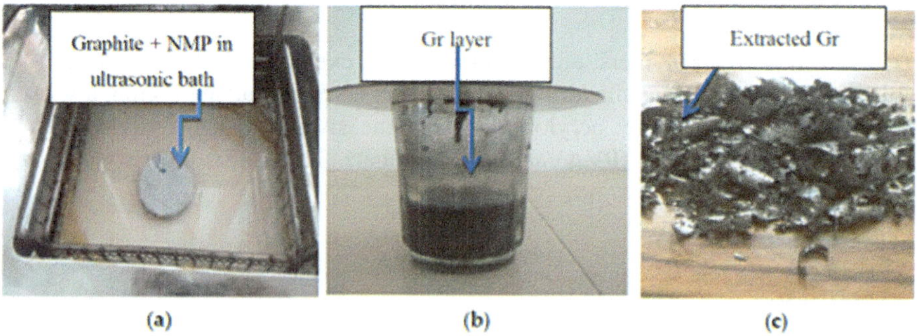


**Fig. 1** Extraction of Gr. (**a**) Sonication of graphite and NMP; (**b**) Formation of Gr layer; and (**c**) Extracted Gr. (Source: Lab Data, Manufacturing Research Lab, Guru Nanak Dev Engineering College, Ludhiana, India)

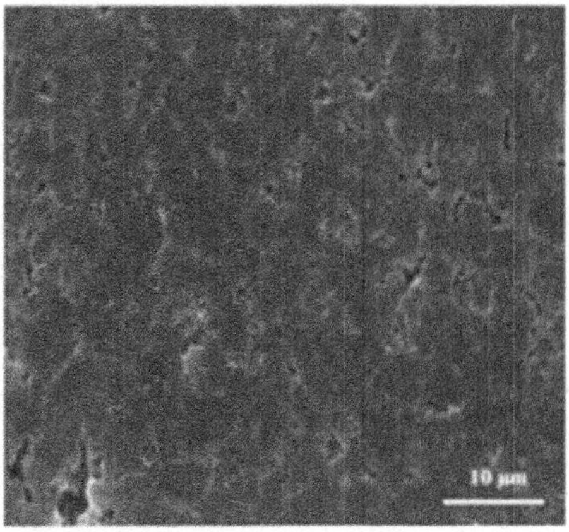


**Fig. 2** Scanning electron microscopy (SEM) image of extracted Gr. (Source: Lab Data, Manufacturing Research Lab, Guru Nanak Dev Engineering College, Ludhiana, India)

oven at 70 °C for about an hour. Figure 2 shows the scanning electron microscope (SEM) image of the extracted Gr.As observed from Fig. 2, uniform 2D structures have been obtained. This means that the functional group of graphene has not been disturbed. The present study highlights the alternative method for exfoliation of graphite at the lab scale. In this work, out of 50 g graphite, around 3.7 g Gr was finally extracted. Further chemical analysis may be conducted for ascertaining whether the exfoliation is complete or not.This was not conducted in the present study. Further the literature highlights that very few has been reported on in-house development of an ABS–Gr feedstock filament (from low cost graphite material) for FDM [12–14]. The ABS reinforced with Gr shows enhancement of its mechanical, optical properties and electrical and thermal stability [15].

# 4 Materials

The commercially-available ABS was procured from local market (Batra Polymers, Ludhiana, India) having melt flow index (MFI) of 2.9 g/10 min as per the ASTM D 1238-73 standard [10]. The graphite powder (thermal conductivity: 2–90 W/m K) was supplied by Bharat Graphite Pvt. Ltd. (Ludhiana, India). The NMP and acetone acted as dissolute and was supplied by Saiteja Chemicals, Hyderabad, India.

# 5 Twin Screw Extrusion

The compounding of thermoplastic materials on a mechanical scale is essentially done in co-rotational twin-screw extruders (TSE). It is a machine with two different single screws with an exceptionally adaptable economic process. The schematic of TSE is shown in Fig. 3.

The commercially available Thermo Scientific HAAKE Mini CTW (Make: Karlsruhe, Germany) was used for feedstock filament preparation. The screw

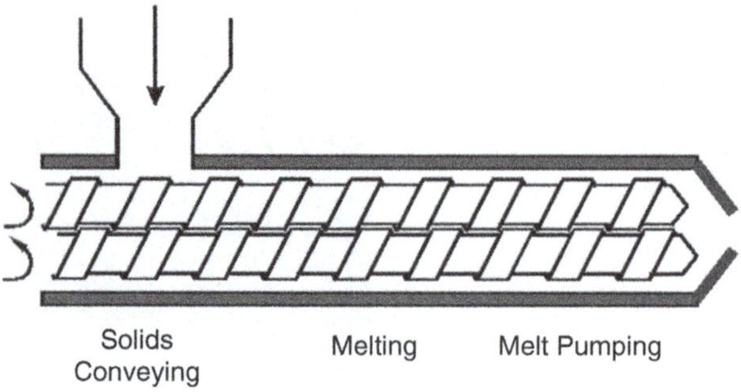


**Fig. 3** Schematic of TSE

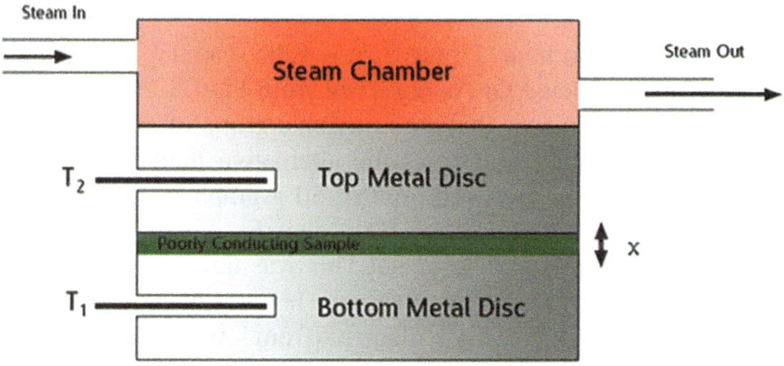


**Fig. 4** Experimental setup for Lee's Disc

rotation speed of 200 RPMs with a load of 10 kg at 180 °C was regulated in order to obtain a final diameter of the extruded filament of 1.75 ± 0.10 mm. In case of mechanical blending, the Gr and ABS were directly fed into the hopper of the TSE. Finally, the polymer composite filament was prepared and fed to FDM for fabrication of functional prototypes [11].

# 6 Testing Techniques for ABS-Gr Blended Feedstock Filament

## *6.1 Lee's Method for Thermal Conductivity*

Lee's Disk technique was utilized to measure the thermal conductivity ofthe developed composite material. Experimental setup for Lee's Disk method (see Fig. 4) has two metal discs, a steam chamber, the specimen to be measured and two thermometers to gauge the temperature inclination. The specimen is set in the middle of the two metal discs (the thermometers are embedded into the metal plates) and the steam chamber is set on top of the top metal circle. During the experiment, water is heated to create steam, which is then pushed through a steam chamber which is straightforwardly over one of the metal circles with a thermometer in it.

This steam makes the metal disc warm up and exchange warmth to the sample that is being tested, thus, exchanges warmth to the base metal disc, which loses warmth to convection. After a specific time, each of the two metal discs achieves consistent temperature. The two temperatures are distinctive and are utilized to calculate the thermal conductivity.

## 6.2 *Measurement of Electrical Conductivity*

The capacity of a material; to convey current is measured in terms of electrical conductivity. It is a characteristic property of a material, indicated by 'σ' while Siemens per meter (S/m) is its SI unit.

In order to calculate electrical conductivity, the famous Ohms law can be used. George Ohm found that, at a steady temperature, the current flowing through a settled straight resistance is specifically corresponding to the voltage, and furthermore inversely proportional to the resistance. This relation between the Voltage, Current and Resistance forms the basis of **Ohms Law** and is demonstrated as follows.

$$Current\ (I) = \frac{Voltage\ (V)}{Resistance\ (R)}\ \text{in Amperes (A)}$$

The other way to calculate the Resistance (R) of the sample is as follows:

$$R = \frac{\rho l}{A}$$

Where,

ρ = Resistivity
l = length of the sample
A = Area of cross section of the sample

It is assumed that the current is uniform over the cross-range of the wire, which is real only for Direct Current.

The reverse of resistivity is called Conductivity. There are settings where the usage of conductivity is more favourable.

$$\text{Electrical conductivity,}\ \sigma = \frac{1}{\rho},\ \text{in Siemens-metre(S-m)}$$

In order to conduct the experiment to measure electrical conductivity, schematic circuit is shown in Fig. 5.

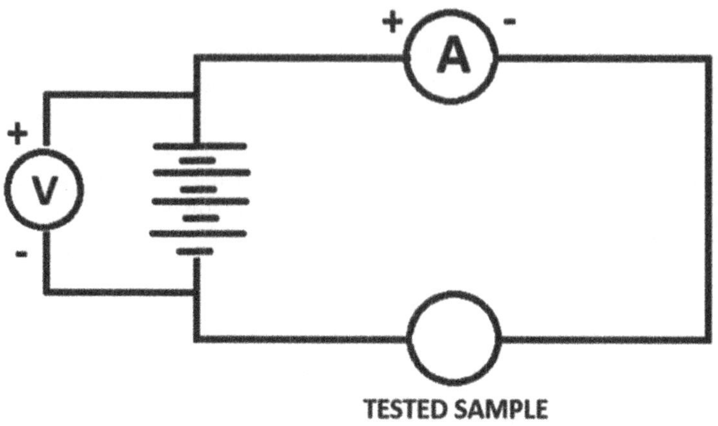


**Fig. 5** Schematic for calculation of electrical conductivity

### *6.3 Measurement of Porosity and Hardness*

Porosity is essentially an undesirable property which arises when any material while cooling or getting changed into solid state from fluid or softening stage catches some air particles into immaculate lattice of material in this way by causing loss of some imperative mechanical properties. The inverted microscope has been used in this study for measurement of porosity. Hardness is another important mechanical property to consider, which measures the material's resistance to plastic deformation. For polymeric materials hardness is normally examined at Shore scale. Scale Shore A is utilized in checking elastomers (like rubbers) and scale Shore D scale is used to identify the hard elastomers (like thermoplastics and thermosets).

## 7 Optimisation Study of Process Parameter of FDM

After the preparation of ABS-Gr composite feedstock filament wire on TSE, the filament wire was fed to FDM. A. *STL* (sterelithography) file of the disc shaped sample was prepared and had been fedto the FDM setup for fabricationof discs (functional prototype, selected as bench mark). Table 1 shows selected FDM process parameters for part fabrication.

Figure 6 shows pictorial view of designed disc, FDM setup and finally prepared disc (selected as bench mark/ functional prototype).

Table 2 shows the control log of experimentation as per Taguchi L8 orthogonal array.

Based upon Table 2, the electrical conductivity, thermal conductivity, Porosity and Shore D Hardness has been measured. The experiment was repeated three times in order to reduce the experimentation error.

**Table 1** FDM process parameter for part fabrication

| Process variable | Operating conditions |
|---|---|
| Nozzle diameter | 0.3 mm |
| Filament diameter | 1.75 mm |
| Layer height | 0.4 mm |
| Perimeter | 3 |
| Solid layers | Top-3, Bottom-3 |
| Fill density | 0.8 |
| Fill pattern | Honeycomb |
| Perimeter speed | 30 mm/sec |
| Infill speed | 60 mm/sec |
| Travel speed | 130 mm/sec |
| Extruder temperature | 240 °C |
| Bed temperature | 55 °C |

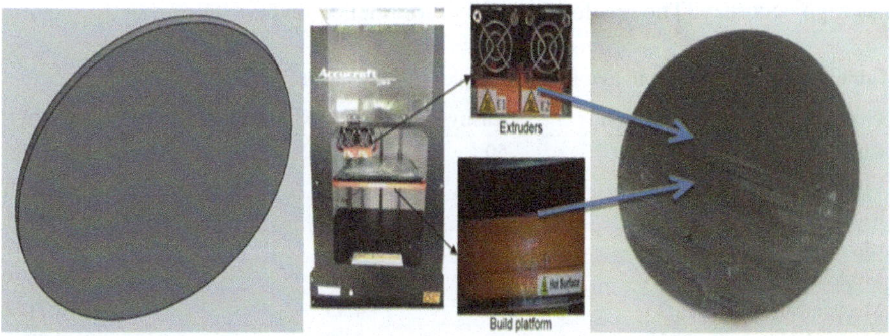


**Fig. 6** FDM setup and the fabricated disc. (Source: Lab Data, Manufacturing Research Lab, Guru Nanak Dev Engineering College, Ludhiana, India)

**Table 2** Control log of experiment

| Blending process | Proportion (by weight) | Infill density (%) |
|---|---|---|
| Chemical | 75:25 | 50 |
| Chemical | 75:25 | 100 |
| Chemical | 90:10 | 50 |
| Chemical | 90:10 | 100 |
| Mechanical | 75:25 | 50 |
| Mechanical | 75:25 | 100 |
| Mechanical | 90:10 | 50 |
| Mechanical | 90:10 | 100 |

## 8 Analysis of Electrical Conductivity Test

The obtained results for electrical conductivity value were calculated for 'larger is better' case using Minitab17. Table 3 gives the utmost value of electrical conductivity 7.29 S/m obtained at experiment condition number 2 at proportion 75:25 and density 100%.

Figure 7 shows the main effects plot for SN ratios of electrical conductivity. As observed from Fig. 7 electrical conductivity value obtained at level 1 of process, Level 1 of proportion and Level 2 of density give the maximum value. This outcome may be credited to the chemical dissolute sample with proportion 75:25 and in-fill density of 100%.

Residual error was obtained as 4.10% (Table 4) with maximum contribution of Proportion (Table 5) i.e. 41.44% predicts the model for electrical conductivity of chemically dissolute sample with proportion 75:25 and in-fill density of 100% value has high accuracy. The results are at 95% confidence level.

**Table 3** Electrical conductivity of tested sample with SN ratios

| Process | Proportion | Density | Electrical conductivity | SNRA1 |
|---|---|---|---|---|
| Chemical | 75:25 | 50 | 4.82 | 13.6609 |
| Chemical | 75:25 | 100 | 7.29 | 17.2546 |
| Chemical | 90:10 | 50 | 3.50 | 10.8814 |
| Chemical | 90:10 | 100 | 5.07 | 14.1002 |
| Mechanical | 75:25 | 50 | 4.30 | 12.6694 |
| Mechanical | 75:25 | 100 | 4.85 | 13.7148 |
| Mechanical | 90:10 | 50 | 2.60 | 8.2995 |
| Mechanical | 90:10 | 100 | 3.63 | 11.1981 |

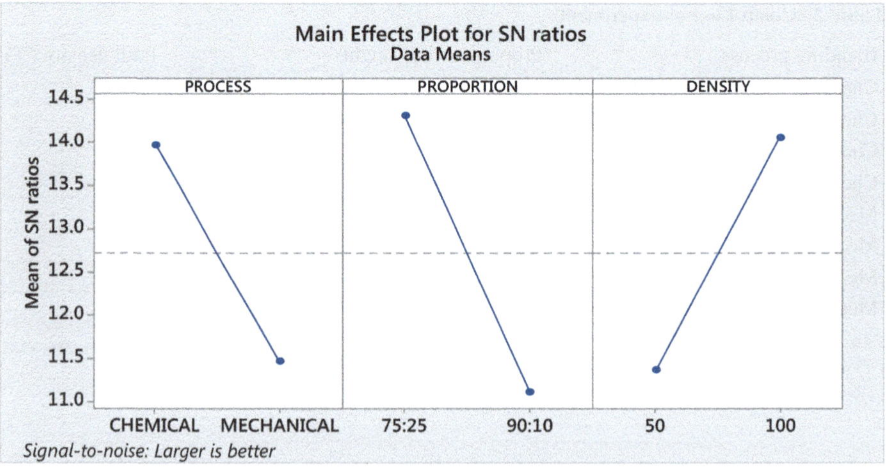


**Fig. 7** Main effects plot for SN ratios of electrical conductivity

**Table 4** Analysis of variance for SN ratios of electrical conductivity

| Source | DF | Seq SS | Adj SS | Adj MS | F | P |
|---|---|---|---|---|---|---|
| Process | 1 | 12.538 | 12.538 | 12.538 | 24.64 | 0.008 |
| Proportion | 1 | 20.546 | 20.546 | 20.5459 | 40.37 | 0.003 |
| Density | 1 | 14.463 | 14.463 | 14.4629 | 28.42 | 0.006 |
| Residual error | 4 | 2.036 | 2.036 | 0.5089 | | |
| Total | 7 | 49.582 | | | | |

*DF* Degree of freedom, *Seq SS* Sequential sums of squares, *Adj SS* Adjusted sum of squares, *Adj MS* Adjustedmean squares

**Table 5** SN ratio response table (larger is better)

| Level | Process | Proportion | Density |
|---|---|---|---|
| 1 | 13.97 | 14.32 | 11.38 |
| 2 | 11.47 | 11.12 | 14.07 |
| Delta | 2.50 | 3.21 | 2.69 |
| Rank | 3 | 1 | 2 |

The optimum value for electrical conductivity has been calculated using formula:

$$\eta_{opt} = m + (m_{A1} - m) + (m_{B1} - m) + (m_{C2} - m)$$

Where, m is overall mean of SN data, $m_{A1}$is mean of SN data for Process at level 1, $m_{B1}$ is mean of SN data for Proportion at level 1, $m_{C2}$ is mean of SN data for Density at level 2.

$y_{opt}^2 = (1/10)^{\eta opt/10}$ for properties, lesser is better.

$y_{opt}^2 = (10)^{\eta opt\ /\ 10}$ for properties, Larger is better

Overall mean of SN ratio (m) was taken from Minitab17

$$m = 12.7$$

Now, signal to noise ratio from response table, $m_{A1} = 13.97$, $m_{B1} = 14.32$, $m_{C2} = 14.07$.

From here,

$$\eta_{opt} = 16.96$$

Now,

$$y_{opt}^2 = (10)^{\eta opt/10}$$

$$y_{opt} = 7.05\ S$$

So, optimum electrical conductivity is 7.05 S while the observed value at these setting is 7.29 S.

## 9 Analysis of Thermal Conductivity Test

The obtained results for thermal conductivity value were calculated for 'larger is better' case using Minitab17.Table 6 gives the utmost value of thermal conductivity 17.60 W/mK obtained at experiment condition number 2 at proportion 75:25 and density 100%.

Figure 8 shows the main effects plot for SN ratios of electrical conductivity. As observed from Fig. 8 the thermal conductivityobtained at level 1 of Process, Level 1 of Proportion and Level 2 of density provides the maximum value of thermal Conductivity. This outcome may be credited to the chemical dissolute sample with proportion 75:25 and in-fill density of 100%.

Residual error was obtained as 1.22% (Table 7) with maximum contribution of process of blending i.e. 20.45% (Table 8) predicts the model for thermal

**Table 6** Thermal conductivity of tested sample with SN ratios

| Process | Proportion | Density | Thermal conductivity | SNRA1 |
|---|---|---|---|---|
| Chemical | 75:25 | 50 | 8.85 | 18.9389 |
| Chemical | 75:25 | 100 | 17.60 | 24.9103 |
| Chemical | 90:10 | 50 | 6.36 | 16.0691 |
| Chemical | 90:10 | 100 | 12.43 | 21.8894 |
| Mechanical | 75:25 | 50 | 2.40 | 7.6042 |
| Mechanical | 75:25 | 100 | 4.65 | 13.3491 |
| Mechanical | 90:10 | 50 | 2.41 | 7.6403 |
| Mechanical | 90:10 | 100 | 3.99 | 12.0195 |

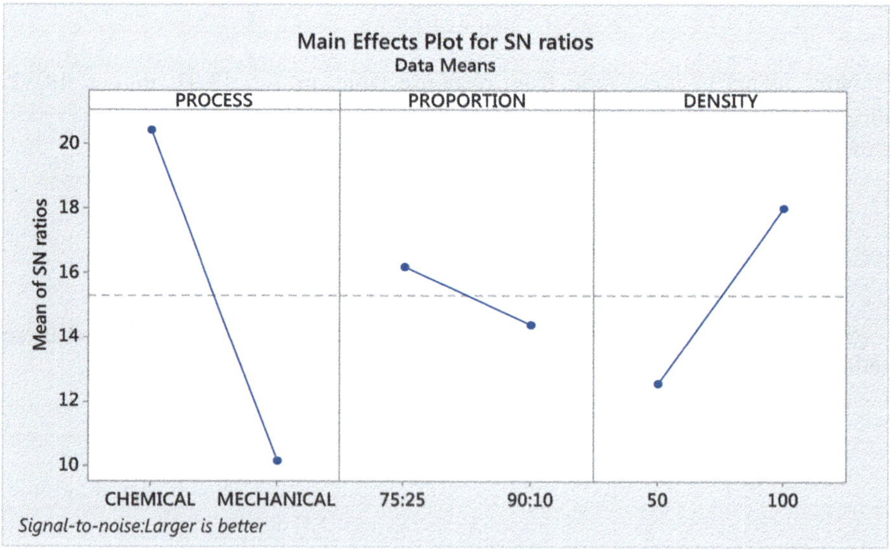


**Fig. 8** Main effects plot for SN ratios of thermal conductivity

**Table 7** Analysis of variance for SN ratios of thermal conductivity

| Source | DF | Seq SS | Adj SS | Adj MS | F | P |
|---|---|---|---|---|---|---|
| Process | 1 | 212.124 | 212.124 | 212.124 | 245.14 | 0.000 |
| Proportion | 1 | 6.451 | 6.451 | 6.451 | 7.46 | 0.052 |
| Density | 1 | 60.037 | 60.037 | 60.037 | 69.38 | 0.001 |
| Residual error | 4 | 3.461 | 3.461 | 0.865 | | |
| Total | 7 | 282.074 | | | | |

**Table 8** SN ratio response table (larger is better)

| Level | Process | Proportion | Density |
|---|---|---|---|
| 1 | 20.45 | 16.20 | 12.56 |
| 2 | 10.15 | 14.40 | 18.04 |
| Delta | 10.30 | 1.80 | 5.48 |
| Rank | 1 | 3 | 2 |

conductivity of chemically dissolute sample with proportion 75:25 and in-fill density of 100% value has high accuracy. The results are at 95% confidence level.

The optimum value for thermal conductivity has been calculated using the formula:

$$\eta_{opt} = m + (m_{A1} - m) + (m_{B1} - m) + (m_{C2} - m)$$

where, m is overall mean of SN data, $m_{A1}$ is mean of SN datafor Process at level 1, $m_{B1}$ is mean of SN data for Proportion at level 1, $m_{C2}$ is mean of SN data for Density at level 2.Overall mean of SN ratio (m) was taken from Minitab17:

$$m = 15.3$$

Now, from response table of signal to noise ratio, $m_{A1} = 20.45$, $m_{B1} = 16.20$, $m_{C2} = 18.04$.

From here,

$$\eta_{opt} = 24.09$$

And,

$$y_{opt} = 16.01 \quad W/mK$$

Hence, Optimum Thermal conductivity is 16.01 W/mK while the observed value was 17.60 W/mK.

## 10 Analysis of Porosity

The obtained results for Porosity value were calculated for 'smaller is better' case using Minitab17. Table 9 gives the least value of porosity 12.05 obtained at experiment condition number 1 at proportion 75:25 and density 50%.

Figure 9 shows the main effects plot for SN ratios of porosity. As observed from Fig. 9 the porosity obtained at level 1 of Process, Level 1 of Proportion and Level 1 of density provides the most favourable value of Porosity. This outcome may be credited to the chemical dissolute sample with proportion 75:25 and in-fill density of 50%.

Residual error was obtained as 5.02% (Table 10) with maximum contribution of process of blending (Table 11) predicts the model for porosity of chemically dissolute sample with proportion 75:25 and in-fill density of 50% value has high accuracy. The results are at 95% confidence level.

**Table 9** Porosity of tested sample with SN ratios

| Process | Proportion | Density | Porosity | SNRA1 |
|---|---|---|---|---|
| Chemical | 75:25 | 50 | 12.05 | −21.6197 |
| Chemical | 75:25 | 100 | 21.76 | −26.7532 |
| Chemical | 90:10 | 50 | 19.02 | −25.5842 |
| Chemical | 90:10 | 100 | 22.26 | −26.9505 |
| Mechanical | 75:25 | 50 | 22.61 | −27.0860 |
| Mechanical | 75:25 | 100 | 31.78 | −30.0431 |
| Mechanical | 90:10 | 50 | 27.58 | −28.8119 |
| Mechanical | 90:10 | 100 | 44.84 | −33.0333 |

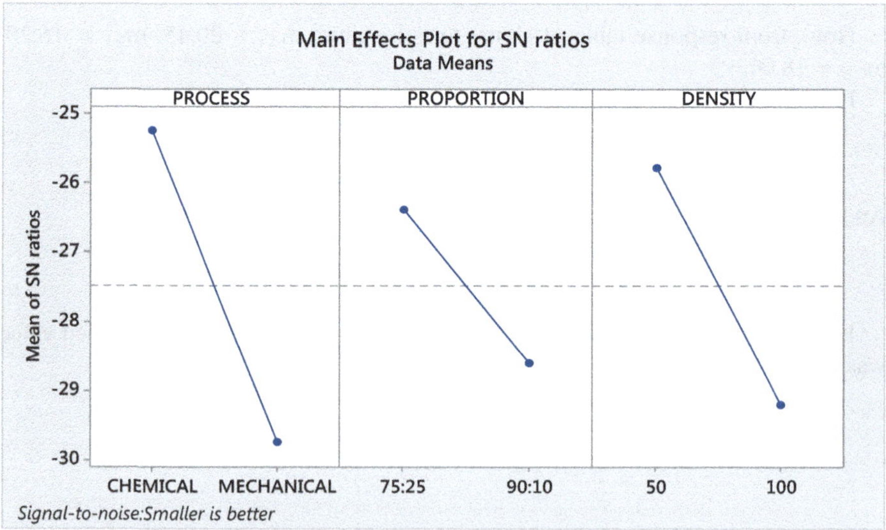


**Fig. 9** Main effects graph for SN ratios of porosity

**Table 10** Analysis of variance for SN ratios of porosity

| Source | DF | Seq SS | Adj SS | Adj MS | F | P |
|---|---|---|---|---|---|---|
| Process | 1 | 40.800 | 40.800 | 40.800 | 40.36 | 0.003 |
| Proportion | 1 | 9.852 | 9.852 | 9.852 | 9.75 | 0.035 |
| Density | 1 | 23.387 | 23.387 | 23.387 | 23.14 | 0.009 |
| Residual error | 4 | 4.043 | 4.043 | 1.011 | | |
| Total | 7 | 78.083 | | | | |

**Table 11** SN ratio response table (Smaller is better)

| Level | Process | Proportion | Density |
|---|---|---|---|
| 1 | −25.23 | −26.38 | −25.78 |
| 2 | −29.74 | −28.59 | −29.20 |
| Delta | 4.52 | 2.22 | 3.42 |
| Rank | 1 | 3 | 2 |

The optimum value for porosity has been calculated using the formula:

$$\eta_{opt} = m + (m_{A1} - m) + (m_{B1} - m) + (m_{C1} - m)$$

Where, m is overall mean of SN data, $m_{A1}$ is mean of SN data for Process at level 1, $m_{B1}$ is mean of SN data for Proportion at level 1, $m_{C1}$ is mean of SN data for Density at level 1.

Overall mean of SN ratio (m) was taken from Minitab17.

$$m = -27.4866667$$

Now, from response table of signal to noise ratio, $m_{A1} = -25.23$, $m_{B1} = -26.38$, $m_{C1} = -25.78$.

From here,

$$\eta_{opt} = -22.4166666$$

and, $$y_{opt} = 13.20$$

Hence, Optimum Porosity is 13.20 while the observed value is 12.05.

## 11 Analysis of Shore Hardness

The obtained results for shore hardness value were calculated for ‘larger is better’ case using Minitab17. Table 12 gives the utmost value of shore hardness 67.4 shore D obtained at experiment condition number 4 at proportion 90:10 and density 100%.

Figure 10 shows the main effects plot for SN ratios ofShore hardness. As observed from Fig. 10 the shore hardness obtained at level 1 of Process, Level 2 of Proportion and Level 2 of density provides the most favourable value of shore

**Table 12** Shore hardness of tested sample with SN ratios

| Process | Proportion | Density | Hardness | SNRA1 |
|---|---|---|---|---|
| Chemical | 75:25 | 50 | 53.6 | 34.5833 |
| Chemical | 75:25 | 100 | 66.3 | 36.4303 |
| Chemical | 90:10 | 50 | 55.7 | 34.9171 |
| Chemical | 90:10 | 100 | 67.4 | 36.5732 |
| Mechanical | 75:25 | 50 | 52.4 | 34.3866 |
| Mechanical | 75:25 | 100 | 63.1 | 36.0006 |
| Mechanical | 90:10 | 50 | 59.6 | 35.5049 |
| Mechanical | 90:10 | 100 | 66.0 | 36.3909 |

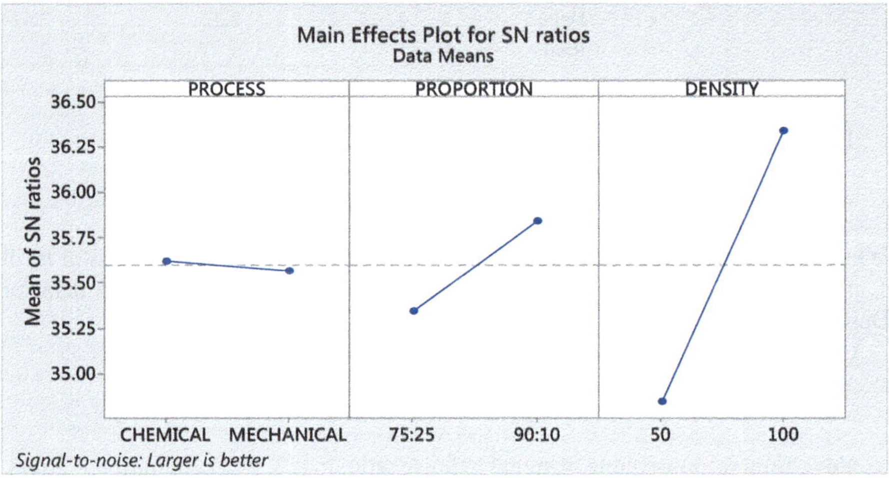


**Fig. 10** Main effectsplot for SN ratios of shore hardness

**Table 13** Analysis of variance for SN ratios of shore hardness

| Source | DF | Seq SS | Adj SS | Adj MS | F | P |
|---|---|---|---|---|---|---|
| Process | 1 | 0.00610 | 0.00610 | 0.00610 | 0.06 | 0.817 |
| Proportion | 1 | 0.49269 | 0.49269 | 0.49269 | 4.92 | 0.091 |
| Density | 1 | 4.50448 | 4.50448 | 4.50448 | 44.99 | 0.003 |
| Residual error | 4 | 0.40049 | 0.40049 | 0.10012 | | |
| Total | 7 | 5.40375 | | | | |

hardness. This outcome may be credited to the chemical dissolute sample with proportion 75:25 and in-fill density of 50%.

Residual error was obtained as 7.4% (Table 13) with maximum contribution of in-fill density (Table 14) predicts the model for porosity of chemically dissolute

**Table 14** SN ratio response table (larger is better)

| Level | Process | Proportion | Density |
|---|---|---|---|
| 1 | 35.63 | 35.35 | 34.85 |
| 2 | 35.57 | 35.85 | 36.35 |
| Delta | 0.06 | 0.50 | 1.50 |
| Rank | 3 | 2 | 1 |

sample with proportion 90:10 and in-fill density of 100% value has high accuracy. The results are at 95% confidence level.

The optimum value for shore hardness has been calculated using the taguchi designed formula as follows:

$$\eta_{opt} = m + (m_{A1} - m) + (m_{B2} - m) + (m_{C2} - m)$$

Where, m is overall mean of SN data, $m_{A1}$ is mean of SN data for Process at level 1, $m_{B2}$ is mean of SN data for Proportion at level 2, $m_{C2}$ is mean of SN data for Density at level 2.

Overall mean of SN ratio (m) was taken from Minitab17

$$m = 35.6$$

Now, from response table of signal to noise ratio, $m_{A1} = 35.63$, $m_{B2} = 35.85$, $m_{C2} = 36.35$.

From here,

$$\eta_{opt} = 36.63$$

$$\text{And,} \quad y_{opt} = 67.84$$

Hence, Optimum Shore D Hardness is 67.84 while the observed value was 67.4.

## 12 Optical Micrograph Observations for Porosity

Porosity is essentially an undesirable property which begins when any material while cooling or getting changed over into strong state from fluid or softening stage ensnares some air particles into immaculate lattice of material in this way by causing loss of some imperative mechanical, electrical and thermal properties. So less is the porosity better will be the mechanical properties of material, better will be the electrical as well as thermal properties.

The samples prepared were examined for porosity through optical microscope at 100X (see Fig. 11) as per ASTM E2015–04 [16].

| | | | |
|---|---|---|---|
| 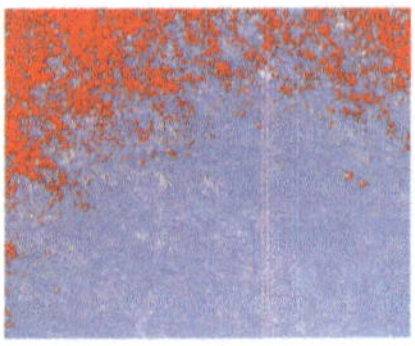 |  | 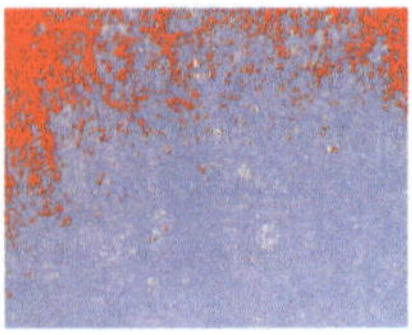 |  |
| Process: Chemical<br>Proportion:<br>75:25<br>Density: 100%<br>Porosity: 21.76 | Process: Chemical<br>Proportion: 75:25<br>Density: 50%<br>Porosity: 12.05 | Process: Chemical<br>Proportion: 90:10<br>Density: 100%<br>Porosity: 22.26 | Process: Chemical<br>Proportion: 90:10<br>Density: 50%<br>Porosity: 19.02 |
| 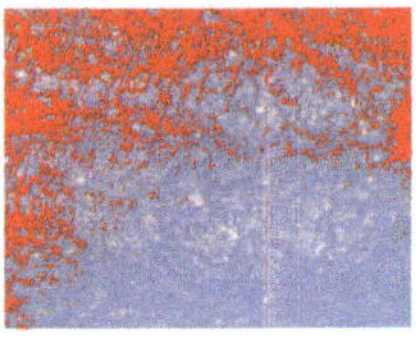 | 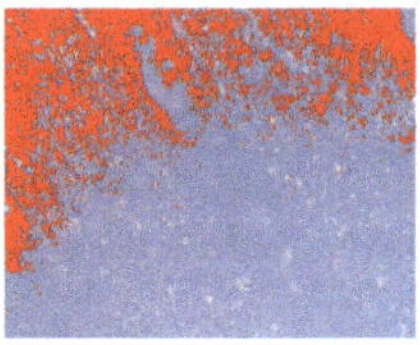 | 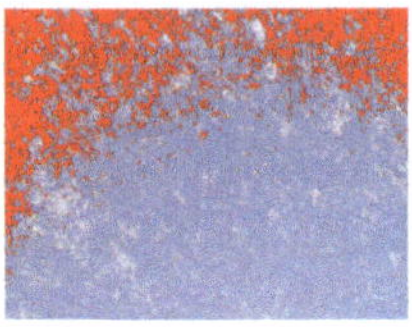 |  |
| Process: Mechanical<br>Proportion: 75:25<br>Density: 100%<br>Porosity: 31.78 | Process: Mechanical<br>Proportion: 75:25<br>Density: 50%<br>Porosity: 22.61 | Process:<br>Mechanical<br>Proportion: 90:10<br>Density: 100%<br>Porosity: 44.84 | Process: Mechanical<br>Proportion: 90:10<br>Density: 50%<br>Porosity: 27.58 |

**Fig. 11** Optical micrographic views for ABS-Graphene chemically and mechanically blended samples. (Source: Lab Data, Manufacturing Research Lab, Guru Nanak Dev Engineering College, Ludhiana, India)

## 13 Differential Scanning Calorimeter

For the present study, METTLER TOLEDO model DSC3 with *STAR*$^e$(SW 14.00) has been used to examine the glass transition temperatures of two different methods (chemical and mechanical). Two successive cycles – heating and cooling has been run on DSC. The temperature range was set from 30 °C to 280 °C at an increasing rate of 10 K/min, in the presence of N2 gas at flow rate 50 ml/min. Figures 12 and 13 shows the DSC graphs for chemically and mechanically blended samples. As shown in Fig. 12, small drop was observed for glass transition in the first heating cycle (104.37 °C), while 105.77 °C during the second heating cycle. The glass transition temperature of virgin ABS is 105 °C. Similarly for mechanically blended sample with proportion 75:25 and 100% in-fill density, a small drop occurred in the curve having value 97.76 °C shows glass transition in the first heating cycle, while 98.25 °C during the second heating cycle (see Fig. 13). The glass transition temperature shows that the observed values in chemically blended samples are much closer to the glass transition temperature of ABS i.e. 105 °C while in case of mechanically blended samples the temperature range is quite far. However these samples are thermally stable and can be successfully recycled.

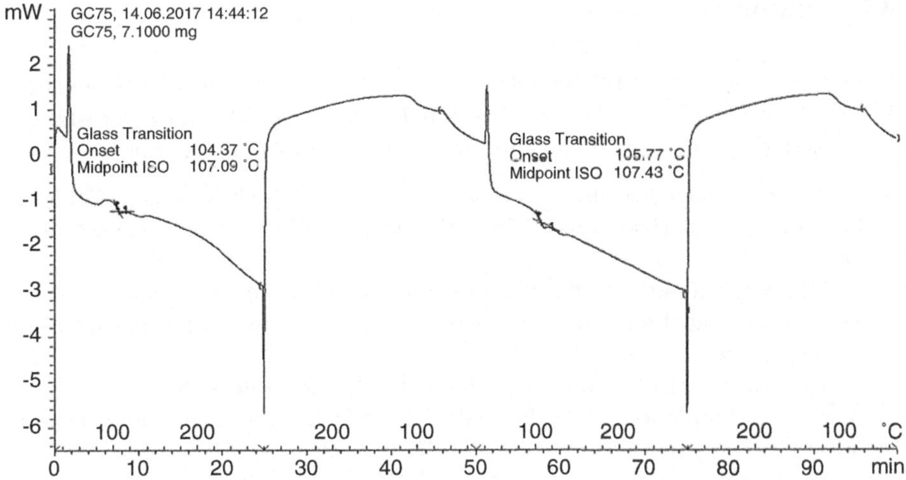


**Fig. 12** DSC graph for chemically blended sample with proportion 75:25 and 100% infill density. (Source: Lab Data, Manufacturing Research Lab, Guru Nanak Dev Engineering College, Ludhiana, India)

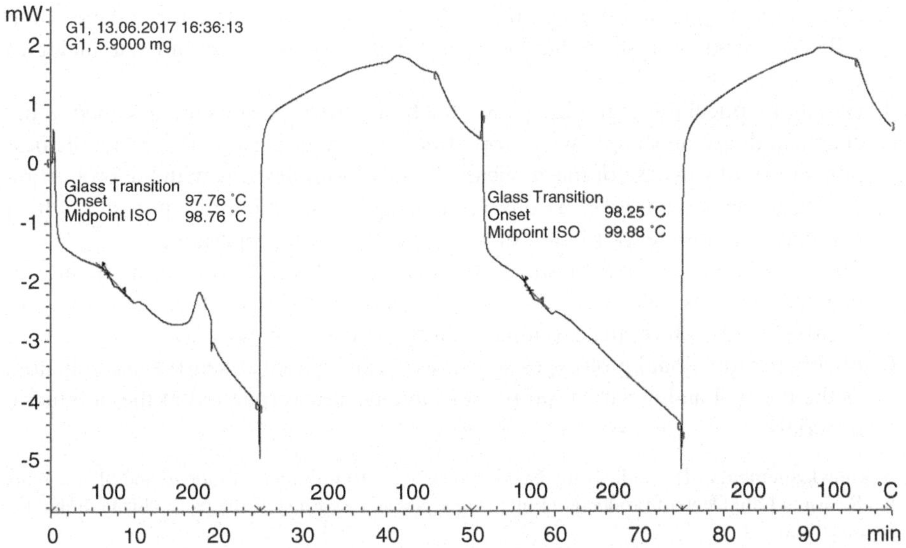


**Fig. 13** DSC graph for mechanically blended sample with proportion 75:25 and 100% infill density. (Source: Lab Data, Manufacturing Research Lab, Guru Nanak Dev Engineering College, Ludhiana, India)

## 14 Summary

In this research study, ABS-Gr filament of FDM has been successfully prepared. Further, investigations for electrical conductivity, thermal conductivity, porosity and shore hardness were made. Following are the conclusions from this study.

1. The best settings for electrical conductivity of ABS-Gr mix are: Chemical Dissolution with Proportion 75% ABS and 25% Gr by weight having in-fill Density 100%

   The best settings for thermal properties of ABS-Gr blend are: Chemical Dissolution with Proportion 75% ABS and 25% Gr by weight having in-fill Density 100%

   The best settings for porosity analysis of ABS-Gr blend are: Chemical Dissolution with Proportion 75% ABS and 25% Gr by weight having in-fill Density 50%

   The best settings for good hardness of ABS-Gr blend are: Chemical Dissolution with Proportion 90% ABS and 10% Gr having in-fill Density 100%
2. For electrical conductivity, proportion of Gr in ABS matrix (by weight) is the significant factor. Similarly for thermal conductivity, process of blending is the significant factor in developing ABS-Gr blend.
3. As regards to porosity, process of blending is the significant factor in developing ABS-Gr blend. For shore hardness, in-fill density is the significant factor in developing ABS-Gr blend.
4. Based on optical micrographic views, the best porosity results are obtained in the chemical dissolute sample with proportion 75:25 and density 50% while the best porosity results in case of mechanically blended samples was with the proportion 75:25 and density of 50%. As observed, sample 2 has minimum porosity, which may have contributed for better thermal and electrical properties.
5. The glass transition temperature shows that the observed values in chemically blended samples are much closer to the glass transition temperature of ABS i.e. 105 °C, therefore, these samples can be successfully recycled.
6. Finally the functional prototypes so prepared can be used as sensors and actuators as the thermal and electrical properties have been incorporated in the polymeric materials.

**Acknowledgements** The authors are thankful to DST (GOI) for financial support and Manufacturing Research Lab (Dept. of Production Engg., Guru Nanak Dev Engg. College, Ludhiana, India) for technical support.

## References

1. Van Krevelen DW, TeNijenhuis K. Properties of polymers: their correlation with chemical structure; their numerical estimation and prediction from additive group contributions. Elsevier; 2009 Feb 9.

2. Shirakawa H, McDiarmid A, Heeger A. Twenty-five years of conducting polymers. Chemical Communications. 2003;2003(1):1–4.
3. Difallah BB, Kharrat M, Dammak M, Monteil G. Mechanical and tribological response of ABS polymer matrix filled with graphite powder. Materials & Design. 2012 Feb 29;34:782–7.
4. Han Z, Fina A. Thermal conductivity of carbon nanotubes and their polymer nanocomposites: a review. Progress in polymer science. 2011 Jul 31;36(7):914–44.
5. Novoselov KS, Geim AK, Morozov SV, Jiang D, Zhang Y, Dubonos SV, Grigorieva IV, Firsov AA. Electric field effect in atomically thin carbon films. science. 2004 Oct 22;306(5696):666–9.
6. Geim AK. Graphene: status and prospects. science. 2009 Jun 19;324(5934):1530–4.
7. Sharma V, Goyal M, Jindal P. Preparation, characterization and study of mechanical properties of graphene/ABS nano-composites. Indian Journal of Science and Technology. 2017 May 5;10 (17).
8. Singh VK, Shukla A, Patra MK, Saini L, Jani RK, Vadera SR, Kumar N. Microwave absorbing properties of a thermally reduced graphene oxide/nitrile butadiene rubber composite. Carbon. 2012 May 31;50(6):2202–8.
9. Bai X, Wan C, Zhang Y, Zhai Y. Reinforcement of hydrogenated carboxylated nitrile–butadiene rubber with exfoliated graphene oxide. Carbon. 2011 Apr 30;49(5):1608–13.
10. ASTM A. D1238-Standard test methods for melt flow rates of thermoplastics by extrusion plastometer. Annual Book of ASTM Standards; American Society for Testing and Materials: Conshohocken, PA, USA. 2013.
11. Singh R, Sandhu GS, Penna R, Farina I. Investigations for Thermal and Electrical Conductivity of ABS-Graphene Blended Prototypes. Materials. 2017 Jul 31;10(8):881.
12. Potts JR, Dreyer DR, Bielawski CW, Ruoff RS. Graphene-based polymer nanocomposites. Polymer. 2011 Jan 7;52(1):5–25.
13. Meng F, Huang F, Guo Y, Chen J, Chen X, Hui D, He P, Zhou X, Zhou Z. In situ intercalation polymerization approach to polyamide-6/graphite nanoflakes for enhanced thermal conductivity. Composites Part B: Engineering. 2017 May 15;117:165–73.
14. Ivanova O, Williams C, Campbell T. Additive manufacturing (AM) and nanotechnology: promises and challenges. Rapid Prototyping Journal. 2013 Jul 26;19(5):353–64.
15. http://librarygndec.blogspot.in/2017/08/new-mtech-thesis-submitted-from.html
16. https://www.astm.org/Standards/E2015.htm

# Investigate the Effects of the Laser Cladding Parameters on the Microstructure, Phases Formation, Mechanical and Corrosion Properties of Metallic Glasses Coatings for Biomedical Implant Application

**Mahmoud Z. Ibrahim, Ahmed A. D. Sarhan, M. O. Shaikh, T. Y. Kuo, Farazila Yusuf, and M. Hamdi**

## 1 Introduction

Metallic glasses (MG) or amorphous metals are the description of non-crystalline metals. Conventionally, metals tend to form crystalline structure during transition from liquid to solid state. For MG, the atoms are randomly arranged in the matrix showing no crystals or no ordered-arrangement of atoms [3]. In Fig. 1, a scanning electron microscope (SEM) image showing the difference between the amorphous and crystalline phases of Zr-based MG sample, and Fig. 2 shows the corresponding X-ray diffraction pattern (XRD) [4]. The MG systems have superior mechanical properties and promising tribological and corrosion resistance ability [5]. These superior properties are owing to the homogeneous single phase structure (amorphous structure), no grain boundaries, no dislocations and other crystal defects [5–7].

M. Z. Ibrahim
Centre of Advanced Manufacturing and Material Processing, Department of Mechanical Engineering, Faculty of Engineering, University of Malaya, Kuala Lumpur, Malaysia

Department of Mechanical Engineering, Southern Taiwan University of Science and Technology, Taiwan, Republic of China

Department of Mechanical Engineering, Faculty of Engineering, Ain Shams University, Cairo, Egypt

A. A. D. Sarhan (✉)
Department of Mechanical Engineering, King Fahd University of Petroleum and Minerals, Dhahran, Saudi Arabia
e-mail: ahsarhan@kfupm.edu.sa

M. O. Shaikh · T. Y. Kuo
Department of Mechanical Engineering, Southern Taiwan University of Science and Technology, Taiwan, Republic of China

F. Yusuf · M. Hamdi
Centre of Advanced Manufacturing and Material Processing, Department of Mechanical Engineering, Faculty of Engineering, University of Malaya, Kuala Lumpur, Malaysia



B. AlMangour (ed.), *Additive Manufacturing of Emerging Materials*,
https://doi.org/10.1007/978-3-319-91713-9_10

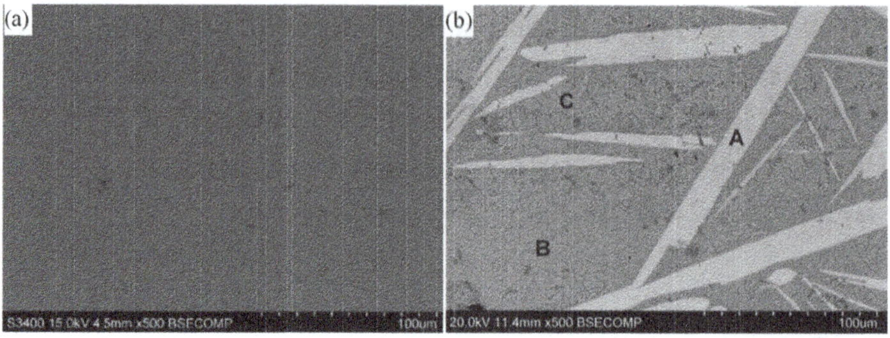


**Fig. 1** SEM image of Zr-based MG (**a**) full amorphous phase, (**b**) Combined amorphous phase (A) with crystalline phases (B,C) [4]

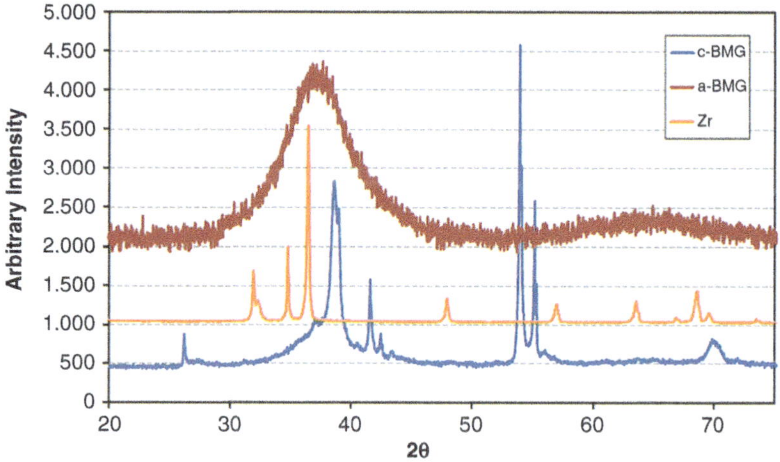


**Fig. 2** XRD pattern of three different Zr alloy sample, Red: full amorphous phase, Blue: combined amorphous and crystalline phases, Orange: full crystalline phase

## *1.1 Evolution of MG*

In 1960, the first MG was proposed. The MG was produced under super cooling rates (cooling rate reached $10^6$ K/s) to preserve the amorphous structure of liquid metal. Because of the challenging requirements of the excessive cooling rates, only thin ribbons (~10 μm) and limited dimensions could be produced [3]. Chen and Turnbull performed detailed investigations on the crystallization process of MG and later they succeeded to extend the critical thickness of the MG castings to 1 mm [8].

Further researcher focused on increasing the critical casting thickness of the MG until Peker and Johnson introduced Vitreloy 1 (Zr41.2Ti13.8Cu12.5Ni10Be22.5) with high glass forming ability (GFA) at relatively low cooling rates (10 K/s or less)

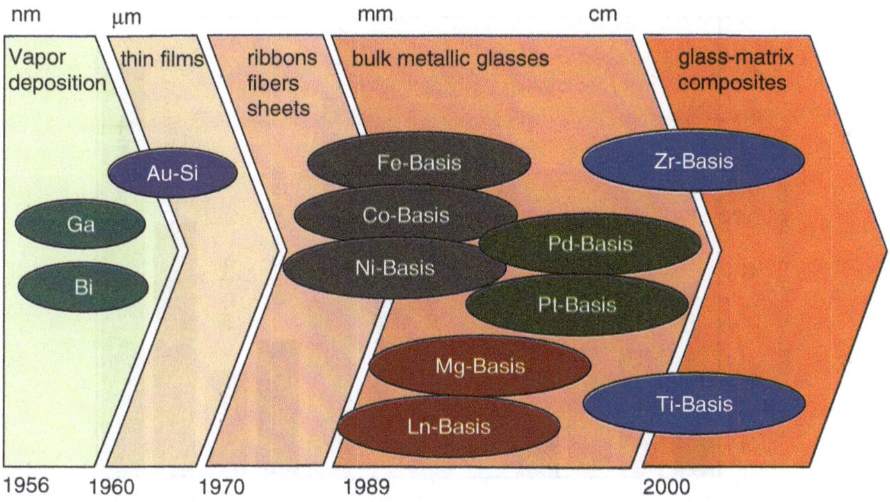


**Fig. 3** Historic advances in MG systems [11]

[9]. Recent researches reported that higher GFA can be achieved (from 0.1 K/s to 100 K/s) by applying the following rules; (1) multicomponent MG system (three or more elements), (2) significant differences in the atomic sizes of the main elements, (3) deep eutectics in the phase diagram of the alloys, (4) large difference between the glass transition temperature ($T_g$) and crystallization temperature ($T_x$) [10]. These rules enabled the fabrication of larger MG objects (up to few centimeters) and emerged them in structural applications. Figure 3 shows the development of MG systems in last few decades.

For various applications and different industries, different metal-based MG systems were proposed. The most common MG systems are the Fe, Zr, Ti, Co, Ni, Pt, Mg, Pd and Cu-based MG. Figure 4 shows the maximum diameter – which can represent the GFA – of different MG systems that can be fabricated with the corresponding discovery year [12].

## *1.2 Mechanical Properties of MG*

As mentioned above, MG exhibits superior mechanical properties and tribological and corrosion characteristics compared to the conventional known crystalline metallic alloys. Because of the excellent corrosion resistance of MG and proved biocompatibility, MG is considered a promising candidate for biomedical application [13]. The mechanical properties of common MG systems in biomedical applications are stated in Table 1. It is noted that MG systems exceeds the mechanical properties of the common metallic alloys. It is worth mentioning that Zr-based MG showed

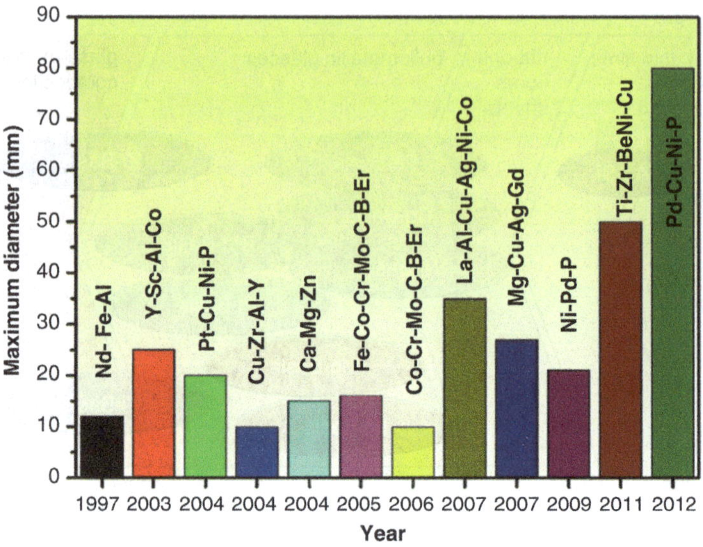


**Fig. 4** Different MG systems with the corresponding maximum diameter and discovered year [12]

**Table 1** Mechanical properties of common biomedical MG systems compared to the most popular biomedical metallic alloys [2, 13, 14]

| Material | Critical diameter, mm | Compressive strength/ (UTS), MPa | Young's modulus | Vickers's hardness, $H_v$ (HRC) |
|---|---|---|---|---|
| Ti-alloy | | (241–896) | 40–115 | (30–39) |
| Co-alloy | | (650–1900) | 200–240 | (40–47) |
| Stainless steel | | (490–1700) | 200–220 | (25–50) |
| Ti-based MG | 2–5 | 1200–2640 | 80–119 | 530–816 |
| Zr-based MG | 3–20 | 1450–2158 | 68–102 | 411–590 |
| Fe-based MG | 2–16 | 2500–4200 | 176–220 | 845–1253 |
| Ca-based MG | 0.9–6 | 364–600 | 19.4–46 | 0.7–1.42 |
| Mg-based MG | 0.025–6 | 550–848 | 35–66 | 2.16–4 GPa |
| Cortical bone | | 50–150 | 1–20 | |

promising GFA which allow fabrication of larger objects, and in the other side, Ti-based MG has the lowest GFA. This results in high potential to apply Zr-based and Fe-based in biomedical implants. However, biodegradable MG (Ca- and Mg-based MG) has lower Young's modulus close to the cortical bone which reduce the

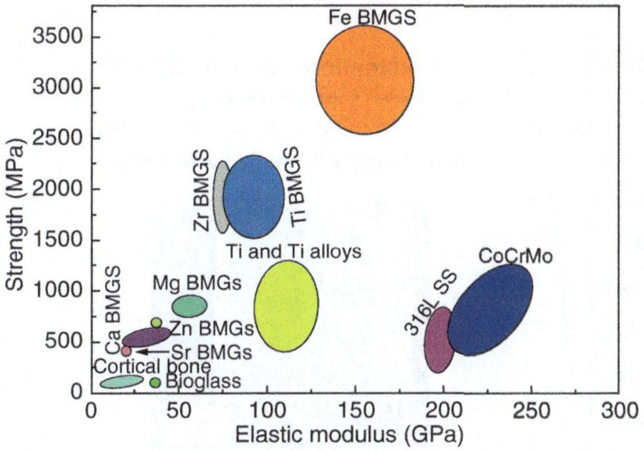


**Fig. 5** Graph comparing the different metallic alloys and MG used in biomedical applications

effect of stress shielding, but they exhibit relatively low strength compared to the other MG systems. This brings such MG to more investigation to ensure durability and acceptable performance when implanted. Figure 5 shows graphical comparison between the different metallic alloys and MG used in biomedical applications.

Generally, all MG suffer from brittleness and low toughness compared to crystalline alloys. This drawback hinders the application of MG and represent challenging for fabricating MG. So, great effort was focused on toughening MG through developing mixed structure (amorphous structure with dendrite crystalline structure). This mixed structure – usually called dendrite reinforced amorphous composite – can enhance the Poisson's ratio and the increase the compressive ductility by 30%, [15]. These advances promote the applications of MG in biomedical implants products (Fig. 6).

## *1.3 Biocompatibility of MG Systems*

According to D. Williams, biocompatibility is defined as "the ability of a biomaterial to perform its desired function with respect to a medical therapy, without eliciting any undesirable local or systemic effects in the recipient or beneficiary of that therapy, but generating the most appropriate beneficial cellular or tissue response in that specific situation, and optimizing the clinically relevant performance of that therapy" [16]. From this definition, a biocompatible material should meet the following requirements;

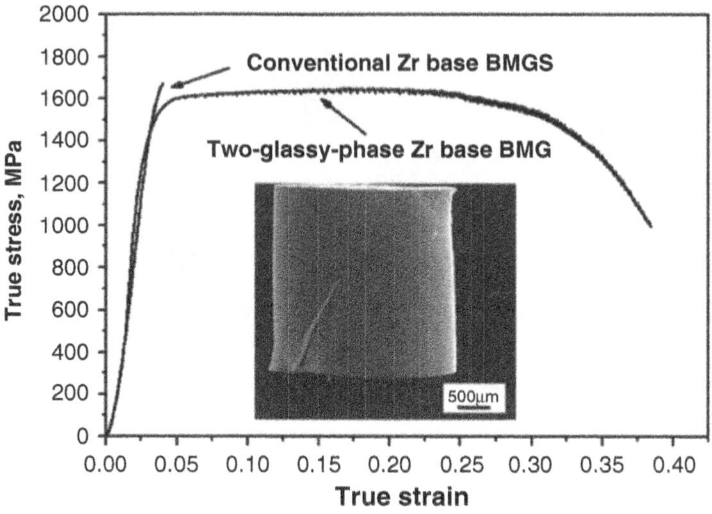


**Fig. 6** Enhancement of compressive strain of Zr-based MG [15]

1. High corrosion resistance, the material should exhibit lower ion release within the hosting body, so reduce the hazardous effect of released ions, and hence, prevent further side-effects of these ions.
2. Non-toxic, the material is not considered as cytotoxic, in other words, the material used is not destroying cells and living tissues in contact
3. Allergy-free, the material should be accepted by the hosting body without irritation or inflammation.
4. Suitable mechanical properties, to be able to withstand the conditions of different loadings during operation.
5. And acceptable osteointegration, in terms of ability to bond with the living tissue.

Generally, metallic alloys were considered to have acceptable biocompatibility. However, all the developed metallic alloys failed to show excellent durability in bone-implants applications. In the other side, MG systems proved to be an excellent substitute of conventional metallic alloys.

Generally, most of the MG systems exhibit high corrosion resistance due to the single glassy phase as mentioned above [7]. X. Lan et al. reported the enhanced tribological properties of Ti-based MG specially after the addition of Sn. The results showed high hardness values which indicate excellent wear resistance [17]. A cytotoxicity evaluation study revealed the excellent biocompatibility of Fe-based MG and revealed the promising potential in bone-implants and dental applications, [18]. Many other studies proved the excellent biocorrosion properties of MG and their potential in different biomedical applications [19–22] (Fig. 7).

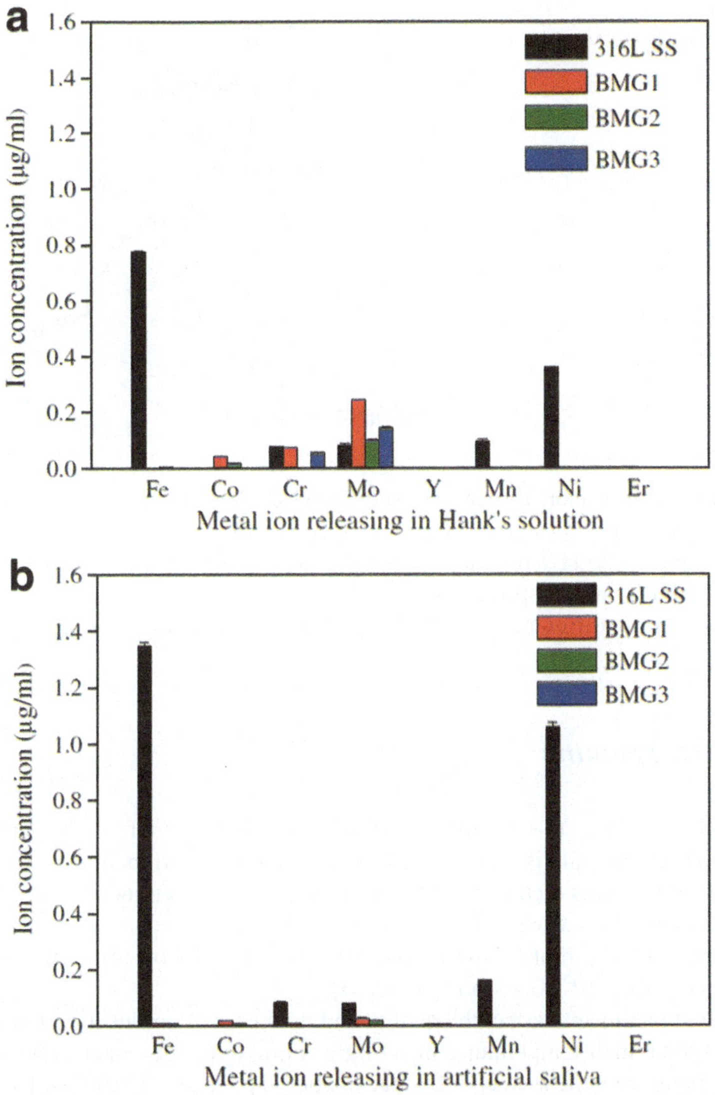


**Fig. 7** Ion-release of three different Fe-based MG and 316 L SS in (**a**) Hank's solution, (**b**) artificial saliva [18]

## 2 Techniques to Fabricate MG

Different techniques were developed to fabricate metallic glass objects. These techniques were subjected to exhibit rapid cooling sufficient to form the amorphous structure. Starting from micron-sized thickness ribbons to cylinders with few

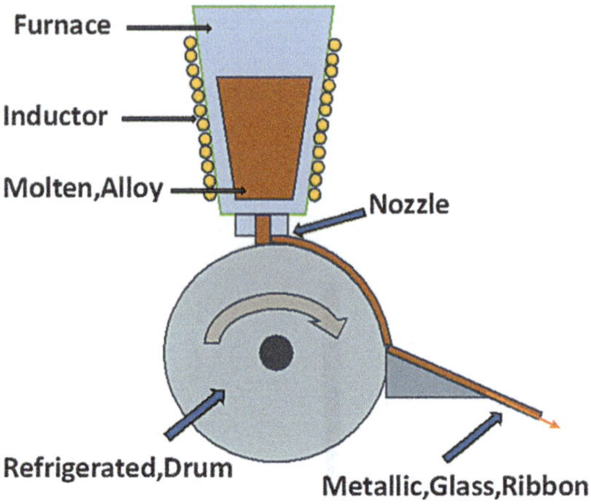


**Fig. 8** Melt spinning technique to produce MG ribbon [25]

centimeters in diameter, researchers succeeded to develop many fabrication techniques to produce MG products. GFA of available MG systems still lack to produce larger size products. This limitation represents a challenging research area for further emerging MG in more applications.

In this section, different available MG production techniques are discussed.

## 2.1 *Melt Spinning*

Melt spinning is the most common technique to produce high quality amorphous ribbons. This technique is able to produce continuous thin ribbon (20–30 μm) with different widths. This technique is based on injecting molten metal through certain nozzle onto rotating wheel. The wheel is made from copper, so it allows rapid solidification of the molten metal forming very thin ribbon. This technique can achieve more than $10^5$ K/s cooling rate [23].

Different parameters affect the quality and thickness of produced ribbons. Wheel rotating speed, melt temperature, injecting gas pressure, and nozzle-wheel gap are the main parameters. It was reported that smaller gap results in better surface quality of ribbons and hence, higher degree of amorphous structure [24] (Fig. 8).

## 2.2 *Casting*

Typical casting of MG using water-cooled copper molds is commonly used to produce MG ingots. The process is based on melting the MG in inert gas atmosphere – to prevent further oxidations – using electric arc, then inject the melt into copper

**Fig. 9** Copper-mold injection casting used for MG production

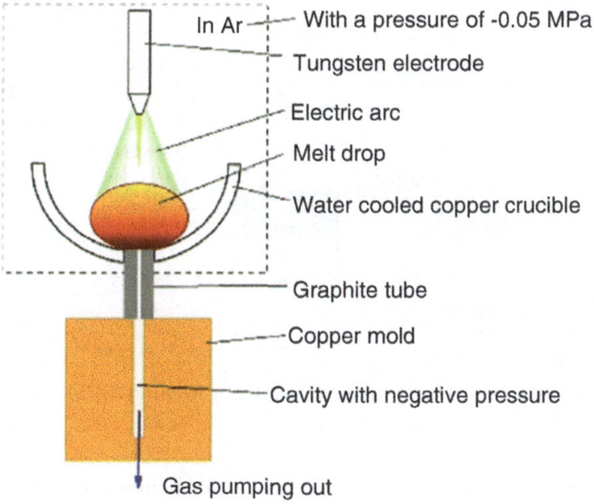


**Fig. 10** Schematic of continuous casting system proposed by T. Zhang et al

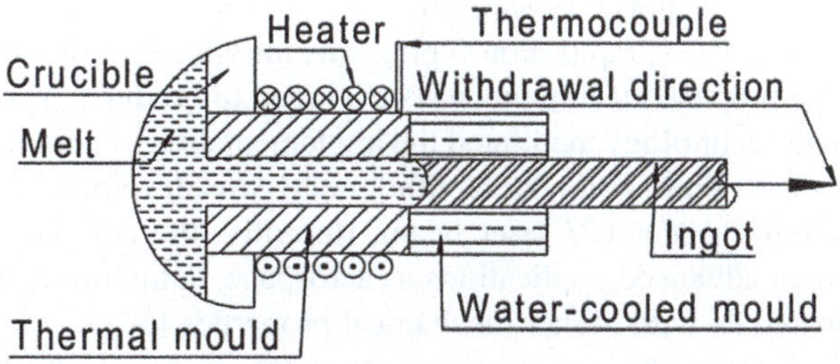


mold through negative gas pressure. The copper mold facilitates high cooling rates sufficient for glassy phase formation [26]. Figure 9 shows schematic of copper-mold casting technique.

T. Zhang et al. developed a novel technique for continuous casting of MG. The authors succeeded to produce Zr-based MG rod with 10 mm diameter rod and several tens centimeters long. The authors proposed horizontal continuous casting system equipped with water-cooled mold at the exit as shown in Fig. 10 [27].

## 2.3 *Additive Manufacturing*

Additive manufacturing (AM) – or 3D printing – propose fabrication technique based on adding successive layers to build objects. The process is categorized into three categories: powder-based, liquid-based and solid-based [28]. The powder and solid-based process can be used for almost all materials, while the liquid-based is limited to certain photosensitive polymers. However, powder-based and liquid-

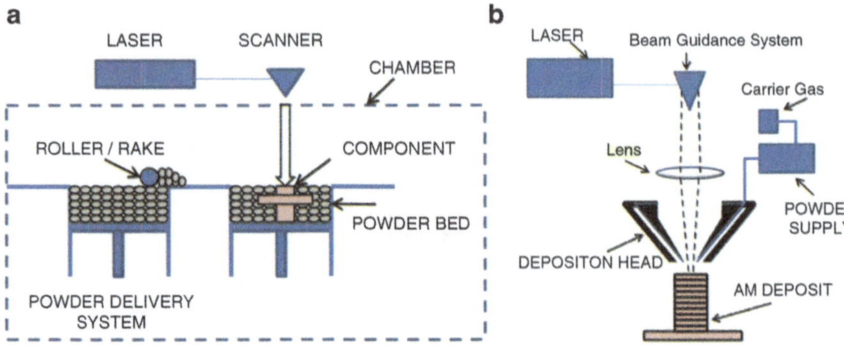


**Fig. 11** Principle of (**a**) Selective Laser Melting (SLM), and (**b**) Laser Engineering Net Shaping [32]

based are attracting more attention because of higher accuracy and precision. AM has become of great interest because of their relatively lower cost – compared to conventional manufacturing techniques – flexibility, and ability to produce complex objects in shorter times [29].

In 1980's, rapid prototyping was introduced as the first form of AM technology. The proposed technique was developed to produce 3D objects using CAD software. This technology facilitated the implementation of complex designs with low cost and in short time. Later, stereolithography was developed employing polymer resin that solidify under UV laser beam. Recently, the AM is developing and extending to more advanced applications as aerospace, automotive, biomedical, etc. that requires materials with higher mechanical properties [30].

Commonly, AM utilizes laser energy to melt and sinter printing material during building the model. Selective laser melting (SLM), laser engineering net shaping (LENS) and electron beam melting (EBM) are the common techniques for AM. These methods can process metals, polymers, ceramics or blend of them.

SLM is based on employing laser beam to melt and sinter powder layer with specified path spread on a piston. The piston movement – upward – determines the layer thickness and excessive powder is removed by the end of process [31]. In the other side, LENS inject material powder into the melting pool, so no removal of excessive powder is needed, Fig. 11 shows the principle of SLM and LENS process.

In EBM method, an electron laser beam powered with high voltage (30–60 kV) in a vacuumed chamber to avoid oxidation. The process uses two magnetic coils to focus and deflect the electron beam respectively (Fig. 12).

### 2.3.1 Laser Cladding Technique for MG Coatings

Laser cladding is another technique of AM where a single or multiple layer of material is developed on the surface of substrate. The key advantage of laser cladding is the strong metallurgical bond built between the successive layers. This

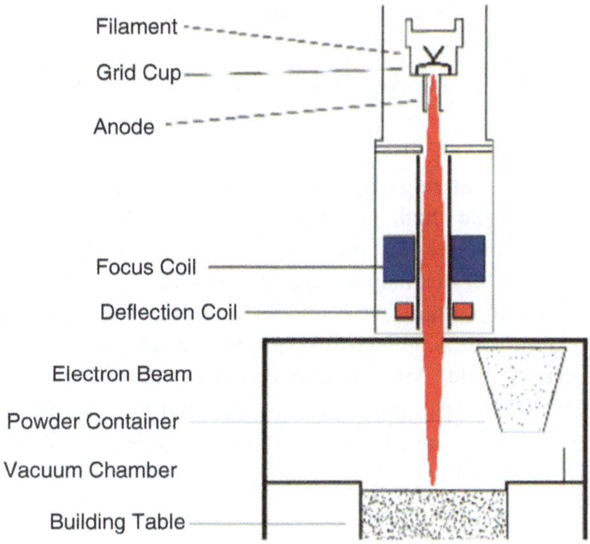


**Fig. 12** Schematic showing the EBM setup and principle [33]

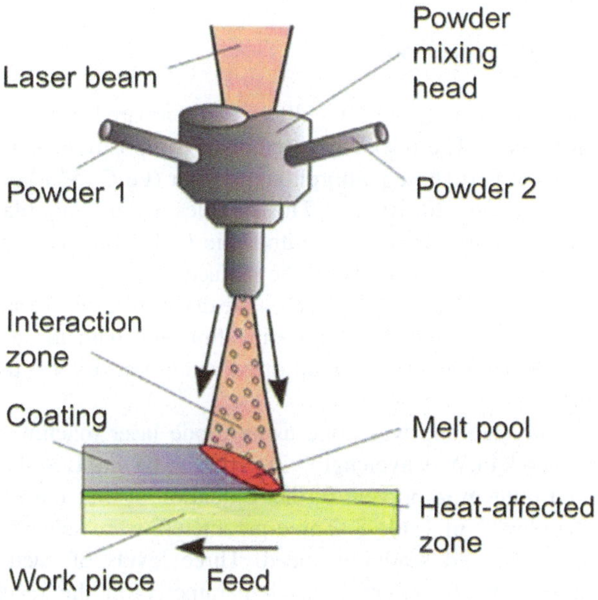


**Fig. 13** Powder-feed laser cladding technique [34]

leads to dense structure of the coating with excellent mechanical properties. The metallurgical bond is built as the laser beam penetrates the layer and the substrate, allowing dilution take place at the interface [34]. In addition, laser cladding offers versatile and productive method, besides it can be utilized to repair expensive metallic parts [35] (Fig. 13).

**Table 2** Elemental composition of ASTM F2229 substrate

| Material | Fe | Cr | Ni | Mo | Mn | Si | Cu | N | C | P | S |
|---|---|---|---|---|---|---|---|---|---|---|---|
| ASTM F2229 | Balance | 20.00 | 0.10 | 21.00 | 1.00 | 0.75 | 0.25 | 0.90 | 0.08 | 0.03 | 0.01 |

Laser cladding can work with powder-bed or powder-feed technique. The results proved that both techniques do not affect the resulted coating layer [36]. The parameters affecting the properties of the coating layer are laser power, scanning speed, laser beam spot size, layer thickness and the overlap percentage [37, 38].

An important factor for the amorphous structure is the cooling rate. A critical cooling rate is defined for each MG system depending on the composition and the stability of the MG. As laser cladding offers high cooling rate, but it is important to control the laser parameters to maintain the amorphous phase. An experiment was performed to investigate the effect of the laser parameters on the properties of laser cladded MG.

### 2.3.2 Experimental Procedures

Sample Preparation

Substrate samples of 30 × 30 mm are prepared from 3 mm ASTM F2229 stainless-steel sheet. The respective elemental composition is listed in Table 2. The coating material is Fe-based amorphous powder (Fe-Cr-Mo-B-C) with normally distributed particle size 20–70 μm. The samples were sand blasted using 40-grit alumina particles, then cleaned in ultrasonic bath using DI-water then ethanol to remove dirt and contaminations on the surface.

The powder was preplaced – the dry powder is spread on the surface then ethanol is added and put the sample in ultrasonic bath to ensure uniform distribution of powder on the surface, then dried using hair dryer – using 0.22 g/cm$^2$ spread rate, Fig. 14.

The cladding was done using diode laser machine – LDF 4000–60 maximum power 4.4 kW, wavelength 978–1025 ± 10 nm, LaserLine, Germany – under Ar gas environment to prevent further oxidation of the coating layer following the parameters levels in Table 3. Three parameters were selected for the study; laser power, spot size and scanning speed. Three levels of each parameter were chosen to investigate the effect of each parameter on the resulted microstructure, phases formation, hardness and the corrosion resistance of the Fe-based MG coating. The selected parameters levels are based on the preliminary experiments held.

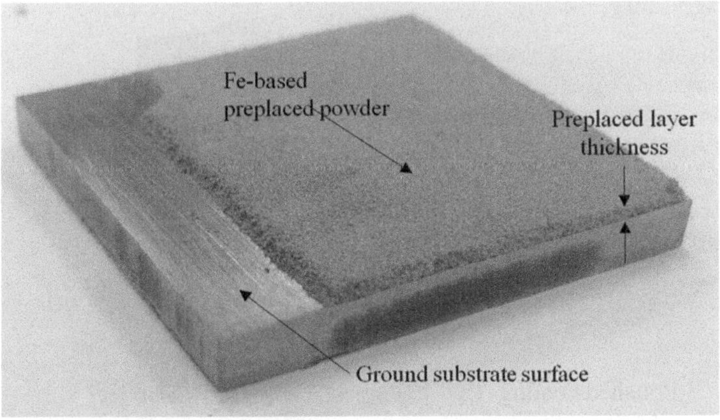


**Fig. 14** Preplaced Fe-based amorphous powder

**Table 3** Selected parameters to fabricate Fe-based MG coatings

| Sample code | Laser power, W | Spot size, mm | Scanning speed, mm/s |
|---|---|---|---|
| A1 | 1500 | 4 × 4 | 15 |
| A2 | 1500 | 4 × 4 | 20 |
| A3 | 1500 | 4 × 4 | 25 |
| B1 | 2000 | 4 × 4 | 40 |
| B2 | 2000 | 4 × 4 | 45 |
| B3 | 2000 | 4 × 4 | 50 |
| D1 | 4000 | 4 × 6 | 30 |
| D2 | 4000 | 4 × 6 | 35 |
| D3 | 4000 | 4 × 6 | 40 |
| E1 | 4000 | 4 × 8 | 20 |
| E2 | 4000 | 4 × 8 | 25 |
| E3 | 4000 | 4 × 8 | 30 |

### Characterization and Testing

For X-ray diffraction (XRD) analysis, a sample 15 × 15 mm was cut from each sample. The surface of the coating layer was flattened using SiC 80-grit sandpaper, then ground using 150-grit to 4000-grit and polished using 40 nm alumina suspension until mirror-like surface is obtained. Then, the samples are cleaned in ultrasonic bath using DI water and ethanol. The analysis was done using Bruker D2 Phaser, Germany using Cu $K_\alpha$ $\lambda = 0.154060$ nm, 2θ range of 0~140° and scanning rate 3° $min^{-1}$.

SEM imaging (Hitachi S-3000N, Tokyo, Japan) was used to analyze the microstructure at the interface and at the coating layer, with ×1.5k magnification. The hardness measurements were taken along the cross-section at 0.1 mm distance

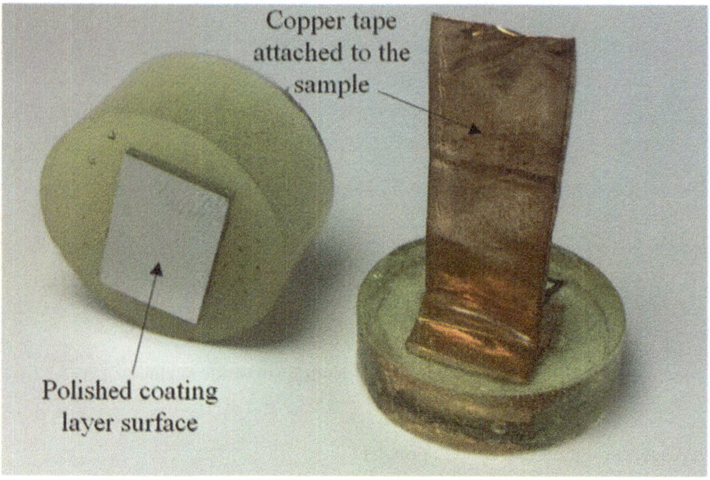


**Fig. 15** Prepared sample for electrochemical corrosion test

between each successive point at 100 g testing load and 10 sec. Dwell time using Vicker's Hardness tester.

An electrochemical corrosion – Potentiodynamic polarization – test was performed in Ringer's solution (pH 7.4) as electrolyte at 37 °C to test the corrosion resistance of the Fe-based BMG coating and compare it to the bare stainless steel (Cronidur 30). The selected samples for corrosion resistance evaluation were chosen based on the best results of surface hardness, which is a measure of amorphous quality. The samples were prepared in epoxy mold as shown in Fig. 15 so the coating surface is only exposed during the test. The sample was set as working electrode (WE), Saturated Calomel Electrode (SCE) as reference electrode (RE) and platinum strip as counter electrode (CE). The open circuit potential was kept for 40–60 min until saturation attained. Then linear potential sweep was applied from −1.5 to 1.5 V with scanning rate 10 mV/s. The corrosion potential and current were obtained by Tafel extrapolation method.

### 2.3.3 Results

XRD Analysis

The obtained XRD patterns of the as received materials and samples are shown in Figs. 16 and 17. The analysis of XRD analysis included the analysis of multiple phases formed and the crystallinity content percentage. The crystallinity content percentage was obtained using HighScore Plus [39]. The XRD patterns showed that all samples exhibited amorphous structure – broad peak at 44° – with existence of

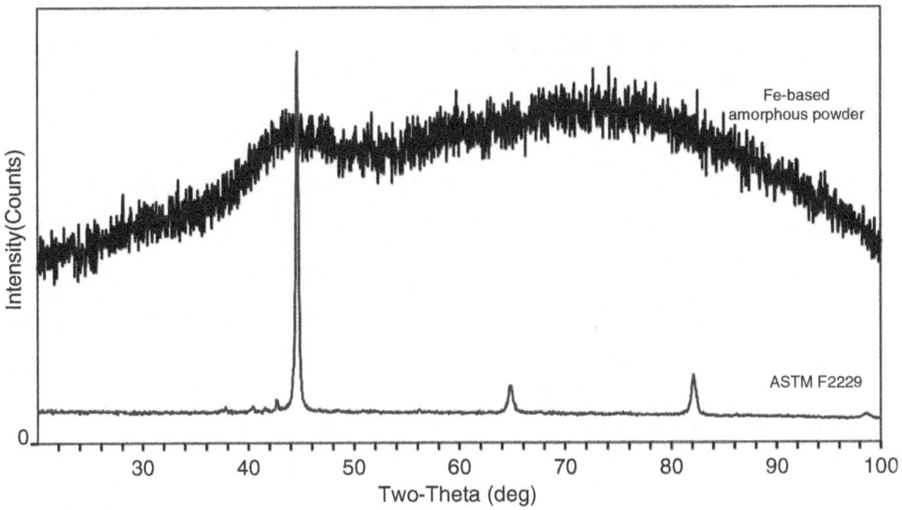


**Fig. 16** XRD pattern of Fe-based amorphous powder and ASTM F2229

crystalline metallic phases as Fe, Cr and ASTM F2229 besides intermetllic compounds as iron boron ($B_6Fe_{23}$), chromium carbide ($Cr_{23}C_6$) and iron chromium carbide. It is noticed in sample E1, the phase iron nitride (FeN) appeared. All recognized intermetallic compounds are featured as hard and stable compounds.

It is noticed that the crystallinity content percentage around 15 $\pm$ 1%, so no significant difference between the examined samples. However, sample E1 was found to have the highest crystallinity content (18.11%). Form previous experiment, it was shown that laser power has the most significant effect on the crystallinity content. Also, the increase in scanning speed would enhance the amorphous structure in coating layer, but over-increase of scanning speed results in unsuccessful coating.

The effect of laser power on the amorphous structure of coating layer can be explained as the laser power represents the total energy transmitted to the sample per unit time, which in turn determines the temperature of the melt pool. In the other side, the scanning speed is a measure of the heating and cooling rate involved during the laser cladding process. Hence, laser power and scanning speed has an interrelation effect on the formed structure within coating layer.

The spot size is inversely proportional to the transmitted energy to the sample. In laser cladding applications, increasing spot size would enhance the coverage efficiency, but in the other side requires more power to compensate the decrease in the energy transmitted. So, power density can describe the interrelation effect between the laser power and spot size described in Eq. 1.

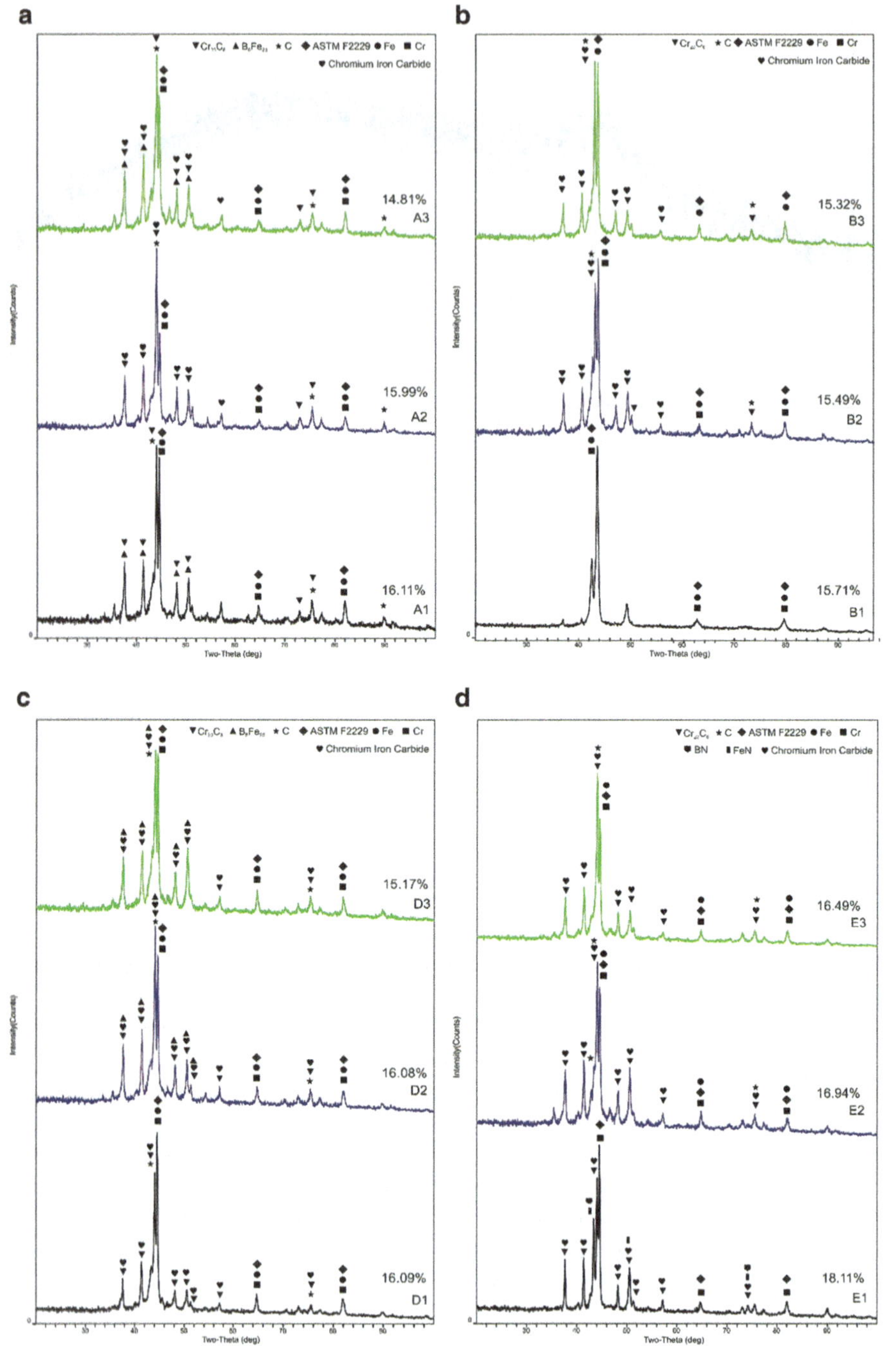


**Fig. 17** XRD pattern (**a**) samples A1, A2, A3, (**b**) samples B1, B2, B3, (**c**) samples D1, D2, D3, (**d**) samples E1, E2, E3

$$\text{Power density}\left(\text{W/mm}^2\right) = \frac{\text{Power (W)}}{\text{Spot size (mm}^2)} \tag{1}$$

Based on that, sample E1 and sample B1 have the same power density with different scanning speed – 20 and 40 respectively. The XRD results revealed that sample B1 exhibited lower crystallinity (15.71%) content and the recognized phases are Fe, Cr and ASTM F2229 only, while sample E1 exhibited higher crystallinity content (18.11%) and the recognized phases are iron-nitride (FeN), $Cr_{23}C_6$ and iron chromium carbide phases. From these findings, we can conclude that scanning speed affect the crystalline phases formed, besides the crystallinity content.

It can be concluded that laser power, scanning speed and spot size have significant effect on the formed phases. In addition, the three parameters have interrelation effect of the fromed phases in the Fe-based MG coating.

## Microstructure Examination

Figure 18 shows the SEM image of the microstructure taken at the interface. As shown, all the samples showed crystalline phases embedded in amorphous structure appeared as plain grey surface. Equiaxed grains appeared in all samples which represents the Cr and Fe phases recognized in XRD. In addition, nano-crystals are recognized. However, in samples D1 and E1, a dentritic structure becomes dominant and clear which is referred to $Cr_{23}C_6$ and Fe-Cr-C. A columnar dendritic structure usually is a result of higher cooling rate of multi-element compounds which is present at the interface. It is noticed that the dentritic structure layer is decreasing with the increasing scanning speed [40].

Figure 19 shows the SEM image of the microstructure taken at the coating layer. The samples showed different microstructure from that obtained at the interface. Beside the dendritic structure – coarse and equiaxed – bright square-shaped crystals are existed. As mentioned before, the equiaxed grains appeared represent the soft crystalline phases (Fe and Cr), while the coarse dendritic crystals appeared in sample E1 correspond to $Cr_{23}C_6$ and Fe-Cr-C. The bright square-shaped crystals phases are corresponded to hard intermetallic compounds ($B_6Fe_{23}$, FeN) [41].

As revealed by SEM pictures, coarse dendritic crystals appeared in sample E1 which exhibited high crystallinity content (18.11%) – as shown in XRD. For samples E2 and E3, the dendritic structure layer becomes smaller and amorphous content increases as the scanning speed is increasing. The samples D1, D2 and D3 showed the same trend as samples E1, E2 and E3.

Although the power density and scanning speed of sample A2 and sample E1 are the same, it is found that sample E1 exhibited higher crystallinity content. This reveals the significance effect of the laser power.

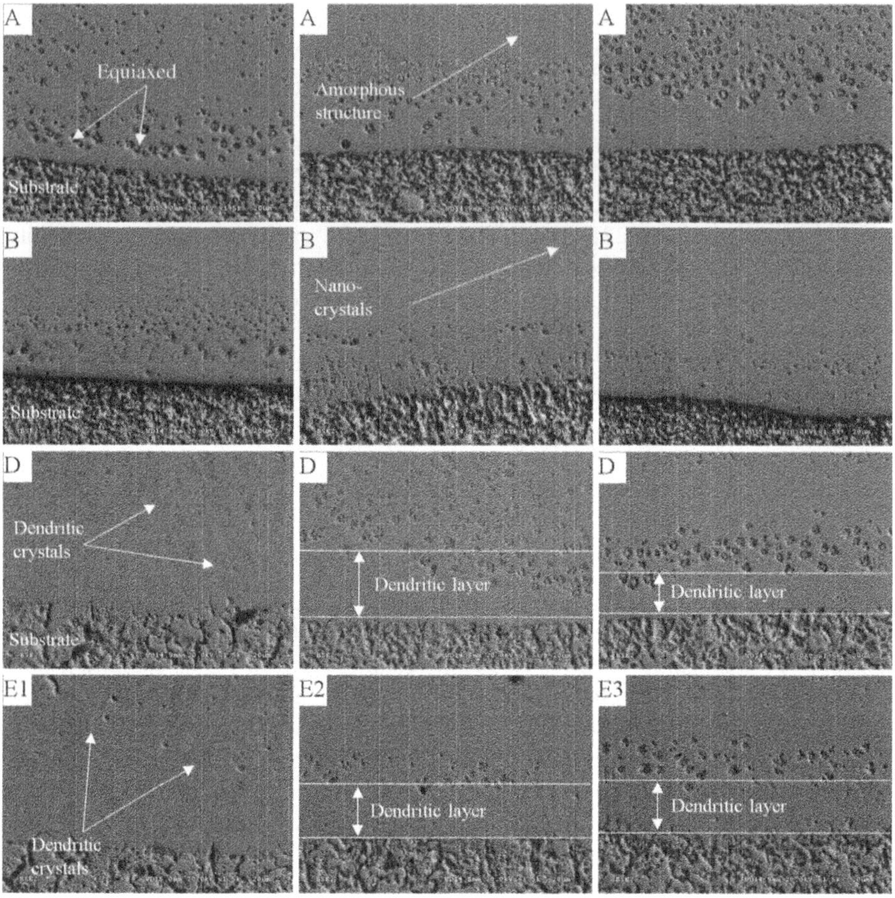


**Fig. 18** SEM image at the interface of fabricated samples

It is clear from the microstructure analysis that the laser power and scanning speed affects significantly the types and the shape of formed structure. The multiple phases formed in the samples exhibited different crystalline structure within the interface or the coating layer.

### Hardness Measurement

The hardness values in HV0. 1 of the surfaces are shown in Fig. 20. The hardness of the coating exceeded 1300 HV0.1 which is more than 5 times the hardness of the substrate (the substrate hardness is 230 HV0.1). The high value of hardness is referred to the amorphous structure at the coating layer. However, the presence of crystalline phase degrades the hardness depending on the formed phases. For example, samples B1 and B2 exhibited formation of Cr and Fe phases, and the

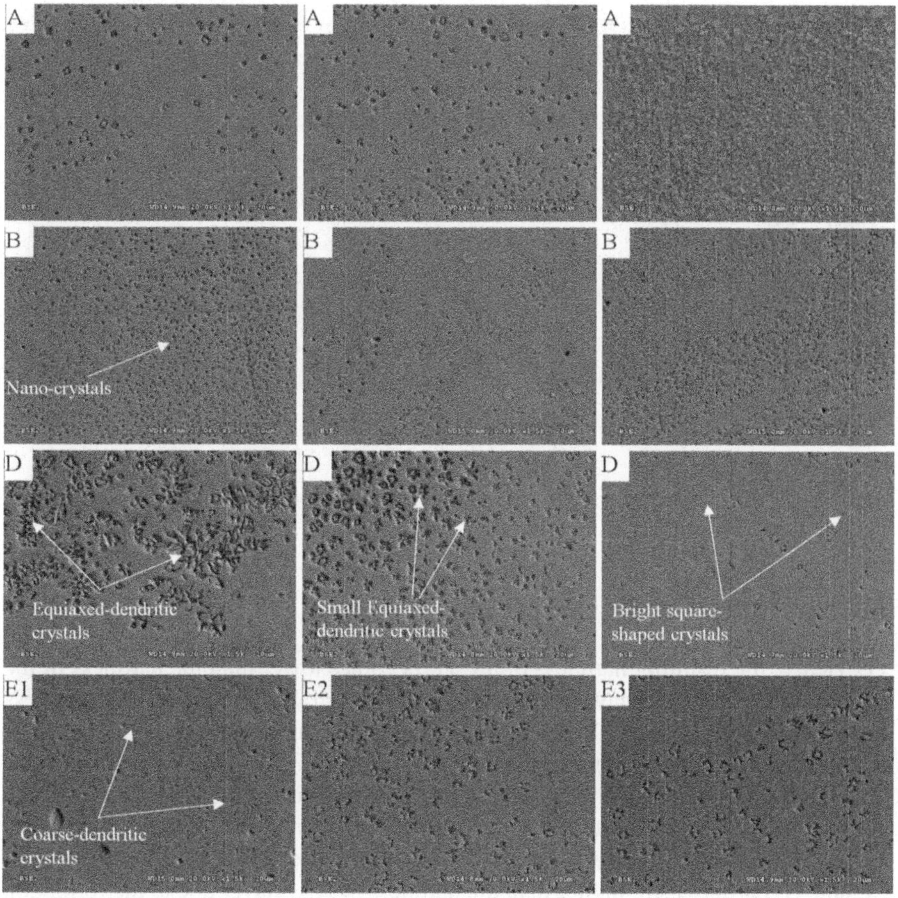


**Fig. 19** SEM image at coating layer of fabricated samples

hardness value is dropped below 600 HV0.1. While in samples D1 and E1, $Cr_{23}C_6$ and Fe-Cr-C was the dominant phases. So, the measured hardness was about 900 HV0.1. this can be understood as the carbides are harder than the pure metal phases as Cr and Fe.

For more investigations, a profile hardness value along the cross-section for each sample is revealed in Fig. 21. From the hardness measurement, four layers are found along the cross-section; coating layer (CL), transition layer (TL), heat affected zone (HAZ) and substrate. The CL is featured with high hardness above 1000 HV0.1 which indicates the amorphous structure existence. It is important to mention that all samples exhibited mixed structure at CL which revealed in the microstructure images. That explains why the hardness value is fluctuating in the CL.

The hardness value is dropped (below 800 HV0.1) in the transition layer because of the higher content of crystalline phases. In addition, hard phases – $Cr_{23}C6$, $B_6Fe_{23}$, Cr-Fe-C, FeN – are found in CL, while Fe and Cr are found in the

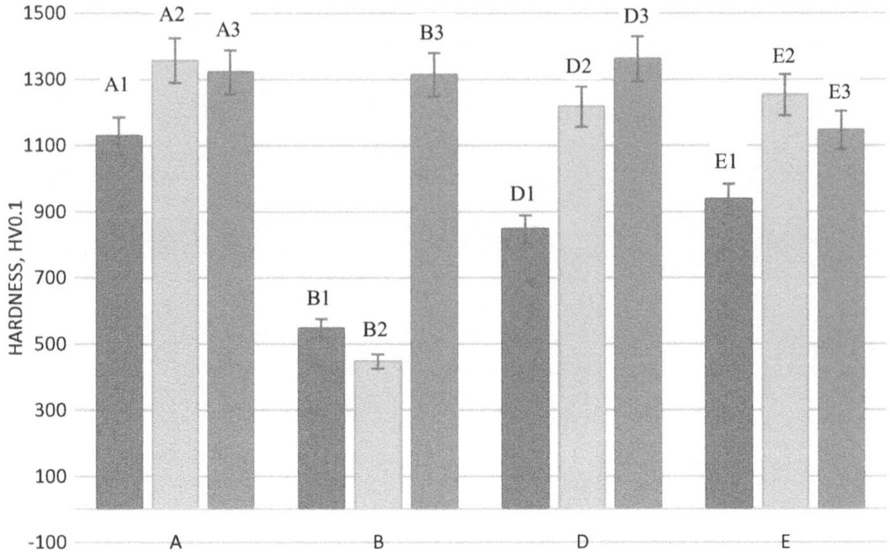


**Fig. 20** Surface hardness of Fe-based MG coating

TL. Hence, the resulting hardness of the TL becomes lower than the hardness of the CL. After the TL, HAZ layer – which is in the substrate – is found with hardness value below 500 HV0.1. The HAZ layer depth are variant according to the laser power density and the scanning speed.

It is noticed that both samples D1 and E1 exhibited almost equal hardness value at the CL and the TL with noticeable fluctuation. This result conforms with the XRD and microstructure investigations that showed higher crystallinity content. However, CL and TL hardness still higher than the hardness of HAZ layer due to the formation of hard compounds as well as the amorphous structure.

## Electrochemical Corrosion Test

The Potentiodynamic polarization curves for the Cronidur 30 substrate and the amorphous coatings are shown in Fig. 22. Furthermore, the values of corrosion potential ($E_{corr}$) and corrosion current ($i_{corr}$) can be estimated from the Tafel slope and are shown in Table 4.

The comparison of the polarization curves indicates that the Fe-based MG coatings have a higher corrosion potential as compared to Cronidur 30. Furthermore, they exhibit a significantly wider passive region. It is noted that samples A2 and D3 exhibited the highest open circuit potential while the sample E2 was the worst compared to the coated samples.

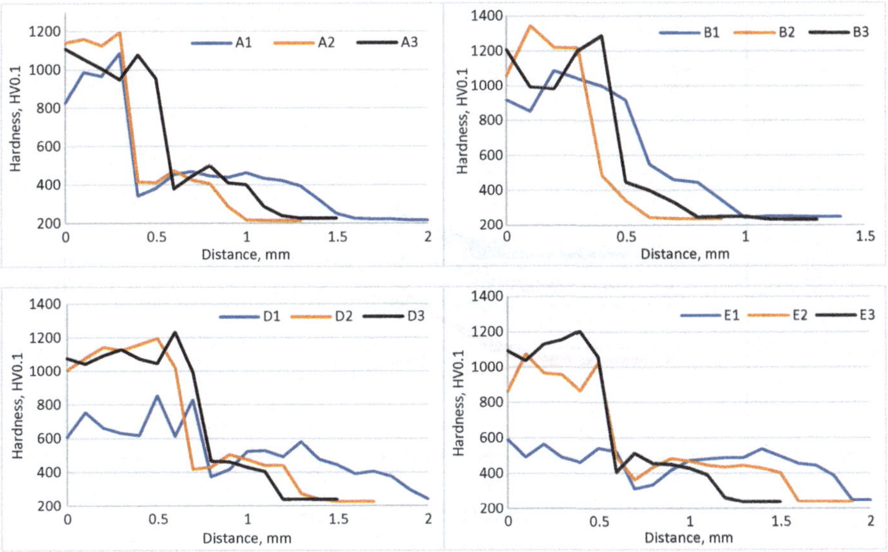


**Fig. 21** Profile hardness measured at the cross-section of the fabricated samples

These results correspond to the evaluation of corrosion resistance of the tested samples, i.e. sample A2 exhibited the best corrosion resistance and E2 exhibited the poorest corrosion resistance while samples B3 and D3 lies in-between. From these findings, it can be concluded that amorphous structure found in the Fe-based MG coating affects directly the corrosion resistance level. The excellent corrosion resistance of MGs is referred to the absence of the structure defects as grain boundaries which are weak and corrode rapidly [42].

As all samples exhibited crystalline phases with different percentages, it is noted that lower crystalline percentage showed better corrosion resistance. However, the type of crystalline phases has a significant effect on the corrosion resistance. For example, sample B3 has lower crystallinity content than sample A2, but sample B3 showed lower corrosion resistance. This can be understood from the XRD findings, which showed that hard compounds – $Cr_{23}C_6$ and Fe-Cr-C – were dominant in sample A2 while soft phases – Fe and Cr – were dominant in sample B3 (Table 4).

### 2.3.4 Conclusion

A successful Fe-based MG layer was developed on stainless-steel substrate using laser cladding technique. The fabricated samples were investigated using XRD, SEM, Micro-hardness and electrochemical corrosion test. The samples showed amorphous structure with different crystallinity content percentages.

The XRD analysis revealed the existence of various compounds as $Cr_{23}C_6$, $B_6Fe_{23}$, Fe-Cr-C besides pure Fe and Cr crystalline phases. Both the microstructure

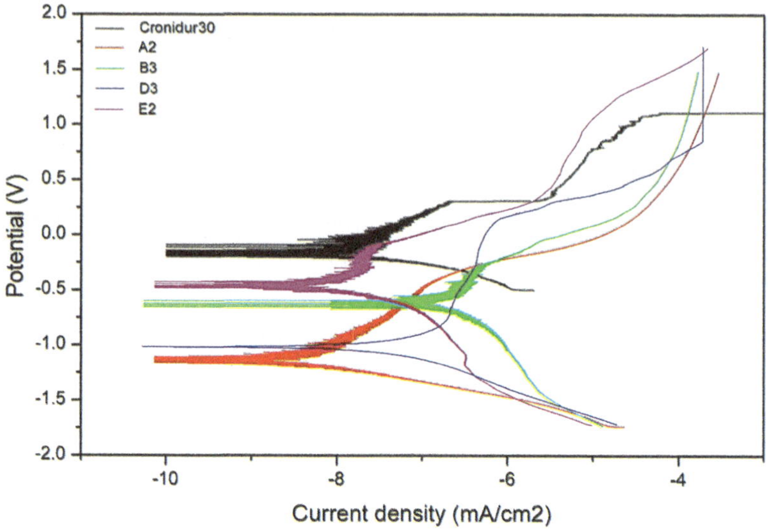


**Fig. 22** Potentiodynamic curves of the Cronidur 30 substrate and Fe-based MG coatings in Ringers solution

**Table 4** Corrosion potential and current for coated samples and uncoated substrate

| Sample | $E_{corr}$ (V) | $i_{corr}$ (mA/cm$^2$) |
|---|---|---|
| Cronidur30 | −0.186 | −7.8 |
| A2 | −0.942 | −6.8 |
| B3 | −0.462 | −6 |
| D3 | −0.846 | −6.3 |
| E2 | −0.407 | −7.2 |

analysis and Micro-hardness measurements confirmed the results obtained from the XRD analysis. The results showed that the amorphous structure as well as the multi-phases recognized affect the corrosion resistance and the hardness. So, the crystallinity content percentage is not the only factor, but also the type of the crystalline phases – hard or soft phases.

It is concluded that the laser power has the strongest effect on the properties of the fabricated coating layer. Also, scanning speed has a significant effect on the crystallinity content percentage. Increasing the scanning speed increases the cooling rate, hence better amorphous structure.

From the hardness measurement and the electrochemical corrosion test results, samples A2 (1500 W, 20 mm/s with spot size 4 × 4 mm) and D3 (4000 W, 40 mm/s with spot size 4 × 6 mm) exhibited the highest hardness value and the best corrosion resistance.

**Acknowledgements** The authors would like acknowledge Nippon Sheet Glass Foundation (Japan) for the grant provided, Energietechnik Essen GmbH for supplying ASTM F2229 (CORINDUR 30) free sample and LiquidMetal® Coatings for supplying Fe-based amorphous powder as free sample. Also, the authors would like to thank Southern Taiwan University for Science and Technology for providing the necessary facilities and resources to carry out the experimental work. Also, the authors would like to thank both University Malaya and King Fahd University of Petroleum & Minerals for providing financial and technical support.

## References

1. Jensen, W. (2018), Production of hip implant by using additive manufacturing Available at: https://www.eos.info/press/case_study/additive_manufactured_hip_implant [Accessed 8 Jun. 2018].
2. M. Z. Ibrahim, A. A. D. Sarhan, F. Yusuf, and M. Hamdi, "Biomedical materials and techniques to improve the tribological, mechanical and biomedical properties of orthopedic implants – A review article," *J. Alloys Compd.*, vol. 714, pp. 636–667, 2017.
3. W. KLEMENT, R. H. WILLENS, and P. DUWEZ, "Non-crystalline Structure in Solidified Gold–Silicon Alloys," *Nature*, vol. 187, no. 4740, pp. 869–870, Sep. 1960.
4. N. Espallargas, R. E. Aune, C. Torres, N. Papageorgiou, and A. I. Muñoz, "Bulk metallic glasses (BMG) for biomedical applications—A tribocorrosion investigation of Zr55Cu30Ni5Al10 in simulated body fluid," *Wear*, vol. 301, no. 1, pp. 271–279, 2013.
5. Y. Waseda and K. T. Aust, "Corrosion behaviour of metallic glasses," *J. Mater. Sci.*, vol. 16, no. 9, pp. 2337–2359, Sep. 1981.
6. A. L. Greer, T. Egami, T. Iwashita, and W. Dmowski, "Mechanical Properties of Metallic Glasses," *Metals (Basel).*, vol. 3, no. 1, pp. 77–113, 2013.
7. S. Wang, "Corrosion Resistance and Electrocatalytic Properties of Metallic Glasses," in *Metallic Glasses - Formation and Properties*, InTech, 2016.
8. J. F. Löffler, "Bulk metallic glasses," *Intermetallics*, vol. 11, no. 6, pp. 529–540, Jun. 2003.
9. A. Peker, W. L. Johnson, and M. Keck, "A highly processable metallic glass: Zr41.2Ti13.8Cu12.5Ni10.0Be22.5," *Cit. Appl. Phys. Lett. Appl. Phys. Lett*, vol. 63, no. 65, pp. 2342–2136, 1993.
10. Y.-L. Gao, J. Shen, J.-F. Sun, G. Wang, D.-W. Xing, H.-Z. Xian, and B.-D. Zhou, "Crystallization behavior of ZrAlNiCu bulk metallic glass with wide supercooled liquid region," 2003.
11. J. Eckert, "Application of amorphous alloys: potential and challenges to overcome."
12. J. Qiao, H. Jia, and P. K. Liaw, "Metallic glass matrix composites," *Mater. Sci. Eng. R Reports*, vol. 100, pp. 1–69, 2016.
13. H. F. Li and Y. F. Zheng, "Recent advances in bulk metallic glasses for biomedical applications.," *Acta Biomater.*, vol. 36, pp. 1–20, May 2016.
14. Q. Chen and G. A. Thouas, "Metallic implant biomaterials," *Mater. Sci. Eng. R Reports*, vol. 87, pp. 1–57, 2015.
15. J. C. Huang, J. P. Chu, and J. S. C. Jang, "Recent progress in metallic glasses in Taiwan," *Intermetallics*, vol. 17, no. 12, pp. 973–987, Dec. 2009.
16. D. F. Williams, "On the mechanisms of biocompatibility," *Biomaterials*, vol. 29, no. 20, pp. 2941–2953, Jul. 2008.
17. X. Lan, H. Wu, Y. Liu, W. Zhang, R. Li, S. Chen, X. Zai, and T. Hu, "Microstructures and tribological properties of laser cladded Ti-based metallic glass composite coatings," *Mater. Charact.*, vol. 120, pp. 82–89, 2016.
18. Y. B. Wang, H. F. Li, Y. F. Zheng, and M. Li, "Corrosion performances in simulated body fluids and cytotoxicity evaluation of Fe-based bulk metallic glasses," *Mater. Sci. Eng. C*, vol. 32, no. 3, pp. 599–606, 2012.

19. J. Schroers, G. Kumar, T. M. Hodges, S. Chan, and T. R. Kyriakides, "Bulk metallic glasses for biomedical applications," *JOM*, vol. 61, no. 9, pp. 21–29, Sep. 2009.
20. J. A. Horton and D. E. Parsell, "Biomedical Potential of a Zirconium-Based Bulk Metallic Glass."
21. L. Huang, D. Qiao, B. A. Green, P. K. Liaw, J. Wang, S. Pang, and T. Zhang, "Bio-corrosion study on zirconium-based bulk-metallic glasses," *Intermetallics*, vol. 17, no. 4, pp. 195–199, 2009.
22. Y. Sun, Y. Huang, H. Fan, Y. Wang, Z. Ning, F. Liu, D. Feng, X. Jin, J. Shen, J. Sun, and J. J. J. Chen, "In vitro and in vivo biocompatibility of an Ag-bearing Zr-based bulk metallic glass for potential medical use," *J. Non. Cryst. Solids*, vol. 419, pp. 82–91, 2015.
23. R. C. Budhani, T. C. Goel, and K. L. Chopra, "Melt-spinning technique for preparation of metallic glasses," *Bull. Mater. Sci.*, vol. 4, no. 5, pp. 549–561, Dec. 1982.
24. T. Gheiratmand and H. R. M. Hosseini, "Finemet nanocrystalline soft magnetic alloy: Investigation of glass forming ability, crystallization mechanism, production techniques, magnetic softness and the effect of replacing the main constituents by other elements," *J. Magn. Magn. Mater.*, vol. 408, pp. 177–192, Jun. 2016.
25. Xue Liang, Jiuhua Chen, Maria Teresa Mora, Jose Fernandez Urdaneta, Qiaoshi Zeng, (2017) Effect of Precipitation on the Hardness of Ternary Metallic Glass. Advances in Materials Physics and Chemistry 07 (06):255–262
26. X. Wang and Xin, "Surface Crystallization in Mg-Based Bulk Metallic Glass during Copper Mold Casting," *Adv. Mater. Sci. Eng.*, vol. 2014, pp. 1–4, May 2014.
27. T. Zhang, X. Zhang, W. Zhang, F. Jia, A. Inoue, H. Hao, and Y. Ma, "Study on continuous casting of bulk metallic glass," *Mater. Lett.*, vol. 65, no. 14, pp. 2257–2260, Jul. 2011.
28. C. K. Chua, C. H. Wong, W. Y. Yeong, C. K. Chua, C. H. Wong, and W. Y. Yeong, "Chapter One – Introduction to 3D Printing or Additive Manufacturing," in *Standards, Quality Control, and Measurement Sciences in 3D Printing and Additive Manufacturing*, 2017, pp. 1–29.
29. B. Dutta, F. H. Froes, B. Dutta, and F. H. Froes, "Chapter 3 – Additive Manufacturing Technology," in *Additive Manufacturing of Titanium Alloys*, 2016, pp. 25–40.
30. K. V. Wong and A. Hernandez, "A Review of Additive Manufacturing," *ISRN Mech. Eng.*, vol. 2012, pp. 1–10, Aug. 2012.
31. I. Yadroitsev, A. Gusarov, I. Yadroitsava, and I. Smurov, "Single track formation in selective laser melting of metal powders," *J. Mater. Process. Technol.*, vol. 210, no. 12, pp. 1624–1631, Sep. 2010.
32. W. E. Frazier, "Metal Additive Manufacturing: A Review," *J. Mater. Eng. Perform.*, vol. 23, no. 6, pp. 1917–1928, Jun. 2014.
33. R. Udroiu, "Powder Bed Additive Manufacturing Systems and its Applications," Acad. J. Manuf. Eng., vol. 10, no. 4, pp. 122–129, 2012.
34. C. Leyens and E. Beyer, "Innovations in laser cladding and direct laser metal deposition," in *Laser Surface Engineering*, Elsevier, 2015, pp. 181–192.
35. R. Liu, Z. Wang, T. Sparks, F. Liou, and J. Newkirk, "13 – Aerospace applications of laser additive manufacturing," in *Laser Additive Manufacturing*, 2017, pp. 351–371.
36. M. Wai Yip, S. Barnes, and A. Aly Diaa Mohmmed Sarhan, "Deposition of a Silicon Carbide Reinforced Metal Matrix Composite (P25) Layer Using CO $_2$ Laser," *J. Manuf. Sci. Eng.*, vol. 137, no. 3, p. 31010, 2015.
37. F. Arias-González, J. del Val, R. Comesaña, J. Penide, F. Lusquiños, F. Quintero, A. Riveiro, M. Boutinguiza, and J. Pou, "Fiber laser cladding of nickel-based alloy on cast iron," *Appl. Surf. Sci.*, vol. 374, pp. 197–205, Jun. 2016.
38. M. Xu, J. Li, J. Jiang, and B. Li, "Influence of Powders and Process Parameters on Bonding Shear Strength and Micro Hardness in Laser Cladding Remanufacturing," *Procedia CIRP*, vol. 29, pp. 804–809, 2015.
39. T. Degen, M. Sadki, E. Bron, U. König, and G. Nénert, "The HighScore suite." Powder Diffraction, p. pp S13-S18, 2014.

40. S. Guo and C. Su, "Micro/nano ductile-phases reinforced Fe-based bulk metallic glass matrix composite with large plasticity," *Mater. Sci. Eng. A*, vol. 707, pp. 44–50, Nov. 2017.
41. M. F. De Carvalho, U. Federal, D. S. Carlos, R. Washington, L. Km, S. P. São, and C. Sp, "LASER CLADDING OF Fe-BASED BULK METALLIC GLASSES," vol. 19, no. April 2017, 2015.
42. R. Li, Z. Li, Y. Zhu, and K. Qi, "Structure and corrosion resistance properties of Ni–Fe–B–Si–Nb amorphous composite coatings fabricated by laser processing," *J. Alloys Compd.*, vol. 580, pp. 327–331, Dec. 2013.

# Fabrication of PLA-HAp-CS Based Biocompatible and Biodegradable Feedstock Filament Using Twin Screw Extrusion

**Nishant Ranjan, Rupinder Singh, I. P. S Ahuja, and Jatenderpal Singh**

## 1 Introduction

FDM deals with additive manufacturing (AM) rule of setting out the material in type of layers/cuts on a fixtureless stage [1–3]. There are few distinct strategies for 3D printing; however, the most generally utilized is a procedure known as FDM. The FDM printers utilize a thermoplastic fiber, which is warmed to its dissolving point and afterward expelled, layer by layer, to make a three dimensional product [4]. FDM is one of the AM processes that build a part of any geometry by ordered deposition of polymeric material in form of feed stock filament on a layer-by-layer basis. The method uses heated thermoplastic filaments which square measure extruded from the tip of the nozzle in an exceedingly prescribed manner in an exceedingly semi-molten state and solidify at chamber temperature. The properties of engineered elements depend upon the setting of various method parameters mounted at the time of fabrication [5]. Feedstock filament is one of the mostly effected input parameters for 3D printing by using FDM [6].

Bone repair or recovery is a typical yet entangled clinical issue in orthopaedic surgery. Despite the fact that autografts what is more, allografts have been generally

N. Ranjan
Department of Production Engineering, Guru Nanak Dev Engineering College, Ludhiana, Punjab, India

Department of Mechanical Engineering, Punjabi University, Patiala, Punjab, India

R. Singh (✉)
Department of Production Engineering, Guru Nanak Dev Engineering College, Ludhiana, Punjab, India

I. P. S Ahuja
Department of Mechanical Engineering, Punjabi University, Patiala, Punjab, India

J. Singh
Department of Defence Production, DGQA, SQAE (A), Jabalpur (M.P), India

 
B. AlMangour (ed.), *Additive Manufacturing of Emerging Materials*,
https://doi.org/10.1007/978-3-319-91713-9_11

utilized as a part of clinical treatment and look into, they both have particular issues. Autografts require auxiliary surgery to get giver bone from the patient's claim body; however the measure of contributor bone is constrained [7, 8]. Bone tissue engineering substitutes are another alternative for treating bone imperfections, and have been proclaimed as an elective system to recover bone [9]. The improvement of biomimetic materials has for some time been a noteworthy objective in the field of bone tissue engineering. Natural bone is a complex organic-inorganic nano-composite material, in which HAp nano-crystallites and collagen fibers are efficient in a various levelled engineering more than a few length scales [10, 11]. Normal polymers and their subordinates have been progressively utilized as a contrasting option to synthetic polymers due to their biodegradability and organic action [12–15].The CS, a characteristic biodegradable polymer, is a low acetyl substituted type of chitin. Inferable from its one of kind properties, including biodegradability, non-poisonous quality, antibacterial impact and bio-similarity, much consideration has been paid to CS-based biomedical materials [16–19]. However, the absence of bone-bonding bioactivity, low mechanical properties and extricating of auxiliary respectability under wet conditions constrain its utilization in bone tissue building. In this way, it is attractive to build up a composite material with the positive properties of both CS and HAp. The outlined composites are relied upon to indicate expanded osteo-conductivity and biodegradation together with adequate mechanical properties, which will be of incredible significance for bone renovating [20–22]. It has been accounted for that Hap and CS composites indicate best biocompatibility and good holding capacity with encompassing host tissues inalienable from HAp [23, 24]. It has additionally been demonstrated that HAp and CS composites can additionally upgrade tissue regenerative adequacy and osteo-conductivity. The methodologies as of now used to get HAp composite materials depend on mechanical blending, co-precipitation [25, 26].

PLA is a nontoxic, biodegradable material with high mechanical strength that is generally utilized as a part of surgery [27–32].With a specific end goal to build the mechanical quality of CS-HAp composite platforms; in this study effort has been made to use PLA into the CS-HAp framework. An advantage of PLA-based biomaterials has been their ability to be fabricated into a variety of structures with the appropriate mechanical properties, topography, geometry, and architecture as required for diverse biomedical applications. One of the oldest methods to fabricate PLA based products has been fiber-spinning from either polymer solution or melt. As PLA is soluble in a wide array of solvents, solution spinning processes has also been widely utilized to fabricate fibers for biomedical applications [33]. Historically, mono and multi filament sutures have been prepared fromPLA-based fibers by spinning; but due to their longer degradation times, other aliphatic polyesters such as poly-glycolic acid (PGA) have now replaced PLA [34]. In addition, woven, knitted, and braided structures produced from spun fibers have found orthopedic applications in bone, ligament, and cartilage regeneration [35, 36]. Further the HAp has stoichiometric Ca/P ratio of 1.67, which is identical to bone apatite [37–41], hence PLA reinforced with Hap can be one of the solution for human implants. But hitherto very less has been reported for preparation of in-house biocompatible and biodegradable feedstock filament comprising of PLA reinforced with CS, HAp for commercial FDM setup. In this study rheological, tensile, thermal and microscopic properties have been ascertained and regression model has been developed to determine the best setting of TSE for drawing the feedstock filament for further printing of functional prototypes. These functional prototypes will act as osteo-conductive and osteo-inductive osteo-allograft by providing attractive surface for new bone cells (osteoblasts).

## 2 Melt Flow Index

Melt flow index (MFI) of any thermoplastic material is defines as a deposition in g/10 min through melt flow tester at certain standard temperature i.e. 210 degree centigrade. The MFI is a determination of glass transition temperature as well as flow-ability of the thermoplastic materials. This is a well-known melting property of thermoplastic material checked by equipment called melt flow tester. Figure 1 shows a schematic of melt flow tester.

The above shown construction (Fig. 1) of melt flow tester is consist of cylindrical barrel which is wrapped and surrounded by a heater coil, heater have an outer

**Fig. 1** Basic construction of melt flow tester

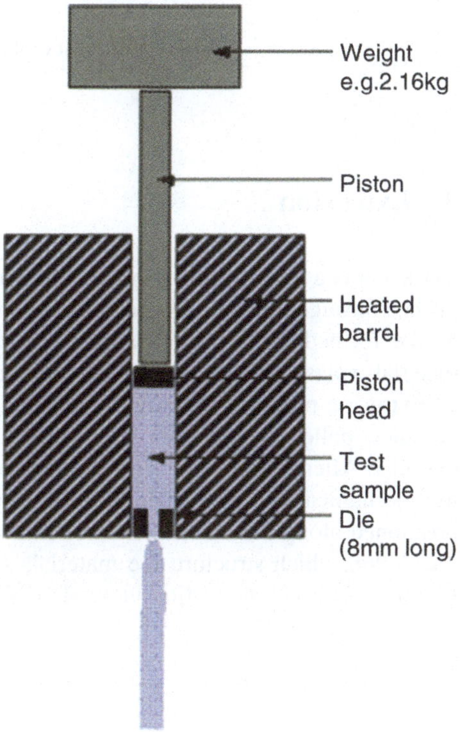

periphery of insulation material to restrict the losses of heat and this heater is controlled automatically by process control box. This set-up is also consisting of a standard weight applied through the piston, a piston head and dies to deposit the materials.

## *2.1 Procedure to Determine Melt Flow Index*

- To determine MFI of certain composition, mixture is poured to the cylinder and a temperature condition is to be set.
- Then left the machine sometime to achieve the predefined temperature.
- After establishing a predefined temperature, pressure is applied with the piston bar into the cylinder by certain weight (according to ASTM D1238) above to the piston bar.
- After placement of plunger, melted plastic come out of the barrel.
- When the time cycle on the digital timer completes, the cutter will automatically cut the sample from the barrel.
- Weigh the melted sample using a digital balance. This weight will be used for calculating the MFI of the sample.

**Formula to calculate MFI:-**

$$\text{MFI} = \text{Weight of sample in g}/10\,\text{min}.$$

# 3 Extrusion

Extrusion is a manufacturing process in which metal or plastic materials are forced/ pushed through a fixed cross-sectional profile or die to produce a continuous ribbon of the formed product or, Extrusion is a manufacturing process in which raw materials (plastic/metal) are melted and made into continual profile.

Extrusion process is starting by feeding of material (plastic/metal) in the form of granules, pellets, powders or flakes from a hopper into the zone of barrel of the extruder. After that, materials are gradually melting by heat, which was produced by the mechanical energy (generating by screw turning) and by the heaters which is positioned along the barrel. The molten materials (plastic/metal) are pushed/forced into a die, which structure the materials into a pipe hard during cooling process. Figure 2 shows cross-section view of single and TSE.

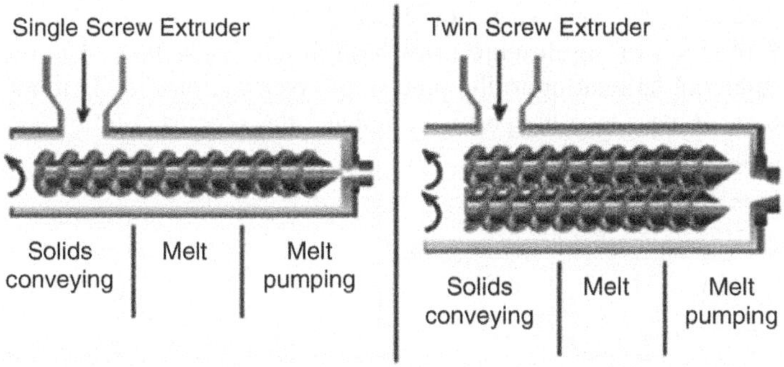


**Fig. 2** Cross-section view of single and twin-screw extruder

### *3.1 Twin Screw Extrusion*

TSE is used extensively for mixing, compounding or reacting polymeric materials. The flexibility of TSE equipment allows this operation to be designed specifically for the formulation being processed. For example, the two screw may be co-rotating or counter-rotating, intermeshing or non-intermeshing. In addition, the configurations of the screws themselves may be varied using forward conveying elements, reverse conveying elements, kneading blocks, and other designs in order to achieve particular mixing characteristics. In the TSE process, the raw materials may be solids (granules, powders & flours), slurries, liquids, and possibly gases.

## 4 Fused Deposition Modeling (FDM)

FDM is an AM technology commonly used for modeling, prototyping, and production applications.

### *4.1 The Technology*

The FDM process constructs three-dimensional objects directly from 3D CAD data. A temperature-controlled head extrudes thermoplastic material layer by layer.

The FDM process starts with importing an STL file of a model into a pre-processing software. This model is oriented and mathematically sliced into horizontal layers varying from +/− 0.127 to 0.254 mm thickness. A support structure is created where needed, based on the part's position and geometry. After reviewing the path data and generating the tool-paths, the data is downloaded to the FDM machine. The system operates in X, Y and Z axes, drawing the model one layer at a time. This process is similar to how a hot glue gun extrudes melted beads of glue. The temperature-controlled extrusion head is fed with thermoplastic modeling material that is heated to a semi-liquid state. The head extrudes and directs the material with precision in ultrathin layers onto a fixtureless base. The result of the solidified material laminating to the preceding layer is a plastic 3D model built up one strand at a time. Once the part is completed the support columns are removed and the surface is finished.

### *4.2 Process*

FDM begins with a software process, developed by Stratasys, which processes an STL file (stereo-lithography file format) in minutes, mathematically slicing and orienting the model for the build process. If required, support structures are automatically generated. The machine dispenses two materials – one for the model and one for a disposable support structure. The thermoplastics are liquefied and deposited by an extrusion head, which follows a tool-path defined by the CAD file. The materials are deposited in layers as fine as 0.04 mm (0.0016") thick, and the part is built from the bottom up – one layer at a time. FDM works on an "additive" principle by laying down material in layers. A plastic filament or metal wire is unwound from a coil and supplies material to an extrusion nozzle which can turn the flow on and off. The nozzle is heated to melt the material and can be moved in both horizontal and vertical directions by a numerically controlled mechanism, directly controlled by a computer-aided manufacturing(CAM) software package. The model or part is produced by extruding small beads of thermoplastic material to form layers as the material hardens immediately after extrusion from the nozzle. Stepper motors or servo-motor are typically employed to move the extrusion head. Several materials are available with different trade-offs between strength and temperature properties. FDM setup shows in Fig. 3.

A "water-soluble" material can be used for making temporary supports while manufacturing is in progress, this soluble support material is quickly dissolved with specialized mechanical agitation equipment utilizing a precisely heated sodium hydroxide solution.

## 5 Differential Scanning Calorimetry (DSC)

### *5.1 Techniqe*

DSC is used to measure melting temperature, heat of fusion, latent heat of melting, reaction energy and temperature, glass transition temperature, crystalline phase transition temperature and energy, precipitation energy and temperature, de-naturization temperatures, oxidation induction times, and specific heat or heat capacity. DSC measures the amount of energy absorbed or released by a sample when it is heated or cooled, providing quantitative and qualitative data on endothermic (heat absorption) and exothermic (heat evolution) processes.

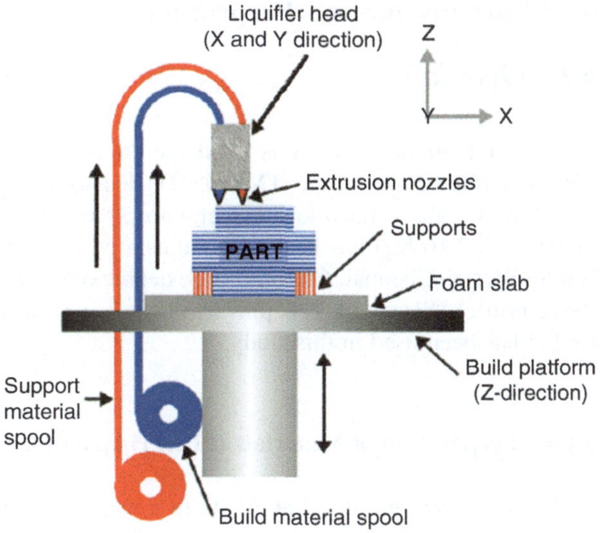


**Fig. 3** Fused deposition modeling process

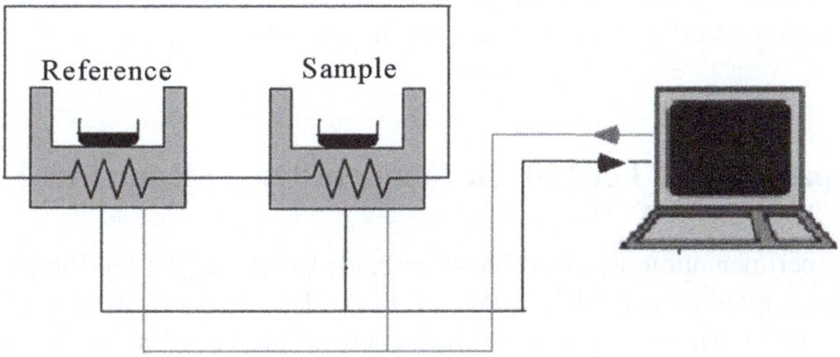


**Fig. 4** Schematic of METTLER TOLEDO DSC

The sample is placed in a suitable pan and sits upon a constantan disc on a platform in the DSC cell with a chromel wafer immediately underneath. A chromel-alumel thermocouple under the constantan disc measures the sample temperature. An empty reference pan sits on a symmetric platform with its own underlying chromel wafer and chromel-alumel thermocouple. Heat flow is measured by comparing the difference in temperature across the sample and the reference chromel wafers. Pans of Al, Cu, Au, Pt, alumina, and graphite are available and need to be chosen to avoid reactions with samples. Figure 4 shows schematic of commercial METTLER TOLEDO DSC.

**Atmospheres** Nitrogen, air, oxygen, argon, vacuum, controlled mixed gases.

# 6 Materials and It's Prepration

## 6.1 Materials

An injection grade of PLA is available from commercial manufacturer, Nature Works (United State), IngeoTM 3052D, density1.24 g/cm$^3$, melting temperature 173–178 °C, glass transition temperature 55–60 °C, melt flow rate 14 g/10 min (at 210 °C, 2.16 Kg) has been selected as a polymer matrix [42] was bought from Nature Works, Chennai. Chitosan (CS, degree of deacetylation>90%), HAp (colorless& brittle) (Size:- 20–100 μm) available from local manufacturer at Ludhiana (India) has been used in this study.

### 6.1.1 Preparation of Materials (PLA, HAp and CS)

PLA pellets were pre-dried at 80 °C in a vacuum oven overnight (8–10 h), while HAp powder and CS were dried at 40 °C for 2 h [43, 44] in vacuum oven so as to remove moisture content. The mixtures of various proportions were added with 2–3 ml of coconut oil and mixed using ball mill at 40 rpm for 4 h.

# 7 Preparation of Feed Stock Filament On TSE for FDM

In pilot experimentation, an effort has been made to prepare the feed stock filament of biocompatible/biodegradable grade of PLA. The first step was to check the possibility of preparation of feed stock filament of PLA reinforced with HAp and CS with varying input parameters. Taking random selection of different composition/ proportions of polymer materials and fillers first check the flowability after that thermal, tensile and microscopic test are to be done. Suitable composition of PLA-HAp-CS for fabrication of feed stock filament on TSE and also making of scaffolds on FDMare to be selected. The feed stock filament has been prepared as per Taguchi L9 (3^3) orthogonal array. For this study commercial, make: HAAKE Mini CTW, Germany has been used.

## 7.1 Rehological Behavior

The feasibility to ascertain the fabrication of feedstock filament of PLA-HAp-CS of varying proportion started with pilot experimentation to access Melt Flow Index (MFI) in accordance with ASTM-D1238. The mixtures of various proportions were added with 2–3 ml of coconut oil and mixed using ball mill at 40 rpm for 4 h. Table 1 show the different composition of PLA-HAp-CS (by weight %age) MFI (g/10 min).

**Table 1** Melt flow index results

| S. No. | Proportions (PLA-HAp-CS) (by weight %) | Observed MFI (g/10 min) | Flow from nozzle | Remarks |
|---|---|---|---|---|
| 1 | 100-0-0 | 13.52 | Continuous | Satisfactory |
| 2 | 84-4-12 | 10.512 | Continuous | Satisfactory |
| 3 | 80-8-12 | 9.015 | Continuous | Satisfactory |
| 4 | 76-12-12 | 3.125 | Discontinuous | Not Satisfactory |
| 5 | 91-8-1 | 12.352 | Continuous | Satisfactory |
| 6 | 90-8-2 | 11.575 | Continuous | Satisfactory |
| 7 | 89-8-3 | 7.474 | Discontinuous | Not Satisfactory |
| 8 | 88-8-4 | 4.465 | Discontinuous | Not Satisfactory |

**Table 2** Results of fabricated feed stock filament on TSE

| S. No | Composition | Output parameters | | |
|---|---|---|---|---|
| | | Flow continuity | Average outside diameter (mm) | Peak strength (MPa) |
| 1 | 84-4-12 | Yes | 1.85 | 2.10 |
| 2 | 80-8-12 | Yes | 1.78 | 2.43 |
| 3 | 91-8-1 | Yes | 1.86 | 3.27 |
| 4 | 90-8-2 | Yes | 1.87 | 3.23 |

As observed from Table 1 the composition/proportion at S. No. 2 (84-4-12), 3 (80-8-12), 5 (91-8-1) and 6 (90-8-2) were selected for further fabrication of feed stock filament. The remaining compositions were not selected, as the flow was not continuous. Sample 1 has been not selected because of in this composition no fillers are to be reinforced. The four compositions/proportions of PLA-HAp-CS as derived from MFI test were extruded on the TSE.

## *7.2 Tensile Behavior*

Four compositions/proportions of PLA-HAp-CS, which derived from MFI tester, are to be drawn by TSE, make: HAAKE Mini CTW, Germany has been used. As shown in Table 2, different combinations of input factors were tried during pilot study. The successful results arrived from the experimentations are listed in Table 2.

According to Table 2, flow continuity of all four compositions/proportions were found suitable and it is expected that the filament will not break at the time of part printing. The experiment was repeated three times in order to reduce the experimental error. It has been observed that peak strength of 91-8-1 and 90-8-2 are maximum and very close to each other while remaining compositions/ proportions have lower peak strength. According to Table 2, it has been concluded that for fabrication of biocompatible/biodegradable feedstock filament best suitable compositions are 91-8-1 and 90-8-2.

## 7.3 Thermal Behavior

Thermal behaviour of the material is one of the major constraints for the use of developed filament in clinical dentistry. For thermal testing of the material in the present study DSC has been used. DSC comprises of glass transition, melting point, enthalpy change, and decomposition of the material with respect to the rising of temperature. Common usage includes investigation, selection, comparison and end-use performance evaluation of materials in research, quality control and production applications.

For thermal analysis METTLER TOLEDO, Model DSC3, Swiss make with $STAR^e$ (SW 14.00) software was used in $N_2$ gas environment. The typical DSC setup determines the behaviour of applied samples by taking references from standard sample, both enclosed in a metallic crucible (Al or platinum). As shown in Fig. 5, DSC sensor uses two crucibles for heating and cooling, one for reference and another for sample.

As shown in Fig. 6, for thermal analysis three heating-cooling cycles are involved. In first cycle, initially heat the sample with comparison to reference from 30 °C to 220 °C at the rate of 10 K/min, after that in the presence of $N_2$ gas (flow rate is 50.0 ml/min) cooling process is done in which matter was cooled from 220 °C to 30 °C at the rate of 10 K/min. After that in the second and third cycle, same process was repeated. The maximum temperature is taken 220 °C because of the material was burnt near 240 °C and lower temperature is generally taken as near to room temperature.

The first cycle of the test is generally not considered for assessment as the sample may contain some foreign substances and may deviate from actual result. The average of second and third cycle was considered for final results. The same are tabulated in Table 3.

Based on Table 3, it has been observed that glass transition temperature lies between 56.5 °C and 57.37 °C, crystallinity of the compositions fall in between 1.20 and 10.50% and the melting point temperature in between 152.46 and 159.87 °C. For fabrication of feedstock filament on TSE and working on FDM glass transition temperature and melting temperature are best suitable range, but for biomedical applications/implants crystallization of material must be low. According to Table 3, it has been concluded that best composition for fabrication of feedstock filament based on crystallization are 91-8-1 and 80-8-12 for further experimentation work. Based on Thermal, Tensile and rheological behaviour it was concluded that two compositions 91-8-1 and 90-8-2 are SEM view performed for selection of best composition.

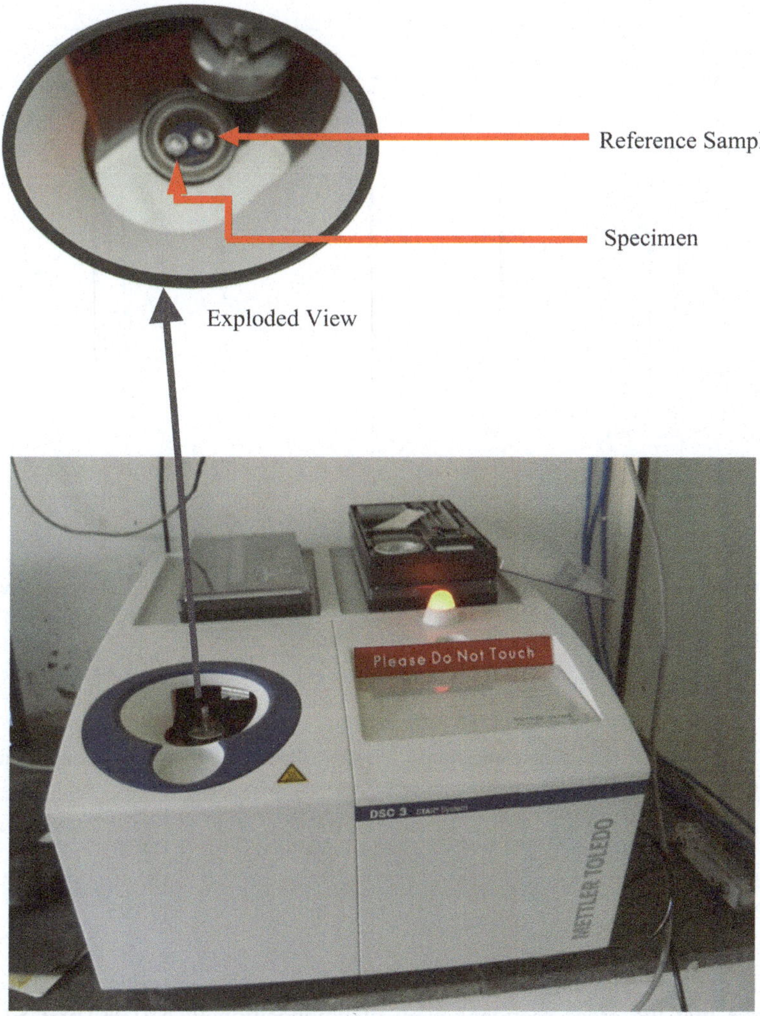


**Fig. 5** Differential scanning calorimetery test

## *7.4 Scanning Electron Microscopic Behavior*

After successful runs of feed stock filament, all samples were subject to SEM analysis (Model no. JEOL JSM-6510LV SEM, Japan) Figs. 7 and 8 shows the cross sectional view of different composition (PLA91-HAp8-CS1 & PLA90-HAp8-CS2) feedstock filament with ×200 magnification, where feedstock filament was broken during the tensile testing. These images show that lots of cavities are present in the cross-section, in whole wire.

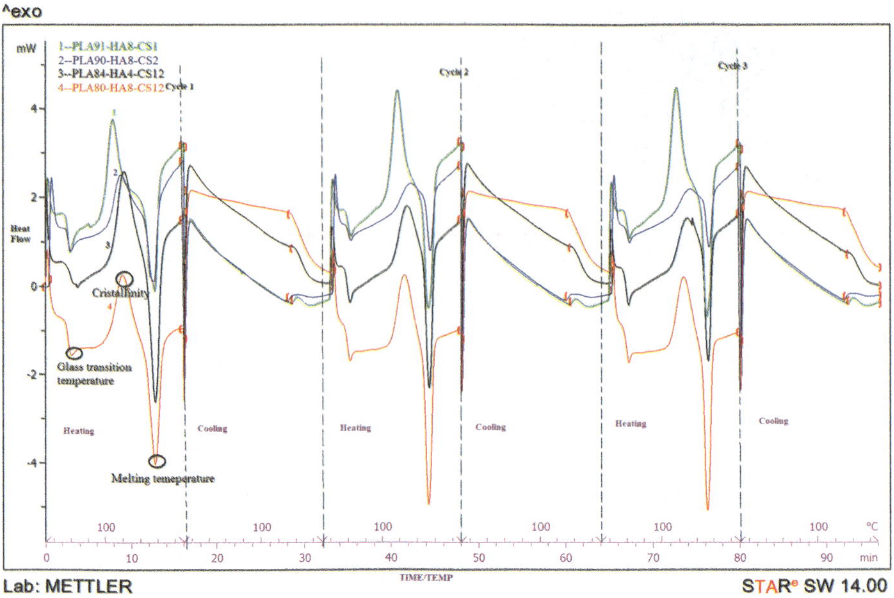


**Fig. 6** Thermal analysis (DSC) curve of feed stock filament

**Table 3** Thermal properties of the specimens

| Compositions | Glass transition temperature (°C) | Crystallization (%) | Melting temperature (°C) |
|---|---|---|---|
| 91-8-1 | 56.50 | 3.13 | 153.17 |
| 90-8-2 | 57.19 | 10.50 | 159.87 |
| 84-4-12 | 56.90 | 7.37 | 152.46 |
| 80-8-12 | 57.37 | 1.20 | 153.95 |

For better understanding higher magnifications images were taken. Figures 9 and 10 shows that the cross sectional view of composition PLA91-HAp8-CS1 with ×400 magnification and PLA90-HAp8-CS2 with ×1600 magnification. As observed from Figs. 7, 8, 9, and 10, the feedstock filament grain structure is similar to open porous structure, which is very useful for cell growth.

## 7.5 *Design of Experiment (DOE)*

The feed stock filament has been prepared as per Taguchi L9 (3^3) orthogonal array. Table 4 shows the control log of experimentation based upon Taguchi L9 (3^3) orthogonal array which was 9 runs of experimentation. A total of 9 sets of specimen

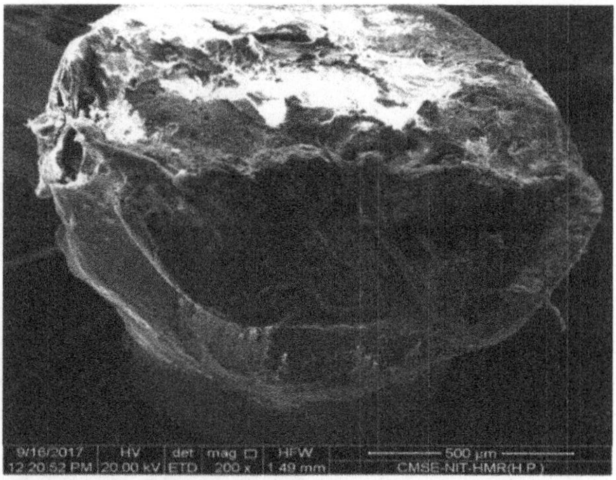


**Fig. 7** Fabricated feedstock filament of composition 91-8-1(×200)

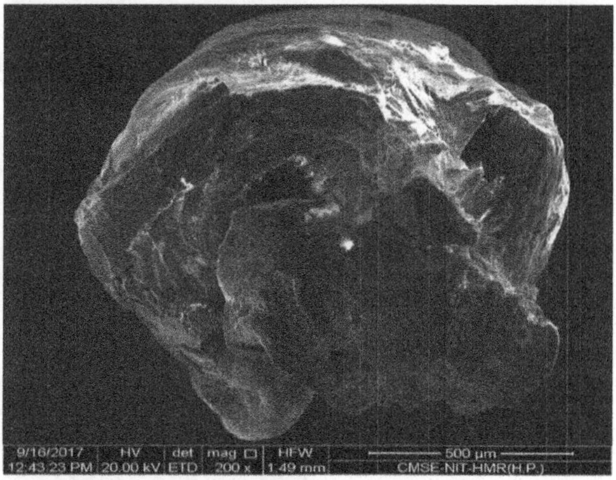


**Fig. 8** Fabricated feedstock filament of composition 90-8-2 (×200)

(feedstock filament) has been prepared by using different controllable parameters (namely; Temperature, Revolution and Dead weight) and total of 27 (9 × 3) experiments were conducted to reduce the effect of environmental, human variations etc. According to previous analysis of compositions/ proportion of material it has been concluded that 91-8-1 (PLA-HAp-CS) are mostly/best suitable for fabrication of biocompatible/biodegradable feedstock filament also development of scaffolds on FDM.

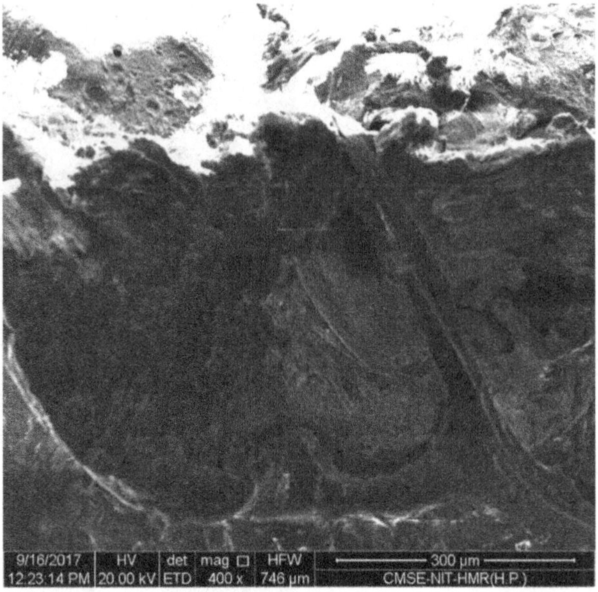


**Fig. 9** Fabricated feedstock filament of composition 91-8-1 (×400)

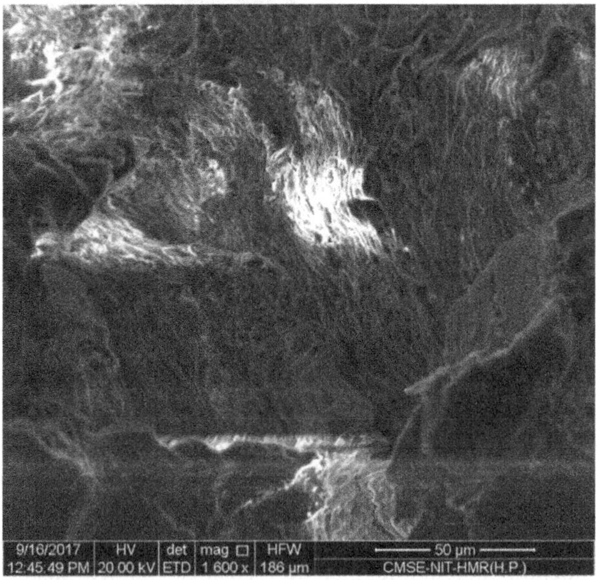


**Fig. 10** Fabricated feedstock filament of composition 90-8-2 (×1600)

## *7.6 Fabrication of Feedstock Filament Based on DOE*

Tensile tests measure the force required to break a materials (polymers/metals) sample specimen and the extent to which the specimen stretches or elongates to that breaking point. Tensile test is the most adoptable test regarding analyse for

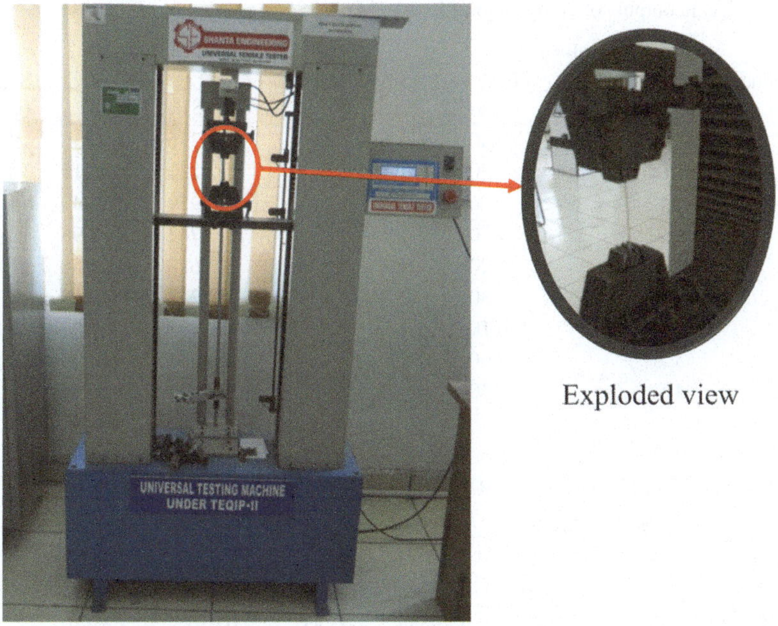


**Fig. 11** Tensile test using Universal Testing Machine

**Table 4** Control log for processing of PLA91-HAp8-CS1

| | Input parameters | | |
|---|---|---|---|
| Experiment run no. | Temperature (°C) | Revolution (rpm) | Dead weight (Kg) |
| 1 | 170 | 100 | 8 |
| 2 | 170 | 140 | 10 |
| 3 | 170 | 180 | 12 |
| 4 | 180 | 100 | 10 |
| 5 | 180 | 140 | 12 |
| 6 | 180 | 180 | 8 |
| 7 | 190 | 100 | 12 |
| 8 | 190 | 140 | 8 |
| 9 | 190 | 180 | 10 |

strength of feedstock filament. Tensile testing is to be done on Universal Tensile Tester (UTT) to check the tensile results at break as well as at peak of feedstock filament. Figure 11 shows the universal tensile tester having capacity up to 5000 N, used only for the plastics/polymers material. Polymerfeedstock filament, polymers strip or flat of any size can be measures by this UTT.

According to Table 4 (DOE), 9 sets of biocompatible/biodegradable feedstock filament wire has been prepared. Further Table 5 shows outputs of tensile test: Strength at peak (PS), Peak load (PL) and Strength at break (BS) and their Signal

**Table 5** Tensile outputs of composition PLA91-HAp8-CS1

| Experiment run no. | Output parameters | | | | | |
|---|---|---|---|---|---|---|
| | Strength at peak (MPa) | Peak load (N) | Strength at break (MPa) | SNRA 1 (PS) | SNRA2 (PL) | SNRA3 (BS) |
| 1 | 3.14 | 6.40 | 2.83 | 9.9386 | 16.1236 | 9.0357 |
| 2 | 3.64 | 7.80 | 3.28 | 11.2220 | 17.8419 | 10.3175 |
| 3 | 4.12 | 9.00 | 3.71 | 12.2979 | 19.0849 | 11.3875 |
| 4 | 3.62 | 7.70 | 3.22 | 11.1742 | 17.7298 | 10.1571 |
| 5 | 4.25 | 9.20 | 3.80 | 12.5678 | 19.2758 | 11.5957 |
| 6 | 3.10 | 6.30 | 2.73 | 9.8272 | 15.9868 | 8.7233 |
| 7 | 4.85 | 10.10 | 4.38 | 13.7148 | 20.0864 | 12.8295 |
| 8 | 3.76 | 7.90 | 3.38 | 11.5038 | 17.9525 | 10.5783 |
| 9 | 4.54 | 9.50 | 4.10 | 13.1411 | 19.5545 | 12.2557 |

**Table 6** Analysis of variance for SNRA 1

| Source | DF | Adj SS | Adj MS | F-value | P-value | % age contribution |
|---|---|---|---|---|---|---|
| Temperature | 2 | 5.2203 | 2.61014 | 36.81 | 0.026 | 36.25 |
| Revolution | 2 | 0.0456 | 0.02280 | 0.32 | 0.757 | 0.32 |
| Dead Weight | 2 | 8.9917 | 4.49584 | 63.41 | 0.016 | 62.44 |
| Error | 2 | 0.1418 | 0.07091 | | | 0.99 |
| Total | 8 | 14.399 | | | | |

**Table 7** Response table for Signal to Noise Ratios larger is better for SNRA 1

| Level | Temperature (T) | Revolution (R) | Dead weight (DW) |
|---|---|---|---|
| 1 | 11.15 | 11.61 | 10.42 |
| 2 | 11.19 | 11.76 | 11.85 |
| 3 | 12.79 | 11.76 | 12.86 |
| Delta | 1.63 | 0.16 | 2.44 |
| Rank | 2 | 3 | 1 |

to Noise (SN) ratio. Based upon Tables 5, 6 and 7 shows percentage contribution of input parameters and their rankings.

As shown in Table 6, percentage error was found to be 0.99%. It shows that model has higher degree of accuracy. Table 7 shows that, dead weight has maximum impact for contributions in SN ratios whereas temperature has minimum contributions.

Further based upon Tables 5, 6, and 7, regression equation for strength at peak has been obtained by using MINITAB software as under:

Regression Equation for Strength at Peak:-

$$SNRA1 = 11.7097 - 0.557^{*}SNRA\_T170 - 0.52^{*}SNRA\_T180 + 1.077^{*}SNRA\_T190 - 0.101^{*}SNRA\_R100 + 0.055^{*}SNRA\_R140 + 0.046^{*}SNRA\_R180 - 1.287^{*}SNRA\_DW8 + 0.136^{*}SNRA\_DW10 + 1.15^{*}SNRA\_DW12$$

Here, SNRA_T170, SNRA_T180, SNRA_T190, SNRA_R100, SNRA_R140, SNRA_R180, SNRA_DW8, SNRA_DW10, SNRA_DW12 has been taken from the Table 7, Obtained from Taguchi analysis. After putting the values form the Table 7 in Regression equation final maximized value has been obtained.

$$SNRA1 = 11.7097 - (0.557^{*}11.15) - (0.520^{*}11.19) + (1.077^{*}12.79) - (0.101^{*}11.61) + (0.055^{*}11.76) + (0.046^{*}11.76) - (1.287^{*}10.42) + (0.136^{*}11.85) + (1.150^{*}12.86) = 16.620(\text{Predicted value})$$

Similarly, for SN ration of strength at peak, final maximized value of the property has been predicted with the help of the following relations.

$$\begin{aligned} \text{SN ratio} &= -10\log X(\text{standard equation for finding SN ratio}) \\ 16.620 &= -10\log X \qquad (1) \\ X &= 10^{-1.6620} \\ X &= 2.1777^{*}10^{-2} \text{ now put value of the x in next equation} \\ X &= 1/n\textstyle\sum 1/y2 \\ 2.1777^{*}10^{-2} &= 1/y^{2}(\text{here n} = 1) \qquad (2) \\ Y^{2} &= 1/2.1777^{*}10^{-2} \\ Y &= 6.776\,(\text{maximized value of the Strength at Peak}) \end{aligned}$$

Where, Y is the value of the property.

Similarly, according to SNRA1 calculate value of SNRA2 and SNRA3.

Regression Equation for Peak Load:-

$$SNRA2 = 18.182 - (0.498^{*}17.68) - (0.518^{*}17.66) + (1.016^{*}19.20) - (0.202^{*}17.98) + (0.175^{*}18.36) + (0.027^{*}18.21) - (1.494^{*}16.69) + (0.194^{*}18.38) + (1.301^{*}19.48) = 23.780\ (\text{Predicted value})$$

Y = 15.432 (maximized value of the Peak Load).

Regression Equation for Strength at Break:-

$$SNRA3 = 10.764 - (0.518^{*}10.247) - (0.606^{*}10.159) + (1.123^{*}11.888) - (0.090^{*}10.674) + (0.066^{*}10.830) + (0.024^{*}10.789) - (1.319^{*}9.446) + (0.146^{*}10.910) + (1.173^{*}11.938) = 15.80\ (\text{Predicted value})$$

Y = 6.166 (maximized value of the Strength at Break)

Based upon Tables 5, 6, 7, 8, and 9 respectively shows the response optimization and best settings predicted from multi factor optimization point of view. The same has been reflected in Fig. 12 as optimization plot, which clearly states that temperature at 190 °C, RPM 140 and dead weight 12 kg are the optimized settings.

**Table 8** Response optimization: SNRA3, SNRA2, and SNRA1

| Response | Goal | Lower | Target | Upper weight | Importance |
|---|---|---|---|---|---|
| SNRA3 | Maximum | 8.7233 | 12.8295 | 1 | 1 |
| SNRA2 | Maximum | 15.9868 | 20.0864 | 1 | 1 |
| SNRA1 | Maximum | 9.8272 | 13.7148 | 1 | 1 |

**Table 9** Best setting obtained with best Signal to Noise Ratios

| Solution | Temperature | Revolution | Dead weight | SNRA3 fit | SNRA2 fit | SNRA1 fit | Composite desirability |
|---|---|---|---|---|---|---|---|
| 1 | 190 | 140 | 12 | 13.1269 | 20.6733 | 13.9918 | 1 |

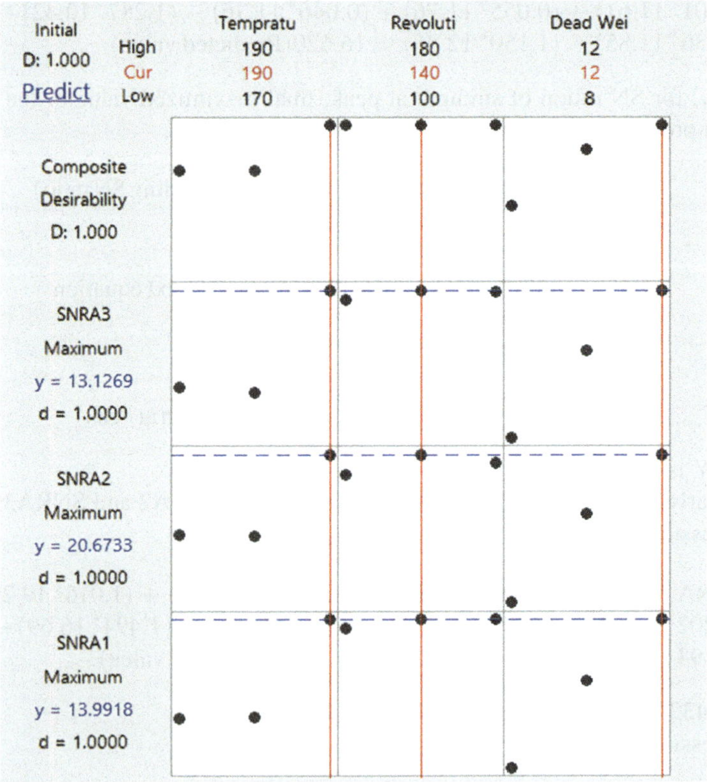


**Fig. 12** Optimization plot of tensile test using UTT

## 8 Summary

In this study, PLA, HAp and CS based composite material has been used for fabrication of biocompatible and biodegradable feedstock filament. Following are the conclusions from the present study:

(a) In MFI testing (according to Table 1) of different compositions it was found that only four composition (84%–4%–12%, 80%–8%–12%, 91%–8%–1%, 90%–8%–2%) flow rate is continuous which is suitable for drawing the feedstock filament.
(b) The peak strength according to Table 2, two compositions/proportion (91%–8%–1%, 90%–8%–2%) is high which is highly suitable to work on FDM for preparation of scaffolds.
(c) In thermal analysis it was concluded that; according to Fig. 6 and Table 3 two compositions (91%–8%–1% and 80%–8%–12%) are less crystallization, which is best suitable for using biomedical applications/implants other compositions are more crystalline.
(d) Based upon SEM analysis (Figs. 7, 8, 9, and 10) concluded that both the compositions/ proportions (91%–8%–1% & 90%–8%–2%) of feedstock filament grain structure is similar to open porous structure. Hence can be gainfully employed for cell growth.
(e) Overall, based on rheological, Thermal and Tensile experimentation it has to conclude that best composition for working on TSE, FDM and for biomedical applications/ implants is "PLA91%-HAp8%-CS1%" because it has continuous and good flowablity, high tensile strength (Peak strength), suitable glass transition and melting temperature and also low crystalline; which has suitable for biomedical applications.
(f) Finally based upon Taguchi analysis and regression model for PLA91%-HAp8%-CS1% based feed stock filament, temperature at 190 °C, RPM 140 and dead weight 12 kg are the recommended optimized settings on TSE. The functional prototypes developed at proposed settings will have better scaffolding which will provides attractive surfaces for new bone cells.

**Acknowledgements** The authors are thankful to SERB (DST), GOI for financial support and Manufacturing Research Lab (Dept. of Production Engineering, Guru Nanak Dev Engineering College, Ludhiana), Punjabi University Patiala for technical support.

# References

1. Anitha, R. and Arunachalam, S. (2001), "Critical parameters influencing the quality of prototypes in fused deposition modeling", *Journal of Materials Processing Technology*, Vol. 113, Issue 1–3, pp. 385–388.
2. Chhabra, M. and Singh, R. (2011), "Rapid casting solutions a review", *Rapid Prototyping Jounral*, Vol. 17, pp. 328–350.
3. Choudhury, A., Chakraborty, D. and Reddy, B. (2007), "Extruder path generation for Curved Layer Fused Deposition Modeling", *Computer-aided design Journal*, Vol. 40, pp. 235–243.
4. Singh, R., Singh, S., Singh, I. P., Fabbrocino, F. and Fraternali, F. (2017), "Investigation for surface finish improvement of FDM parts by vapour smoothing process", *Composites Part B: Engineering*, Vol. 111, pp. 228–234.
5. Kumar, R., Singh, R., Hui, D., Feo, L. and Fraternali, F. (2017), "Graphene as biomedical sensing element: State of art review and potential engineering applications", *Composites Part B: Engineering*, https://doi.org/10.1016/j.compositesb.2017.09.049

6. Singh, R., Kumar, R., Feo, L. and Fraternali, F. (2016), "Friction welding of dissimilar plastic/ polymer materials with metal powder reinforcement for engineering applications", *Composites Part B: Engineering*, Vol. 101, pp. 77–86.
7. Block, J. E. and Poser, J. (1995), "Does xenogeneic demineralized bone matrix have clinical utility as a bone graft substitute?",*Med Hypotheses*, Vol. 45, pp. 27–32.
8. Sasso, R. C., Williams, J. I., Dimasi, N., Meyer Jr. P. R. (1998), "Postoperative drains at the donor site of iliac-crest bone grafts. A prospective, randomized study of morbidity at the donor site in patients who had a traumatic injury of the spine", *J Bone Joint Surgical Am*, Vol. 80, pp. 631–635.
9. Rose, F. R. A. J. and Oreffo, R. O. C. (2002), "Bone tissue engineering: hope vs hype", *BiochemBiophys Res Commun*, Vol. 292, pp. 1–7.
10. Du, C., Cui, F. Z., Zhang, W., Feng, Q. L., Zhu, X. D. and Groot, K. D. (2000), " Formation of calcium phosphate/collagen composites through mineralization of collagen matrix", *J Biomed Mater Res A*, Vol. 50, pp. 518–27.
11. Kikuchi, M., Itoh, S., Ichinose, S., Shinomiya, K. and Tanaka, J. (2001), "Self organization mechanism in a bone-like hydroxyapatite/collagen nanocomposite synthesized in vitro and its biological reaction in vivo", *Biomaterials*, Vol. 22, pp. 1705–11.
12. Burg, K. J. L., Porter, S. and Kellam, J. F. (2000), "Biomaterial developments for bone tissue engineering", *Biomaterials*, Vol. 21, pp. 2347–2359.
13. Chang, M. C. and Tanaka, J. (2002), "XPS study for the microstructure development of hydroxyapatite-collagen nano-composites cross linked using glutaraldehyde", *Biomaterials*, Vol. 23, pp. 3879–85.
14. Grodzinski, J. J. (1999), "Biomedical application of functional polymers", *React Function Polymer*, Vol. 39, pp. 99–138.
15. Hutmacher, D. W. (2000), "Scaffolds in tissue engineering bone and cartilage", *Biomaterials*, Vol. 21, pp. 29–43.
16. Ishihara, M., Ono, K., Saito, Y., Yura, H., Hattori, H. and Matsui, T. (2001), "Photo cross-linkable chitosan: an effective adhesive with surgical applications", *Int Cong Ser*, Vol. 1223, pp. 251–257.
17. Madihally, S. V. and Matthew, H. W. T. (1999), "Porous chitosan scaffolds for tissue engineering", *Biomaterials*, Vol. 20, pp. 1133–1142.
18. Mi, F. L., Tan, Y. C., Liang, H. F. and Sung, H. W. (2002), "In vivo biocompatibility and degradability of a novel injectable-chitosan-based implant", *Biomaterials*, Vol. 23, pp. 181–91.
19. Wang, L. S., Khor, E., Wee, A. and Lim, L. Y. (2002), "Chitosan-alginate PEC membrane as a wound dressing: assessment of incisional wound healing", *J Biomed Mater Res B*, Vol. 63, pp. 610–618.
20. Li, Z. S., Ramay, H. R., Hauch, K. D., Xiao, D. M. and Zhang, M. Q. (2005), "Chitosan-alginate hybrid scaffolds for bone tissue engineering", *Biomaterials*, Vol. 26, pp. 3919–3928.
21. Sarasam, A. and Madihally, S. V. (2005), "Characterization of chitosan-polycaprolactone blends for tissue engineering applications", *Biomaterials*, Vol. 26, Issue 5, pp. 500–508.
22. Shanmugasundaram, N., Ravichandran, P., Reddy, P. N., Ramamurty, N., Pal, S. and Rao, K. P. (2001), "Collagen-chitosan polymeric scaffolds for the in vitro culture of human epidermoid carcinoma cells", *Biomaterials*, Vol. 22, pp. 1943–1951.
23. Andrew, C. A. W., Eugene, K. and Garth, W. H. (1998), "Preparation of a chitin–apatite composite by in situ precipitation onto porous chitin scaffolds", *J Biomed Mater Res A*, Vol. 41, pp. 541–548.
24. Kawakami, T., Antoh, M., Hasegawa, H., Yamagichi, T., Ito, M. and Eda, S. (1992), "Experimental study on osteoconductive properties of chitosan-bonded hydroxyapatite self-hardening paste", *Biomaterials*, Vol. 13, pp. 759–763.
25. Yamaguchi, I., Tokuchi, K., Fukuzaki, H., Koyama, Y., Takakuda, K. and Monma, H. (2001), "Preparation and microstructure analysis of chitosan/hydroxyapatite nanocomposites", *J Biomed Mater Res A*, Vol. 55, pp. 20–27.

26. Yin, Y. J., Zhao, F., Song, X. F., Yao, K. D., Lu, W. W. and Leong, J. C. (2000), "Preparation and characterization of hydroxyapatite/chitosan–gelatin network composite", *J ApplPolymSci*, Vol. 77, pp. 2929–2938.
27. Ariyapitipun, T., Mustapha, A. and Clarke, A. D. (1999), "Microbial shelf life determination of vacuum-packaged fresh beef treated with polylactic acid, lactic acid, and nisin solutions", *Journal of food protection*, Vol. 62, Issue 8, pp. 913–920.
28. Kale, G., Auras, R., Singh, S. P. and Narayan, R. (2007), "Biodegradability of polylactide bottles in real and simulated composting conditions", *Polymer Testing*, Vol. 26, Issue 8, pp. 1049–1061.
29. Fini, M., Giannini, S., Gioradano, R., Giavaresi, G., Grimaldi, M. and Aldini, N. N. (1995), "Resorbable device for fracture fixation: in vivo degradation and mechanical behavior", *Int J Artif Organs*, Vol. 18, pp. 772–776.
30. Shen, X. Y., Tong, H., Jiang, T., Zhu, Z. H., Wan, P. and Hu, J. M. (2007), "Homogeneous chitosan/carbonate apatite/citric acid nano-composites prepared through a novel in situ precipitation method", *Compos Sci Technology,* Vol. 67, pp. 2238–2245.
31. Taddei, P., Monti, P. and Simoni, R. (2002), "Vibrational and thermal study on the in vitro and in vivo degradation of a poly (lactic acid) based bio-absorbable periodontal membrane", *J Mater Sci Mater Med,* Vol. 13, pp. 469–475.
32. Shen, X. Y., Tong, H., Zhu, Z. H., Wan, P. and Hu, J. M. (2007), "A novel approach of homogenous inorganic/organic composites through in situ precipitation in poly-acrylic acid gel", *Mater Lett*, Vol. 61, pp. 629–634.
33. Gupta, B., Revagade, N. and Hilborn J. (2007), "Poly (lactic acid) fiber: an overview", *Prog. Polym.Sci.*, Vol. 32, pp. 455–482.
34. Ge, Z., Goh, J.C., Wang, L., Tan, E.P. and Lee, E.H. (2005),"Characterization of knitted polymeric scaffolds for potential use in ligament tissue engineering", *J. Biomater. Sci. Polym. Ed.*, Vol. 16, pp. 1179–1192.
35. Ouyang, H.W., Goh, J.C., Thambyah, A., Teoh, S.H. and Lee, E.H. (2003),"Knitted polylactide-co-glycolide scaffold loaded with bone marrow stromal cells in repair and regenera-tion of rabbit Achilles tendon", *Tissue Eng.*,Vol. 9,pp. 431–439.
36. Freeman, J.W., Woods, M.D. and Laurencin, C.T. (2007),"Tissue engineering of the anterior cruciate ligament using a braid–twist scaffold design", *J. Biomech.*, Vol. 40, pp. 2029–2036.
37. Singh, R., Kumar, R., Ranjan, N., Penna, R. and Fraternali, F. (2017), "On the recyclability of polyamide for sustainable composite structures in civil engineering", *Composite Structures*, https://doi.org/10.1016/j.compstruct.2017.10.036
38. Kalita, S.J., Bhardwaj, A. and Bhatt, H.A.(2007), "Nanocrystalline calcium phosphate ceramics in biomedical engineering",*Material Science and Engineering: C*, Vol. 27, Issue 3, pp. 441–449.
39. Mostafa, N.Y. and Brown, P.W. (2007),"Computer simulation of stoichiometric hydroxyapatite: Structure and substitutions",*Journal of Physics and Chemistry of Solids*, Vol. 68, Issue 3, pp. 431–437.
40. Teixeira, S., Rodriguez, M.A., Pena, P., De Aza, A.H., De Aza, S. and Ferraz, M.P. (2009),"Physical characterization of hydroxyapatite porous scaffolds for tissue engineering",*Material Science and Engineering: C*, Vol. 29, Issue 5, pp. 1510–1524.
41. Guo, L., Huang, M. and Zhang, X.(2003),"Effects of sintering temperature on structure of hydroxyapatite studied with Rietveld method", *Journal of Materials Science: Material in Medicine*, Vol. 14, Issue 9, pp. 817–822.
42. Auras, R., Harte, B. and Selke S. (2004), "An overview of polylactides as packaging materials", *Macromol. Biosci.*, Vol. 4, pp. 835–864.
43. Ravindra, R., Krovvidi, K.R. and Khan, A.A. (1998), "Solubility parameter of chitin and chitosan", *Carbohydr. Polym.*, Vol. 36, pp. 121–127.
44. Bonilla, J., Fortunati, E., Vargas, M., Chiralt, A. and Kenny, J.M. (2013), "Effects of chitosan on the physicochemical and antimicrobial properties of PLA films", *J. Food Eng.*, Vol. 119, pp. 236–243.

# Index

**A**
ABS–Gr feedstock filament, 281
ABS-P400, 230
Acrylonitile butadiene styrene (ABS), 279
compositions, MFI, 252
DMA curves, 265
higher molecular mobility, 266
prototypes, 251
Adaptive slicing techniques, 228
Addictive manufacturing (AM), 1, 120, 126–155, 165–167, 183, 184, 251, 279, 307, 325
AMCs, 115–116
ASTM-recognized AM methods, 111
for biomedical application, 180–183
EBM, 183, 184
SLM (*see* Selective laser melting (SLM))
CAD software, 308
companies, 57, 58
composites and complex materials, 111
computer control, 180
*vs.* conventional manufacturing
advantages, 59, 62
cell structures, 60
design, 59
disadvantages, 62
DLD, 59, 60
FGMs, 59
ISO/ASTM 52900 standard, 63
multi-directional composite structures, 59
reinforcement, 61
*vs.* conventional methods, 180
copper-matrix composites, 119–120
DED, 64–66
as 3D printing, 180
EOS, 57
fabrication technique, 307
hardness values, 316
iron-based matrix composites, 120–122
laser cladding, 308, 310
laser energy, 308
laser power, 313
mechanical properties (*see* Mechanical properties, AM processed MMCs)
metal materials, 184
microstructural evolution (*see* Microstructural evolution, additively manufactured MMCs)
microstructure examination, 315–316
mixed powder systems, 120 (*see also* Mixed powder systems, AM techniques)
nickel-based matrix composites, 118–119
part quality and surface integrity, 149–154
PBF, 63, 64
porous implants, 180
quality and surface integrity
applied energy density, 150–152
characteristics, mixed powder system, 152–154
sample preparation, 310
SLM and LENS process, 308
titanium alloy, biomedical (*see* Titanium alloys)
TMCs, 116–117
wear behavior
applied energy density, 167
coefficient of friction (COF), 165
extrinsic factors, 165



B. AlMangour (ed.), *Additive Manufacturing of Emerging Materials*,
https://doi.org/10.1007/978-3-319-91713-9

Addictive manufacturing (AM) (*cont.*)
interfacial bonding, 167
size and volume fraction, reinforcements, 165–167
XRD patterns, 312
Additive manufacturing (AM) process, 2
Additive manufacturing technology, 1
Additively manufactured lightweight MMCs
applications, 97–99
distribution pattern, 96, 97
size, 96
volume fraction, 94, 95
Al matrix composites (AMCs), 67
Al matrix nano-composites (AMnCs), 75, 76
Al/5wt.%$Fe_2O_3$ composites, 5
AlSi10Mg
chemical composition, 35, 37
SLS electrode, 42
SLS parameters, 37
SLS tool, 44, 51
Aluminum (Al), 55, 67, 76
AMCs, 115, 116
hardness, 115
Aluminum matrix composites (AMCs), 115–116, 138, 152, 155
Al-$Fe_2O_3$, 5–6
$AlSi_{10}Mg$-SiC, 7–9
alumina and silicon carbide, 5
SLM processing, 7
Analysis of variance (ANOVA), 42, 43, 45, 47, 49, 234, 240, 287
Autografts, 326

**B**
Biomedical applications, 179
Biomedical implants, 302, 303
Bode phase angle diagram, 203
Bone repair, 325
Bone tissue engineering, 326
Buckingham Pi theorem
dimensional accuracy, 242
dimensional analysis, 240
input parameters, 240
surface hardness, 241, 242
surface roughness, 240, 241

**C**
Chromel-alumel thermocouple, 331
Coarse dendritic crystals, 315
Coating layer (CL), 317
CoCr-based alloys, 198
Commercially pure titanium (CP-Ti), 185
Composite material
ABS material test, 259
damping property, 259
feedstock filament, 251–260
rheological behavior, 253
suitability, 256
tensile properties, 257
Young's modulus, 261
Computer-aided design (CAD) model, 198
Computer-aided manufacturing (CAM) software package, 330
Conductivity, 283
Constant phase element (CPE), 203
Control log, 339
Control log of experiment, 285
Copper (Cu)-matrix composites, 119, 120
Copper-mold casting technique, 307
Corrosion behaviors, 198–216
biomaterials, 197
biphasic structure, 198
bone diseases, 197
Co-Cr based alloys, 216–218
EBM-produced Ti-6Al-4V
AM technique, 209
Bode phase diagram, 210
cathodic polarization region, 212
corrosion current density, 211
EBSD analysis, 209
EIS test, 211
fitting data, 211
grain boundary, 213
lamellar α phase, 209
1 M HCl solution, 211
micro-galvanic couple, 213
passive current density, 211
SEM image, 210, 212, 213
SLM technique, 208
volume fraction, 209
XRD patterns, 208
homogeneous laid, 197
metal AM technologies, 197
SLM-produced 316L stainless steel, 218–220
SLM-produced Ti-6Al-4V
acicular α' phase, 200
AM techniques, 198
Bode phase angle diagram, 203
Bode plot diagram, 202
columnar prior β phase, 198
commercial grade 5 alloy, 198
corrosion rate, 206
EIS fitting, 203

EIS test, 202
electrochemical tests, 200
HCl solution, 204
melt powder, 199
morphology, 207, 208
NaCl solution, 204
Nyquist diagram, 202
OCP test, 201
optical microstructure, 199
oxide film, 200, 203
polarization curves, 204, 206
polarization resistance, 204
potentiodynamic curves, 205
potentiodynamic polarization curves, 204
SEM image, 207
semi-equiaxed β grains, 200
the α'-Ti phase, 200
titanium alloys, 198
XRD patterns, 200
SLM-produced Ti-TiB
absorption peak, 216
anode region, 215
CP-Ti, 214
fitted polarization resistance, 214
Nyquist plot, 214
polarization curves, 214
potentiodynamic polarization curve, 215
$TiB_x$ regions, 215
and stainless steel, 221
Young's modulus, 197, 198
Cyclic potentiodynamic polarization, 219

**D**

Design of experiment (DOE), 39
feed stock filament, 336, 338
techniques, 266
Differential scanning calorimetry (DSC), 244–246
atmospheres, 331
melting temperature, 330
test, 335
Dimensional accuracy
control log of experimentation, 233
fabrication, 228
head diameter and neck thickness, 232
mathematical model, 242
pre-processing parameters, 228
shrinkage, 238
smoothing time and orientation angle, 235
and surface finish, 238
Dimensional analysis, 240
Direct energy deposition (DED), 64–66
Direct laser deposition (DLD), 59, 65
Direct metal deposition (DMD), 56
Directed energy deposition (DED), 111
Distribution pattern, 88, 89, 91–93
homogenous, 90, 91
microstructures
3DQCN, 93
Marangoni flow, 92
MMCs, 91
ring-like structure, 92, 93
solidification, 93
non-homogenous
agglomeration/clustering, 88, 89
bimodal, 88
Dynamic viscosity, 79, 80

**E**

Electrical conductivity, 280, 283, 284, 286
Electrical discharge machining (EDM), 37, 38
AISI H13 tool steel, 31
AlSi10Mg, 37
aluminium, silicon and magnesium, 30
AM process, 31, 35
average surface roughness, 40–41, 44–47, 51
chemical composition
AlSi10Mg, 37
Ti-alloy, 38
conventional machining process, 29
copper and graphite electrodes, 37
corundum and silicon carbide, 31
direct production, 31
electrical discharge coating process, 39
electrical discharge machining, 32
EOS GmbH-Electro optical systems, 37
experiment, 38
FA, 34–36
GRA, 30, 33–34, 52
GRGi, 47–49
MATLAB R2014a, 47
MRR, 39–40, 42–43, 51
nontraditional machining, 30
optimization techniques, 32
parameters, 39, 42
parametric setting, 32
rapid prototyping process, 32
SCD, 50, 51
SLS process, 30, 31
Ti6Al4V, 32, 51
Ti-alloy, 30
titanium, 30

Electrical discharge machining (EDM) (*cont.*)
- TWR, 40, 44, 45, 51
- VIKOR method, 32
- work piece and tool electrode, 29

Electro optical systems (EOS), 57
Electrochemical corrosion test, 318–319
Electrochemical impedance spectroscopy (EIS), 202, 203
Electron back scattered diffraction (EBSD), 143, 144
Electron beam melting (EBM), 190–192, 197, 198, 208–213, 221, 308
- beam energy input and chamber environment, 183
- densification rate and microstructural homogeneity, 183
- implants, 183
- titanium alloys
  - advantages, 190
  - defect pore, formation, 192
  - microstructure, 190
  - porous structure model, 191
  - Ti-6Al-4V alloy, 190
  - Young's modulus, 192
- working principle and process, 183, 184

Extrusion process, 328
- TSE, 329

**F**

Fabricated feed stock filament, 333, 337, 338
Fe-based MG coating, 315
Feed stock filament, 332–341
- alternative material, 259
- composite material, 251–260

Firefly algorithm (FA)
- EDM process, 32
- flow chart, 36
- GRA, 32
- grey relational analysis, 47–50
- light intensity, 35
- rhythmic flashlight, 34
- rules, 34

Functionally graded composite material (FGCM), 114
Functionally graded materials (FGMs), 59, 111
Functionally graded metal matrix composites (FGMMCs), 56
Fused deposition modeling (FDM), 56
- ABS, 251
- academic research, 273
- Al and $Al_2O_3$, 254
- $Al_2O_3$ content, 275
- aluminum powder, 253
- AM technology, 325, 329
- ANNOVA method, 275
- autografts, 326
- bone-bonding bioactivity, 326
- composite material, 253
- compositions, MFI, 252, 255
- compositions, weight proportion, 255
- design of experiments, 266–271
- 3D CAD data, 329
- DMA, 259, 260
- 3D printing, 325
- 3D structures, 251
- empirical relationship, 270–271
- experimental study, 251
- fabrication, 258–259
- feedstock filament development, 252
- glass transition temperature, 275
- Gujrat state fertilizer, 252
- head extrudes, 330
- industrial applications, 273
- inspection and testing, 256–257
- liquefier head, 252
- loss factor/tan $\delta$, 264–265
- loss modulus, 263
- mechanical properties, 261
- mechanical testing, 257, 258
- melt flow test, 254
- nylon6-Al-$Al_2O_3$ composite, 273
- optimum parameters, 269–270, 275
- parameter, 285
- printers, 325
- process capability study, 271–274
- process parameters, 269
- properties, filler materials, 252
- rheological behavior, 253–255
- rheological properties, 260
- RT applications, 275
- setup, 285, 330
- single screw extruder, 255–260
- software process, 330
- STL file, 329
- storage modulus, 261–262, 274
- Taguchi approach, 267, 268, 275
- viscoelastic behavior, ABS material, 265–266
- viscoelastic behavior, composite materials, 261–265

**G**

Glass forming ability (GFA), 300
Graphene (Gr)
- chemical analysis, 281
- extraction, 280

extraction, 280–281
polymer, 280
Graphite powder, 280, 281
Grey relational analysis (GRA)
EDM process, 32
FA, 31, 32
grey relational co-efficient, 33
grey relational grade, 33, 34
GST, 33
optimization, 47–50
scale value, 33

**H**
Hall-Petch effect, 161
Hardness, 284
Hardness value, 317

**I**
Inconel
AM techniques, 119
Inconel 625 matrix, 136, 153, 158
Inconel 625 matrix composites, 135
Inconel 690 matrix, 145
Inconel 718 matrix, 135–139
Inconel 738LC alloy, 143
Ni-based superalloys, 118
TiC reinforced, 127
In-situ metal matrix composites, 4, 14
In-situ reactions
aluminum matrix composites, 5–6
nickel matrix composites, 9–14
Ti64/TNZT-N, 22
Ti-Mo-N, 16–22
titanium matrix composites, 14–16
Iron-based matrix composites, 120–122

**L**
Laminated object manufacturing (LOM), 56
Laser additive manufacturing, 1
Laser additive processing, 4
Laser cladding, 308, 310
Laser cladding applications, 313
Laser consolidation (LC), 65
Laser energy density, 81–84
in-situ reaction
amount, 81
morphology, 83, 84
size, 82
Laser engineered net shaping (LENS™), 2, 10, 15, 18, 65, 308
dendritic microstructures, 22
friction coefficient, 13
laser gas nitriding, 14
microstructure, 15
Ti-Nb-Zr-Ta-TiB composites, 15
Ti-TiB composites, 15
Laser-induced melt pools
dynamic viscosity, 79, 80
Marangoni flow, 78
Rayleigh-Benard Convection, 79
recoil pressure, 78
surface tension, 76, 78
Laser metal deposition (LMD), 65, 112–114, 127, 139, 140
Lee's Disk method, 282
Lee's Disk technique, 282
Light-weight, 55
MMCs (*see* Metal matrix composites (MMCs))

**M**
Marangoni flow, 78
Material removal rate (MRR), 30, 39, 42, 51, 52
Mechanical properties, AM processed MMCs, 155–165
categories, 155
direct strengthening
as "classical" strengthening, 155
discontinuously reinforced MMCs, 156
rapid solidification rates, 156
reinforcement size, 158–159
reinforcement type, 158
volume fraction, reinforcements, 157–158
yield strength, for LENS, 156, 157
indirect strengthening
density of dislocations, in matrix, 161, 162
grain refinement, matrix, 160, 161
mechanisms, described, 160
by microstructural modifications, 162, 163
solid solution strengthening mechanism, 163, 164
Ti6Al4V matrix, in TMCs, 160
strengthening, MMCs, 155
stress-strain curves, 155
weakening mechanisms, 164

Mechanical properties (*cont.*)
densification level, 164
microstructural inhomogeneity, 165
microstructure coarsening, 164
Melt flow index (MFI), 281, 332
construction, 327
determination, 328
formula, 328
glass transition temperature, 327
outcomes, 333
Melt flow tester, 327
Melt spinning, 306
Metal matrix composite (MMC), 67–76, 111
AM processes, 56, 66
conventional manufacturing, 114, 115
deposition techniques, 56
ex-situ and in-situ, 3
in-situ processes, 56
liquid phase, 55
LMD processes, 114
mechanical and physical properties, 4
metals and alloys, 3
properties, 3, 4
role, 3
solid phase, 55
solid/liquid two-phase, 56
types
ex-situ reinforced, 67, 68
hybrid ex-situ/in-situ reinforced, 69
in-situ reinforced, 70–73
MMnCs, 74–76
TMCs, 73, 74
Metal matrix nano-composites (MMnCs), 56, 74–76
Metallic glasses (MG), 299
AM, 308
applications, 303
biocompatibility, 303–304
biomedical implants, 303
casting, 306
cytotoxicity evaluation study, 304
development, 301
evolution, 300–301
fabrication, 301
mechanical properties, 299, 301–303
melt spinning, 306
metallic alloys, 303, 304
multicomponent, 301
systems, 301
Microstructural evolution, additively manufactured MMCs, 129–141, 143–149
characteristics, mixed powder system, 127
distribution pattern, reinforcements, 129
energy density, 131
microstructural characterization, 129
scanning speeds, 129, 130
solidification mode and growth velocity, 129
ex-situ reinforced MMCs, 127, 128
factors, 126
in-situ reinforced MMCs, 128–129
laser and scanning parameters, 126
matrix reinforcements
dislocations, in matrix, 148–149
as heterogeneous nucleation sites, 145
LENS process, 144
microstructural refinement, 139–141
and phase transformation, 145, 146
A356 powder and A356/SiC powder mixture, 144, 145
supersaturated matrix, 147, 148
texture, matrix, 141, 143, 144
microstructural features, 127
reinforcement/matrix reactions
in-situ reaction products, 137, 138
interfacial reaction, 133–137
reaction mechanisms, 132
size and morphology, reinforcements, 127
size and volume fraction, 131, 132
volumetric energy density, 126
MINITAB 17 software, 267, 271, 275
MINITAB software, 340
Mixed powder systems, AM techniques
category, 124
characteristics, 152–154
commercially pure (CP)-Ti and irregular-shaped $TiB_2$ powders, 123
direct mechanical mixing, 123
flowability, 122
irregular-shaped water atomized high speed steel powders, 123
laser beam interaction, 125
milling time, 123
morphology, powder particles, 122
size, powder particles, 124
Ti and $TiB_2$ powder mixtures, ball milling, 124
Ti6Al4V and BN powder particles, 125, 126
Multi-response optimization
constraints and goals set, 243
FDM-VS operations, 244, 245
low density settings, 243
90° orientation angle, 244
SN ratios, 243, 244

**N**
Naphthalene serves, 280
Nickel-based matrix composites, 118–119
Ni-Ti-C composites, 11, 12
Ni-TiC-C composites, 10
N-methyl pyrrolidone (NMP), 280
Nyquist plot, 214

**O**
Ohms Law, 283
Open circuit potential (OCP) curve, 200, 201
Optical microscope, 293
Optomec design company, 2
Orthogonal array (OA), 243

**P**
Pipetting, 280
PLA-HAp-CS
  composition, 332
  compositions/proportions, 333
  feedstock filament, 332
  preparation of materials, 332
Plasma deposition manufacturing (PDM), 65
Poly-glycolic acid (PGA), 326
Polymerfeedstock filament, 339
Polymers, 279
Porosity, 284, 290, 291, 293
Potentiodynamic curves, 205
Potentiodynamic polarization curves, 204, 318, 319
Potentiodynamic polarization data, 217
Potentiostaic testing, 220
Powder bed fusion (PBF), 63, 64, 111, 113
Powder mixture
  size of, 85
  volume fraction, 85–88
Power density, 315

**R**
Rayleigh-Benard Convection, 79
Regression equation, 341
Reinforcement, 4, 81, 85–88
  distribution (*see* Distribution pattern)
  laser energy density, 81–84
    in-situ reaction, 81
  powder mixture
    size of, 85
    volume fraction, 85–88
Reinforcement materials, 3
Residual error, 286, 290

**S**
Saturated Calomel Electrode (SCE), 312
Scanning electron microscope (SEM), 281
Screw extrusion process, 256
Selective laser melting (SLM), 5, 56, 185, 187–189, 308
  Al composite, 8
  Al matrix composites, 7, 23
  AM process, 2
  chemical compositions, 9
  components, 181
  computer controls, 181
  cross-sectional and longitudinal sections, 14
  goal, 181
  in situ precipitates, 6
  laser energy density, 181
  laser melting processing parameters, 182
  layer thickness, 182
  mechanical properties, 9
  microstructural views, 6
  microstructures, 16
  optical microscopy image, 6
  parameters, 181, 182
  scanning strategy, 182, 183
  thermite interactions, 5
  titanium alloys
    conventional processing methods, 189
    corrosion resistance, 187
    CP–Ti samples, 188
    potentiodynamic measurements, 187
    relative density and laser power, 185, 187
    Ti–6Al–4V, 187
  XRD patterns, 7
Selective laser melting process, 6
Selective laser sintering (SLS), 56
  AlSi10Mg composite, 30, 37
  AM process, 30
  Copper Phosphite composite tool electrode, 31
  EDM electrode, 35
  EDM process, 37
  EOSINT M 280, 36
  RP process, 32
  ZrB2-CuNi electrode, 31
Shore hardness, 291–293, 296
Shore scale, 284
Signal to noise ratios, 342
Single screw extruder, 256
SN ratio response table, 289, 293
SN ratios, 292

Steel
H13, tool steel, 120
mixed powder systems and AM techniques, 121
stainless, 179
stainless steels, 121
TiB2 reinforced 316L stainless steel matrix, 133
TiB2/316L stainless steel mixed powder systems, 153, 154
Stepper motors or servo-motor, 330
Stereolithography (SLA), 56
Surface finishing techniques, 227–229
classification, 228
post-processing, 229
dimensional and geometrical stabilities, 229
fabrication, 229
VS (*see* Vapour smoothing (VS))
pre-processing
adaptive slicing techniques, 228
angle of orientation/deposition, 227
CAD model, 227
parameters, 228
Surface hardness
Durometer, 231
mathematical model, 241, 242
orientation angle and post-cooling time, 232
post-cooling, 239
Surface roughness
mathematical model, 240, 241
Surface tension, 76, 78

**T**
Tafel extrapolation method, 312
Tafel slope, 318
Taguchi approach, 275
Taguchi $L_{18}$ design of experimentation (DOE), 232, 240
Temperature-controlled extrusion head, 329
Tensile test, 338, 342
Thermal analysis (DSC) curve, 336
Thermal behaviour, 334
Thermal conductivity, 288
Thermal properties, 336
Thermo Scientific HAAKE Mini CTW, 281
Thermoplastic materials, 281
Thermoplastics, 279, 330
Three-dimensional printing, 1
Ti matrix nano-composites (TMnCs), 75, 76
Ti-alloy matrix, 14
Ti-B and Ti64-BN, 14–16
Titanium (Ti), 55, 61, 71, 81, 87, 93, 95
TiC reinforced, 133
TMCs, 116–117
Titanium alloy (Ti6Al4V)
aerospace and biomedical industries, 32
properties, 38
work piece, 30
Titanium alloys (Ti-alloy), 184–192
AM biomedical
EBM, 190–192
mechanical properties, 186
non-toxic property, 185
properties, 184
SLM, 185, 187–189
Ti-6Al-4V, 184
TiNb-based alloys, 185
β phase, 185
composites, 25
conventional, 179
in situ nitridation, 22
machining, 33
manufacturing methods, 179
mechanical properties, 179
and oxygen, 179
toughness and damage resistance, 22
tribological properties, 18
work piece, 40
Titanium carbide (TiC), 9
Titanium-matrix composites (TMCs), 116–117, 123, 134, 140, 146, 150, 151, 155, 156, 160
Ti-TiB composites, 14
Tool wear rate (TWR), 30, 39, 40, 44, 51, 52
Twin screw extrusion (TSE), 281, 329

**U**
Universal tensile tester (UTT), 339

**V**
Vapour smoothing (VS), 230–233, 235, 238
ABS thermoplastic layers, 236
cooling chamber, 229, 230
dimensional accuracy
control log of experimentation, 233
head diameter and neck thickness, 232
shrinkage, 238
smoothing time and orientation angle, 235
and surface finish, 238
hip implant

benchmark component, 231
dimensional accuracy, 230
surface hardness, 230
surface roughness, 230
input parameters, 231
mathematical models, 230
optimum parameter settings, 239
over-lapping, 238
post-cooling, 239
SEM micrograph view, 237
shrinkage, 238
smooth roughness, 239
smoothing chamber, 229, 230
SN ratio plots, 232, 234, 235
surface hardness
control log of experimentation, 233
Durometer, 231
orientation angle and post-cooling time, 232, 235
surface roughness
control log of experimentation, 233
Taguchi $L_{18}$ DOE, 232
uPrint FDM apparatus, 231
Vicker's Hardness tester, 312
Vickers microhardness values, 11
Volume fraction, 80, 85–87, 94, 95

**W**
Water-soluble material, 330
Wear
behavior, AM, 165
coefficient of friction (COF), 165
extrinsic factors, 165
inter-particle spacing (λ), 166
Pin-on-Disc, 165
resistance, AM, 166, 167
volume fraction, reinforcements, 165
Working electrode (WE), 312

**X**
X-ray diffraction (XRD) analysis, 299, 311

**Y**
Young's modulus, 252, 261

The manufacturer's authorised representative in the EU is Springer Nature Customer Service Centre GmbH, Europaplatz 3, 69115 Heidelberg, Germany. If you have any concerns regarding our products, please contact ProductSafety@springernature.com

Printed and bound by CPI Group (UK) Ltd, Croydon, CR0 4YY

30/09/2026

02225230-0010